The Gulf of Guinea Large Marine Ecosystem

Environmental Forcing & Sustainable Development of Marine Resources

Large Marine Ecosystems

Series Editor: Kenneth Sherman

Director, Narragansett Laboratory and Office of
Marine Ecosystem Studies
NOAA-NMFS, Narragansett, Rhode Island, USA
and Adjunct Professor of Oceanography
Graduate School of Oceanography, University of Rhode Island,
Narragansett, Rhode Island, USA

On the cover:

The satellite image of SeaWiFS chlorophyll is focused on the Gulf of Guinea. The image is a composite average of SeaWiFS chlorophyll for the period September 1997 through August 2000.

The color-enhanced scale depicts a shaded gradient of chlorophyll concentration from a high level of 30 mg/m3 shown in reddish to orange color spectra to lesser concentrations in yellow and light blue down to an ecosystem low value of 0.1 mg/m3 depicted in the darker blue.

Credit for providing the image is given to: NASA, SeaWiFS Project and ORBIMAGE.

The Gulf of Guinea Large Marine Ecosystem

Environmental Forcing & Sustainable Development of Marine Resources

Edited by

Jacqueline M. McGlade
Department of Mathematics & CoMPLEX
University College London
London, United Kingdom

Philippe Cury
Institut de recherche pour le developpement (IRD) Research Associate
University of Cape Town, X2,
Rondebosch 7700, South Africa

Kwame A. Koranteng
Marine Fisheries Research Division
Ministry of Food & Agriculture
Tema, Ghana

&

Nicholas J. Hardman-Mountford
The Laboratory, Citadel Hill,
Plymouth, United Kingdom

2002

ELSEVIER

Amsterdam • Boston • London • New York • Oxford • Paris • San Diego
San Francisco • Singapore • Sydney • Tokyo

ELSEVIER SCIENCE B.V.
Sara Burgerhartstraat 25
P.O. Box 211, 1000 AE Amsterdam, The Netherlands

First edition 2002

Library of Congress Cataloging in Publication Data
A catalog record from the Library of Congress has been applied for.

British Library Cataloguing in Publication Data
A catalogue record from the British Library has been applied for.

ISBN: 0 444 51028 1
ISSN: 1570 0461

♾ The paper used in this publication meets the requirements of ANSI/NISO Z39.48-1992 (Permanence of Paper).
Printed in The Netherlands.

Contents

III Fish & Fisheries

Contributors

Editors

Jacqueline M. McGlade Department of Mathematics & CoMPLEX, University College London, London WC1E 6BT, United Kingdom.

Philippe Cury IRD Research Associate, University of Cape Town, X2, Rondebosch 7700, South Africa.

Kwame A. Koranteng Marine Fisheries Research Division, Ministry of Food & Agriculture, PO Box BT-62, Tema, Ghana.

Nicholas J. Hardman-Mountford The Laboratory, Citadel Hill, Plymouth PL1 2PB, United Kingdom.

Other Authors

Joana D. Akrofi Directorate of Fisheries, Ministry of Food & Agriculture, PO Box 630, Accra, Ghana

Angora Aman Université Cocody, UFR SSMT – 22 BP 582, Abidjan 22, Côte d'Ivoire.

Emiliana R. Anang Marine Fisheries Research Division, Ministry of Food & Agriculture, PO Box BT-62 Tema, Ghana.

Emory D.Anderson National Oceanographic & Atmospheric Administration, National Sea Grant College Program, Silver Spring, Maryland 20910, USA.

Robert Arfi IRD, Centre IRD de Bel Air, BP 1386, Dakar, Senegal.

Sonia D. Batten Sir Alister Hardy Foundation for Ocean Science, 1 Walker Terrace, The Hoe, Plymouth, PL1 3BN, United Kingdom.

Ellen Bortei-Doku Aryeetey PO Box LG 490 Legon, Accra Ghana.

Marc Bouvy IRD, Centre IRD de Bel Air, BP 1386, Dakar, Senegal.

Hervé Demarcq IRD Research Associate MCM, Private Bag X2, Rogge Bay 8012, Cape Town, South Africa.

Mamaa Entsua-Mensah Water Research Institute, CSIR, PO Box AH 38, Achimota, Accra, Ghana.

Pierre Fréon MCM, Private Bag X2, Rogge Bay 8012 Cape Town, South Africa.

Julien Hoepffner IRD, centre de Recherches Océanologiques, 29 rue des pêcheurs, BP V18, Abidjan, Côte d'Ivoire.

Chidi A. Ibe Centre de Recherches Oceanologiques, 29 rue des pêcheurs, BP V 18 Abidjan, Cote d'Ivoire

Tapé Joanny IRD, centre de Recherches Océanologiques, 29 rue des pêcheurs, BP V18, Abidjan, Côte d'Ivoire.

Anthony W.G. John Sir Alister Hardy Foundation for Ocean Science, 1 Walker Terrace, The Hoe, Plymouth, PL1 3BN, United Kingdom.

Jacques Konan IRD, centre de Recherches Océanologiques, 29 rue des pêcheurs, BP V18, Abidjan, Côte d'Ivoire.

Pierre Le Loeuff 20 rue Arégnaudeau, 44100 Nantes, France.

Alan Lovell Charles Bruneau Cancer Center, Sainte Justine Hospital, 3175 Cote-Ste Catherine, Montreal, Quebec, H3T 1C5 Canada.

Frédéric Menard IRD-Centre de Recherche Halieutique Méditerranéenne et Tropicale, rue Jean Monnet, BP171, 34203 Sète Cedex, France.

Martin A. Mensah PO Box SK-540, Sakumono, Tema, Ghana

Cornelia Nauen CEC, DG Research, rue de la Loi 200, Brussels B-1049 Belgium

Viveca Nordström IRD, BP 570, Victoria, Iles Seychelles

Daniel Pauly Fisheries Centre, 2204 Main Mall, University of British Columbia, Vancouver, B.C. V6T 1Z4 Canada.

Samuel N.K. Quaatey Fisheries Department, PO Box BT 62, Tema, Ghana

Chris P. Reid Sir Alister Hardy Foundation for Ocean Science, 1 Walker Terrace, The Hoe, Plymouth, PL1 3BN, United Kingdom.

Claude Roy University of Cape Town, Rondebosch 7700, South Africa.

Peter A.G.M. Scheren Royal Haskoning, Barbarossastraat 35, 6522 DK Nijmegen, The Netherlands.

Kenneth Sherman National Marine Fisheries Service, Narragansett Laboratory, 28 Tarzwell Drive, Narragansett, RI 02882 USA.

Soko Guillaume Zabi IRD, centre de Recherches Océanologiques, BP V18, Abidjan, Côte d'Ivoire.

Series Editor's Introduction

Coastal ocean waters around the globe continue to be degraded from habitat alteration, eutrophication, toxic pollution, aerosol contamination, emerging diseases, and non-sustainable fishing practices. The global community of nations, at the 1992 UN Conference on Environment and Development (UNCED), focused on the impacts of growing world population and industrialization on the environment and called on coastal countries to support actions to: (1) prevent, reduce, and control degradation of the marine environment to maintain and improve its life support and productive capacities; (2) develop and increase the potential of marine living resources to meet human nutritional needs as well as social, economic, and development goals; and (3) promote integrated management and sustainable development of coastal areas and the marine environment. No single international organization has been empowered to monitor and assess the changing states of coastal ocean ecosystems on a global scale, and to reconcile the needs of individual nations to those of the community of coastal nations for taking appropriate mitigation and management actions. Post-UNCED actions have, however, begun to move coastal nations toward more sustainable ecosystem-based resource assessment and management than has been generally practiced during the last half century. At present, 59 coastal countries in Asia, Africa, Latin America, and eastern Europe are planning and implementing marine ecosystem-based projects to promote more sustainable uses of the global marine environment and its resources. Many of these projects are linked to similar ecosystem-based efforts underway in western Europe and North America.

The present volume in the new Elsevier Science series on Large Marine Ecosystems will be bringing forward the results of UNCED-related and other ecosystem-based studies for marine scientists, educators, students, and resource managers. The volumes will be focused on LMEs and their productivity, fish and fisheries, pollution and ecosystem health, socioeconomics, and governance. This volume in the new LME series, "The Gulf of Guinea Large Marine Ecosystem: Environmental Forcing and Sustainable Development of Marine Resources," is focused on the results of a broad spectrum of ecosystem-based assessment activity in the Gulf of Guinea. The volume provides the results of a multi-disciplinary assessment of changing multi-decadal environmental conditions in the Gulf of Guinea and the impact of those changes on the fisheries resources and people of the region. The co-editors of the volume, Professor Jacqueline McGlade, Dr. Kwame Koranteng, Dr. Philippe Cury and Dr Nicholas Hardman-Mountford, are scientists with extensive experience in west African marine science and resource management issues. Their volume provides a comprehensive synthesis of information on the present state of marine resources of the region and the socioeconomic implications of findings reported by the authors. Production of this volume was supported, in part, by a grant from the United Nations Industrial Development Organization, Vienna.

Kenneth Sherman, Series Editor
Narragansett, Massachusetts

Preface

The Editors

During the past century, there has been a gradual decline in the abundance of a number of key marine fish species around the world. The belief is that these declines are the result of a mixture of overexploitation and environmental pollution, which have sometimes led to irreversible changes in the trophic structure of the marine ecosystem. Over the past four decades, changes in ecosystem structure and species abundance have also been observed in the Gulf of Guinea. However, the physical system of the Gulf of Guinea is variable in space and time and its dynamics are complex. For example, the large-scale spatial heterogeneity, lead Tilot and King (1993) to divide the system into three subsystems: the Sierra Leone and Guinea Plateau (SLGP), the Central West African Upwelling (CWAU) and the Eastern Gulf of Guinea (EGOG), each exhibiting marked seasonality. And on a smaller scale, the spatio-temporal variability is dominated by upwelling off Ghana and Côte d'Ivoire which occurs seasonally, with weak upwelling around January to March (minor season) and intense upwelling from July to September (major season) (Roy 1995) but which also exhibits a marked interannual variability (Houghton and Colin 1986; Roy *ibid.*). The upwelling has also been seen to propagate westward along the coast, from Ghana to Côte d'Ivoire and is not of a classical Ekman-type (i.e. wind-driven); rather it appears to be controlled by several processes, some of which occur in regions remote from the Gulf of Guinea, and may be related to equatorial upwelling processes (e.g. Moore *et al.* 1978; Servain *et al.* 1982; Houghton and Colin 1986). Thus a better understanding of the interactions between different marine species and their environment must be obtained, before the basis of any long-term changes in the ecosystem can be properly determined.

A number of researchers have been active in the region throughout this period, and through their efforts and the support of their institutions, a number of critical data sets have been collected (Cury and Roy 1991; Bard and Koranteng 1995). The researchers involved in the projects presented here, have extended some of these data sets up to the present and as a result have been able to gain a number of fundamental insights about the effects of long-term changes in the oceanographic conditions within the Gulf of Guinea.

The volume is divided into four sections. The first concerns the research programmes which led to the main results; the second concerns environmental forcing and productivity, the third, fish and fisheries, and the fourth section ecosystem health and the human dimension. The major new results are the outcome of an European Union funded research programme (EU-INCO-DC) on "The impacts of environmental forcing on marine biodiversity and sustainable management on artisanal and industrial fisheries in the Gulf of Guinea" (Chapter 1). The framework within which the results have been set is that of the Large Marine Ecosystem approach, developed by Sherman and colleagues (Chapter 2) and formed the basis for a large Global Environment Facility funded programme in the Gulf of Guinea (Chapter 3). The final contribution to the section (Chapter 4) presents a new way of approaching problems in marine ecology that of consilience or the "jumping together" of different disciplines.

In the first part of the next section, entitled 'Environmental Forcing', Hardman-Mountford and McGlade examine this spatio-temporal variability in two papers using remotely sensed SST (Chapters 5 and 6). The first looks at the spatial structure and seasonal evolution of environmental features, such as the Ghana/Côte d'Ivoire coastal upwelling and a southerly extension of cold water from the Senegalese upwelling, as well as the interannual variability in these features and SST in general. The second uses Principal Components Analysis (PCA) to classify the spatio-temporal variability of the system and to define system and subsystem boundaries. Demarcq and Aman (Chapter 7) also use remotely sensed SST, combined with *in situ* measurements, to assess spatio-temporal variability in the Ghana/Côte d'Ivoire coastal upwelling. The paper by Koranteng and McGlade (Chapter 8) focuses on interannual to decadal scale variability in a number of physico-chemical parameters measured on the continental shelf of Ghana, while the contribution from Arfi, Bouvy and Ménard (Chapter 9) looks at both seasonal and interannual variability in environmental parameters in Côte d'Ivoire.

In the second part of this section, entitled 'Productivity', the focus shifts from forcing functions to responses. The paper by Roy, Cury, Fréon and Demarcq (Chapter 10) sets the context for this section by linking environmental fluctuations with resource variability in the Gulf of Guinea and north-west African region, as well as discussing tools for this type of analysis. John, Reid, Batten and Anang (Chapter11) describe the use of the continuous plankton recorder in the Gulf of Guinea and show a marked seasonal response in phytoplankton colour associated with the Ghana/Côte d'Ivoire coastal upwelling, as well as evidence of interannual variability. Le Loeuff and Zabi (Chapter 12) detail the spatio-temporal response of benthic communities to daily and seasonal cycles, as well as environmental perturbations, and Entsua-Mensah (Chapter 13) discusses the important role of coastal lagoons and marine catchment basins in maintaining ecosystem productivity.

Results from the papers in this section serve to advance both the knowledge base and level of support for a number of existing theories such as Moore *et al.*'s (1978) remote forcing theory (Hardman-Mountford and McGlade Chapter 5) and Marchal and Picaut's (1977) cape effect (Hardman-Mountford and McGlade Chapter 5, Demarcq and Aman). Other findings provide consensus on existing observations, such as the timing of the upwelling seasons (*ibid.*, Arfi *et al.*, John *et al.*), subsystem structure (Hardman-Mountford and McGlade Chapter 6, Le Loeuff and Zabi) and interannual variability (Hardman-Mountford and McGlade Chapter 5, Koranteng and McGlade, Roy *et al.*, John *et al.*). An environmental shift in 1987 was noted by both Hardman-Mountford and McGlade (Chapter 5) and Koranteng and McGlade. There are also some contradictions, such as the position of the minor upwelling (Hardman-Mountford and McGlade Chapter 5, Demarcq and Aman).

The first part of the next section on 'Fish and Fisheries' highlights the fact that the Gulf of Guinea large marine ecosystem is a fascinating system from a fish assemblage perspective. The ecosystem has been the subject of drastic changes during the last few decades that strongly affected the entire functioning of the ecosystem and consequently fisheries. The changing aspects of the ecosystems are described and analysed in a set of papers written by Koranteng (Chapter 14), and Joanny and Ménard (Chapter 15). In these papers two different fish assemblages are described based on scientific surveys and catch statistics, and it is shown how the demersal fish communities are structured, and how they have changed through time. Population structure and reproductive ecology of commercially important species is important for demersal communities, and Lovell and McGlade (Chapter 16) emphasize the need to link individual behaviour to population ecology.

In the second part of this section, the pelagic and demersal fisheries off Ghana and Côte d'Ivoire are presented in three companion papers written by Mensah and Quaatey (Chapter 17), Cury and Roy (Chapter 18), and Koranteng (Chapter 19). These papers show how the abundance of demersal fish stocks could be strongly related to large-scale environmental changes and how they were affected by the sudden emergence of *Balistes capriscus*. Accessibility to data bases is often an important and neglected issue in fisheries; Ménard, Nordström, Hoepffner and Konan (Chapter 20) present the processing and structure of data from trawl fisheries of Côte d'Ivoire (this data base and software are available on CD-ROMs at the Centre de Recherches Océanographiques d'Abidjan, B.P. V18 Abidjan). Finally a trophic model of a small West African coastal lagoon is presented by Pauly (Chapter 21) using the Ecopath and Ecosim approach to establish the changes in trophic structure that have taken place through time.

In the last section, Scheren and Ibe (Chapter 22) review the state of the environment of the Gulf of Guinea large marine ecosystem in relation to pollution, which is largely generated by human activities in the coastal zone. The socio-economics of artisanal marine fisheries and the role of women in marketing is examined by Bortei-Doku Aryeetey (Chapter 23); this work and that of Akrofi (Chapter 24) shows the critical nature of fish processing and the fish 'mammy' system in generating income and controlling the transport of marine fish to markets inland. In the last part of this section Nauen (Chapter 25) examines some of the key institutional drivers outside and inside the region that can help to support fisheries management. This is extended by an analysis of research and agricultural extension work in Ghana by Quaatey (Chapter 26).

References

Bard, F.X. and K.A. Koranteng (eds) 1995 Dynamique et usage des ressources en sardinelles de l'upwelling côtier du Ghana et de la Côte d'Ivoire. ORSTOM édition, Paris.

Cury, P. and C. Roy (eds) 1991 Pêcheries ouest-africaines: variabilité, instabilité et changement. Editions de l'ORSTOM, Paris. 525p.

Houghton, R.W. and C. Colin 1986 Thermal Structure Along 4°W in the Gulf of Guinea During 1983-1984. Journal of Geophysical Research 91(C10):11,727-11,379.

Marchal, E. and J. Picaut 1977 Répartition et abondonce évaluéespar échointégration des poissons du plateau continental ivoiro-ghanéen en relation avec les upwellings locaux. J. Res. Océanogr. 2: 39-57.

Moore, D., P. Hisard, J. McCreary, J. Merle, J. O'Brien, J. Picaut, J-M. Verstraete and C. Wunsch 1978 Equatorial Adjustment in the Eastern Atlantic. Geophysical Research Letters 5(8):637-640.

Roy, C. 1995 The Côte d'Ivoire and Ghana Coastal Upwellings: Dynamics and Changes. In: F.X. Bard and K.A. Koranteng (eds). Dynamique et usage des ressources en sardinelles de l'upwelling côtier du Ghana et de la Côte d'Ivoire. ORSTOM édition, Paris.

Servain, J., J. Picaut and J. Merle 1982 Evidence of Remote Forcing in the Equatorial Atlantic Ocean. Journal of Physical Oceanography 12: 457-463.
Tilot, V. and A. King 1993 A Review of the Subsystems of the Canary Current and Gulf of Guinea Large Marine Ecosystems. IUCN Marine Programme.

> *"Beware and take care of the Bight of Benin,*
> *There's one comes out for forty goes in"*
>
> *18th Century Sailor's Rhyme*

Acknowledgements

The editors are indebted to the sponsors of the research presented in this volume, in particular the European Union, the United Nations Industrial Development Organization and the Global Environment Facility. Such large institutions ultimately rely on individuals to provide help at critical times and to smooth out the bureaucratic hiccups that invariably occur in large projects such as those described in this volume. Four individuals stand out in this regard: Professor Tielak Viegas and Dr Cornelia Nauen from the European Commission, Dr Zoltan Csizer from UNIDO and Dr Ken Sherman from NOAA. They consistently acted to ensure the very best outcome for the major projects described in this volume, and for this we would like to extend our very great appreciation. In addition, we would like to thank the many individuals who have supported the researchers during the course of their projects and especially those who helped make the final workshop in Accra so successful.

We are pleased to acknowledge the interest and financial support of the U.S. National Oceanographic and Atmospheric Administration and UNIDO which have enabled this volume to be published and distributed widely in the region.

Without contributors, there would have been no Gulf of Guinea volume. As editors, we wish to thank them all again for their efforts and willingness to take time to write papers that will allow us to push forward the thinking about the Gulf of Guinea Large Marine Ecosystem.

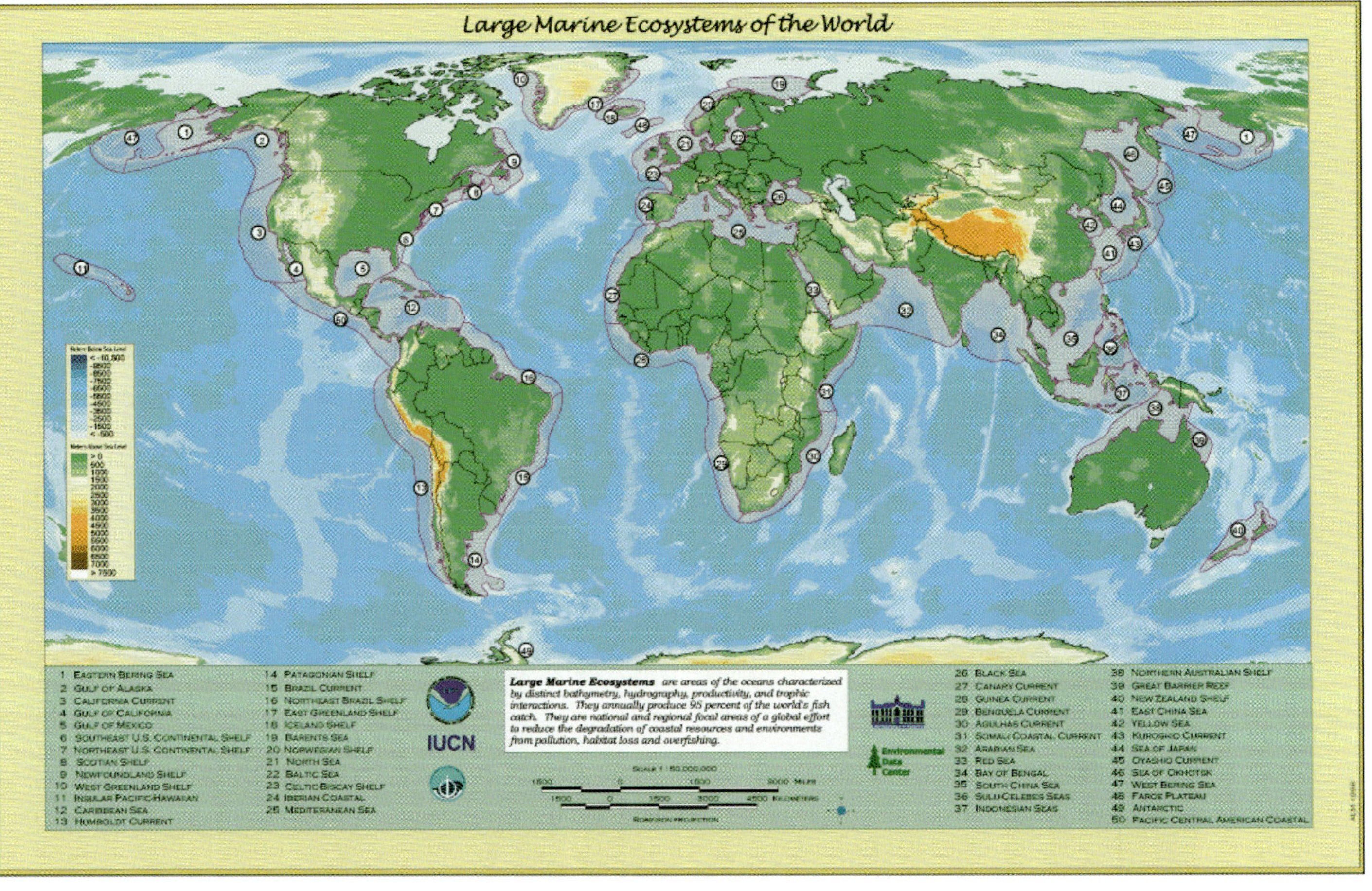

Map 1 Map showing the boundaries of the 49 large marine ecosystems

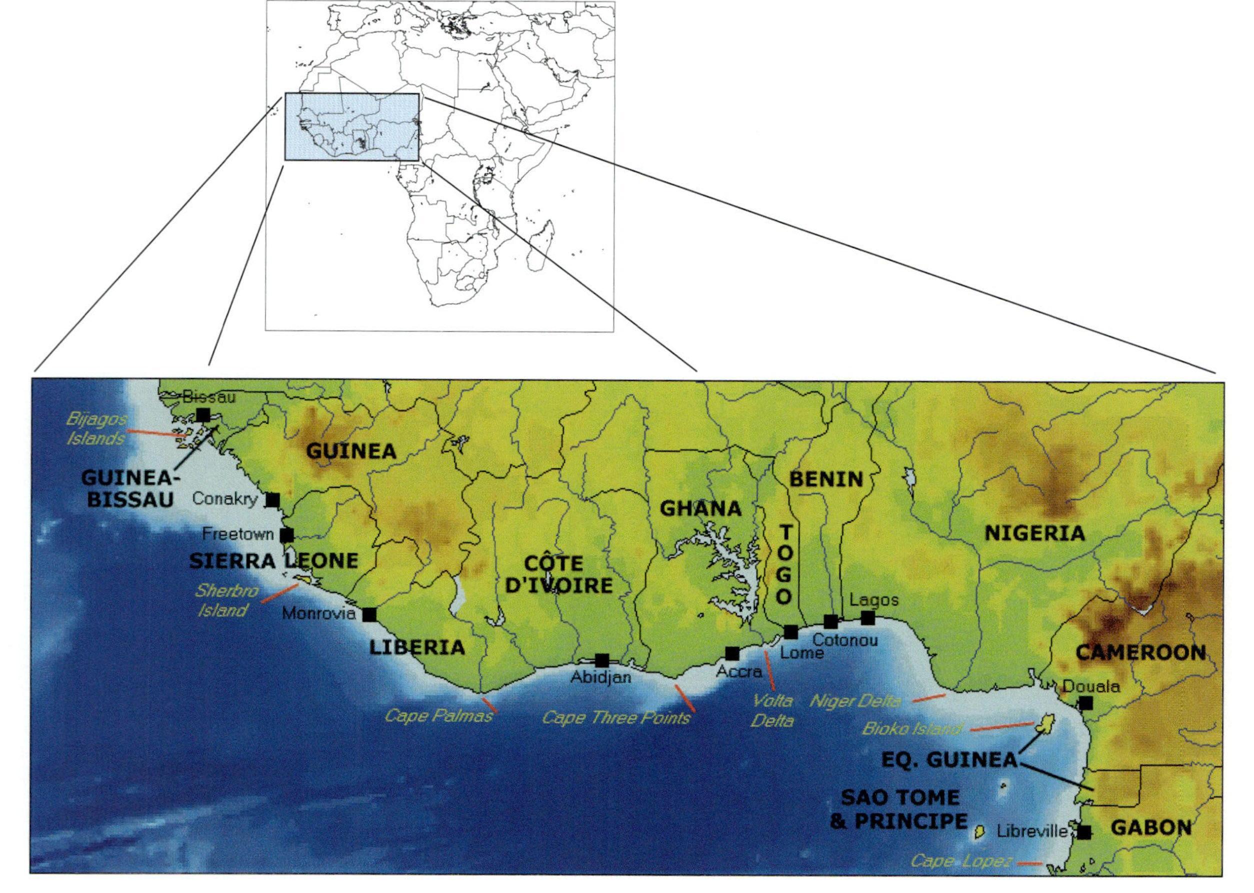

Map 2 Map showing the Gulf of Guinea Large Marine Ecosystem, as defined by Binet and Marchal (1993), including bathymetry, topography, major rivers, country names and boundaries, some major coastal cities and prominent geophysical features.

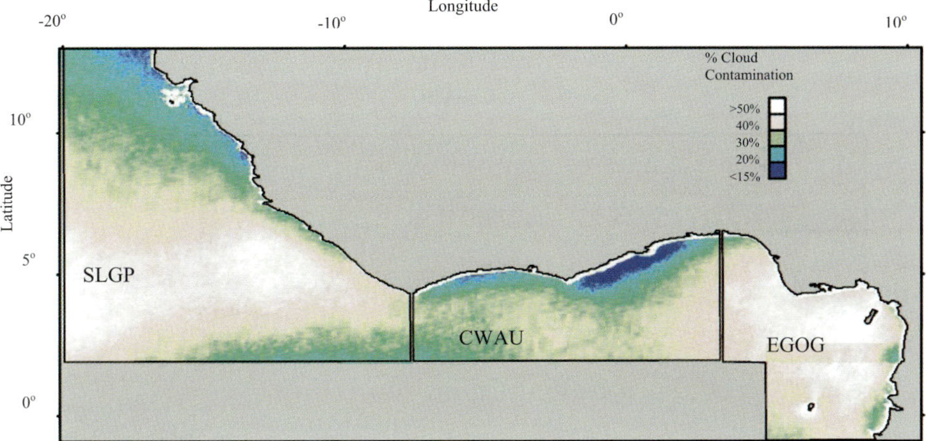

Figure 5-1a Mean cloud contamination of images, averaged over the whole time series. Cloud contamination describes cloud pixels which were recoded to a have a DN value of zero by the JRC during pre-processing.

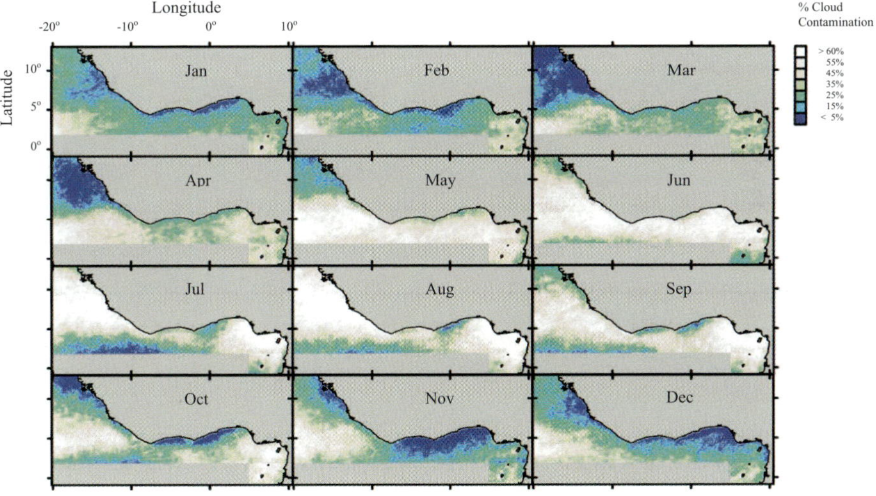

Figure 5-1b Monthly mean cloud contamination from the CORSA-AVHRR data set, calculated from weekly composite images.

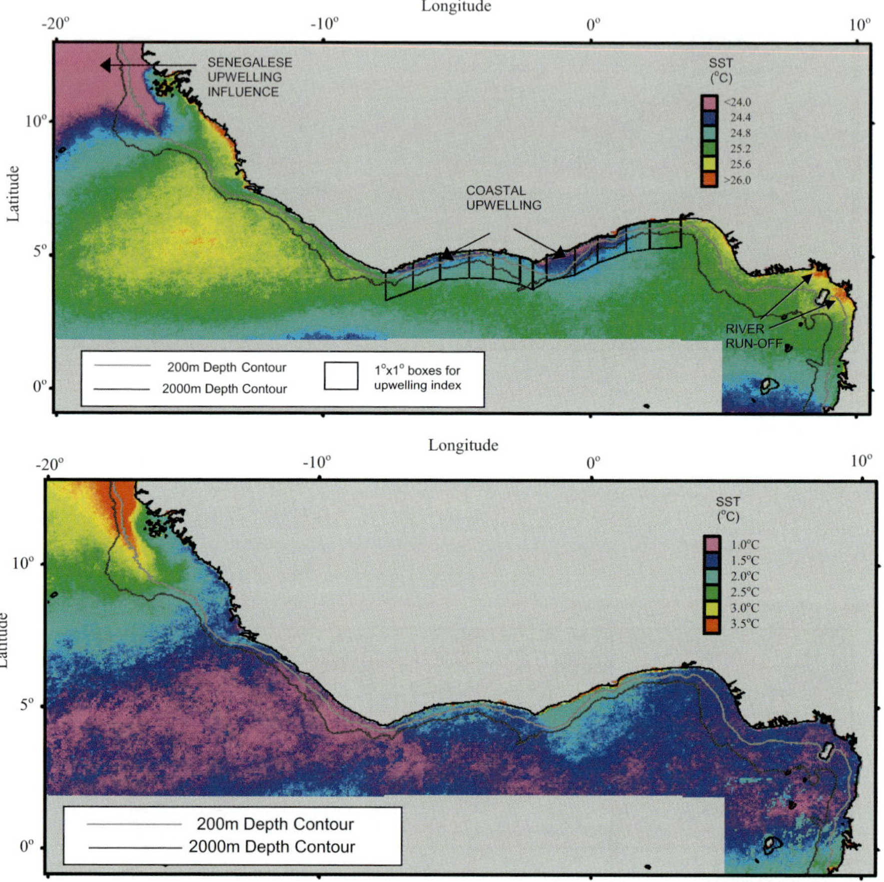

Figure 5-4 a Satellite derived mean SST derived from CORSA-AVHRR weekly SST composites for the period July 1981 to December 1991, annotated to show major oceanographic features: **b** Standard Deviation of SST for the same period.

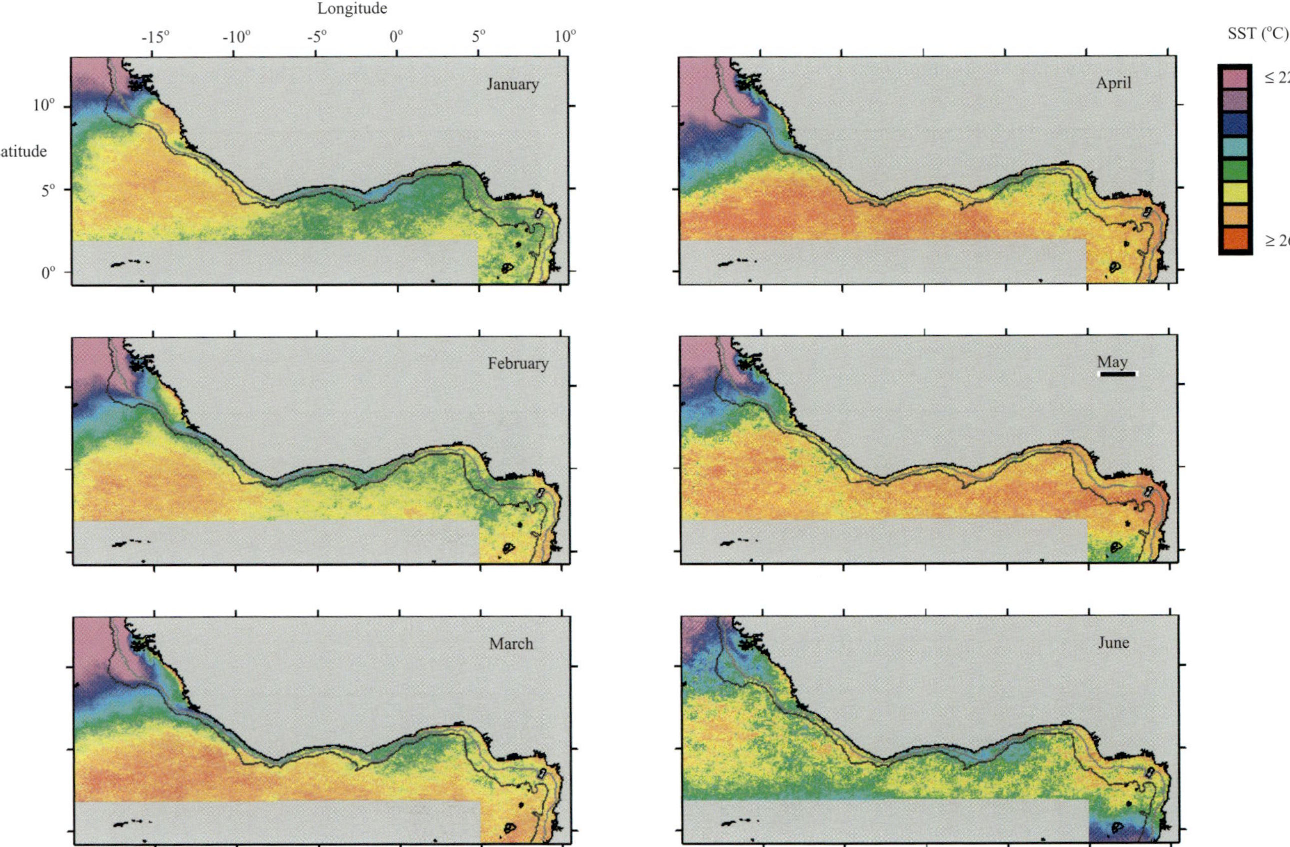

Figure 5-5 Monthly mean SST for the Gulf of Guinea from CORSA-AVHRR SST data. Each monthly mean is averaged from weekly SST composites for each month over the period July 1981 to December 1991.

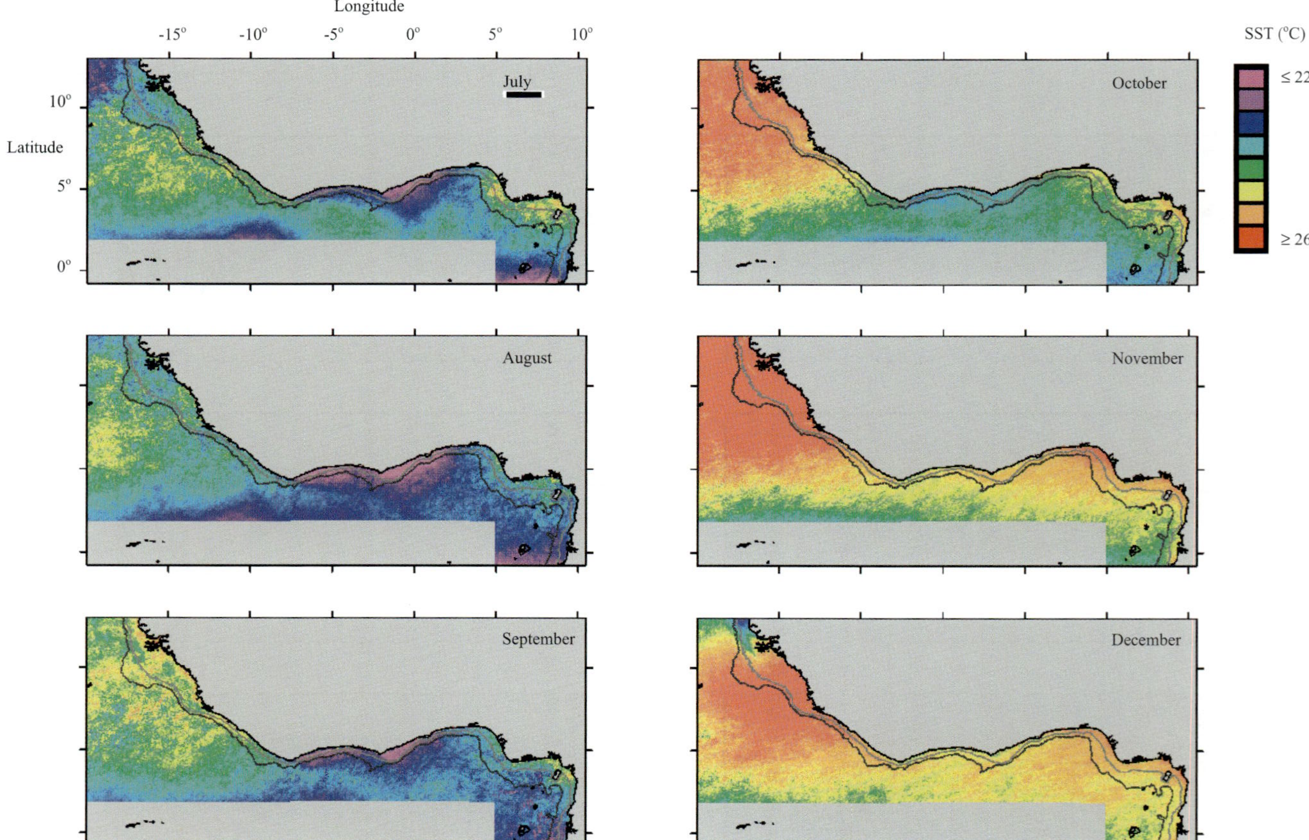

Figure 5 –5 cont. Monthly mean SST for the Gulf of Guinea from CORSA-AVHRR SST data. Each monthly mean is averaged from weekly SST composites for each month over the period July 1981 to December 1991.

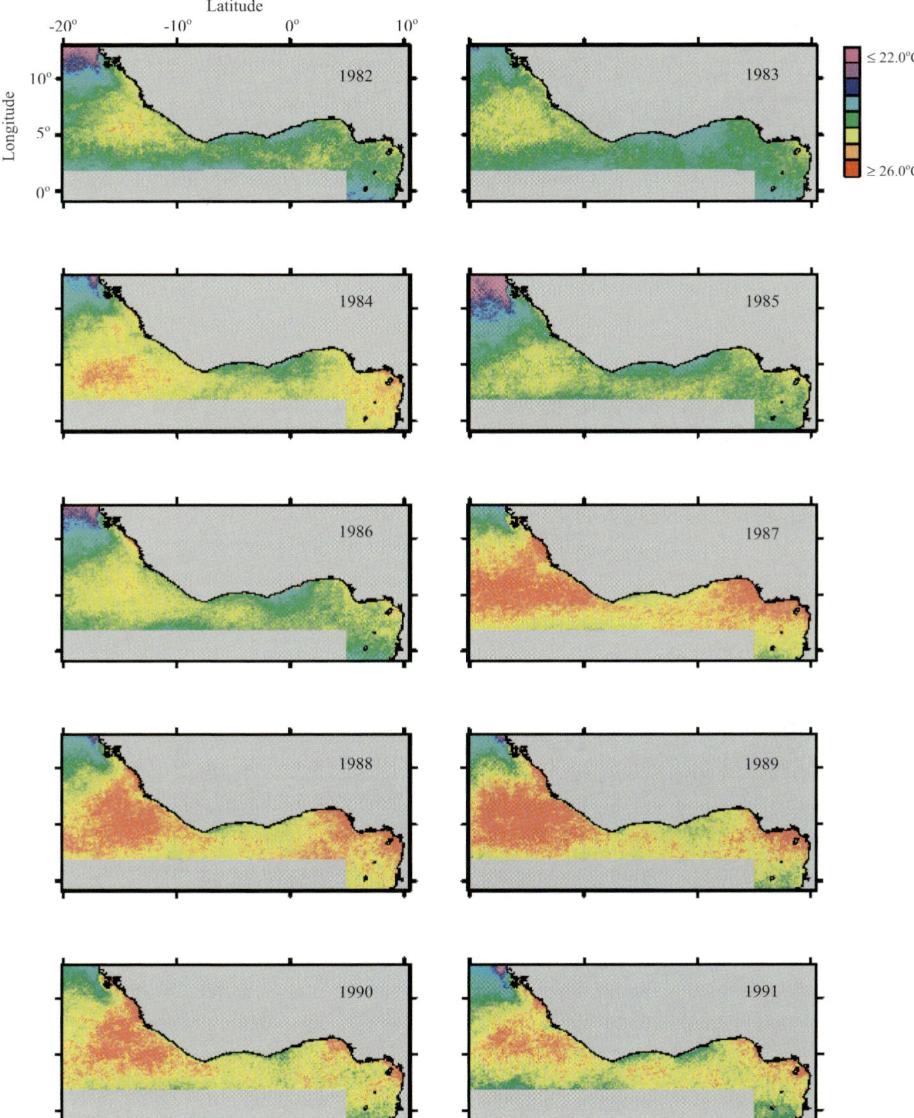

Figure 5-6 Annual mean SST calculated from CORSA-AVHRR SST data. Each annual mean is averaged from weekly composite images for each year of the time series July 1981 to December 1991.

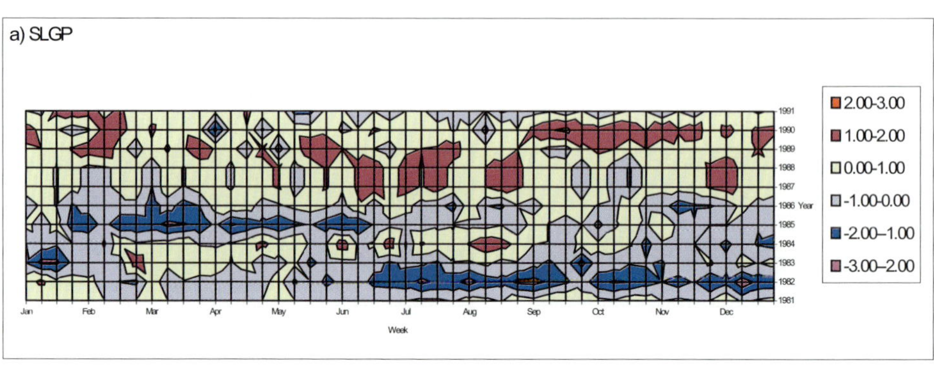

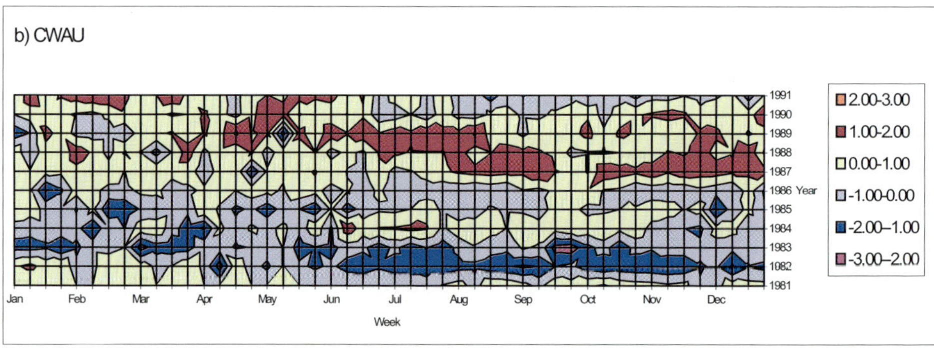

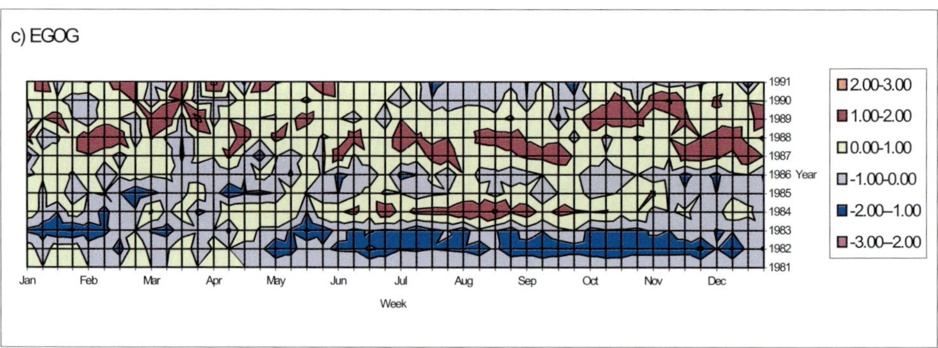

Figure 5-8 Charts showing the spatially averaged interannual SST anomaly for each week of the CORS-AVHRR time series for each subsystem: **a** SLGP, **b** CWAU and **c** EGOG.

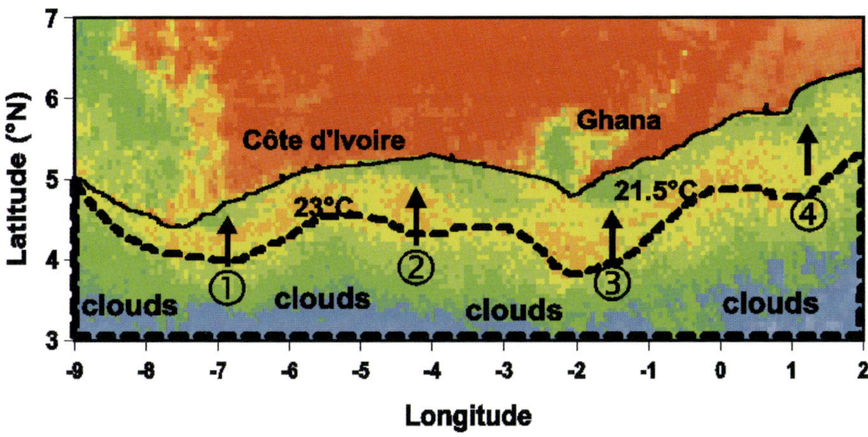

Figure 7-4 Upwelling source regions as identified with METEOSAT during the minor upwelling season on 8 January 1993.

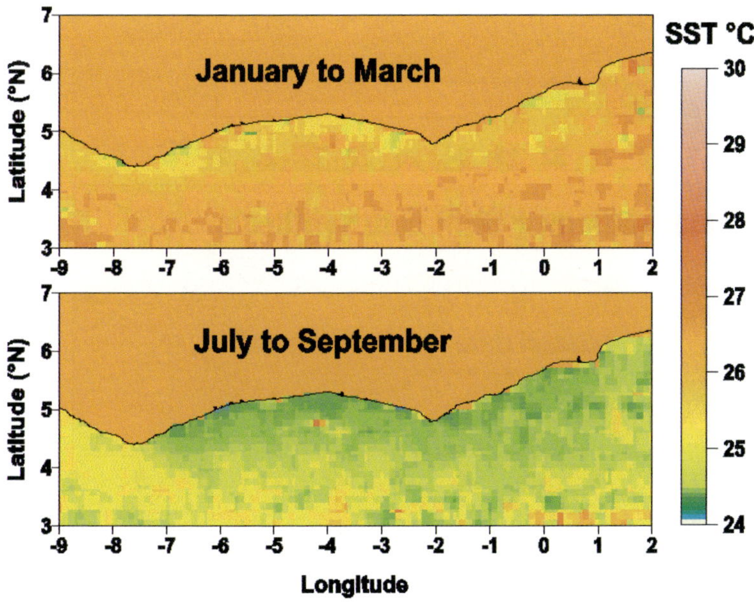

Figure 7-5 Spatial structures for both upwelling seasons as described by the SHIP data compilation from 1984 to 1997

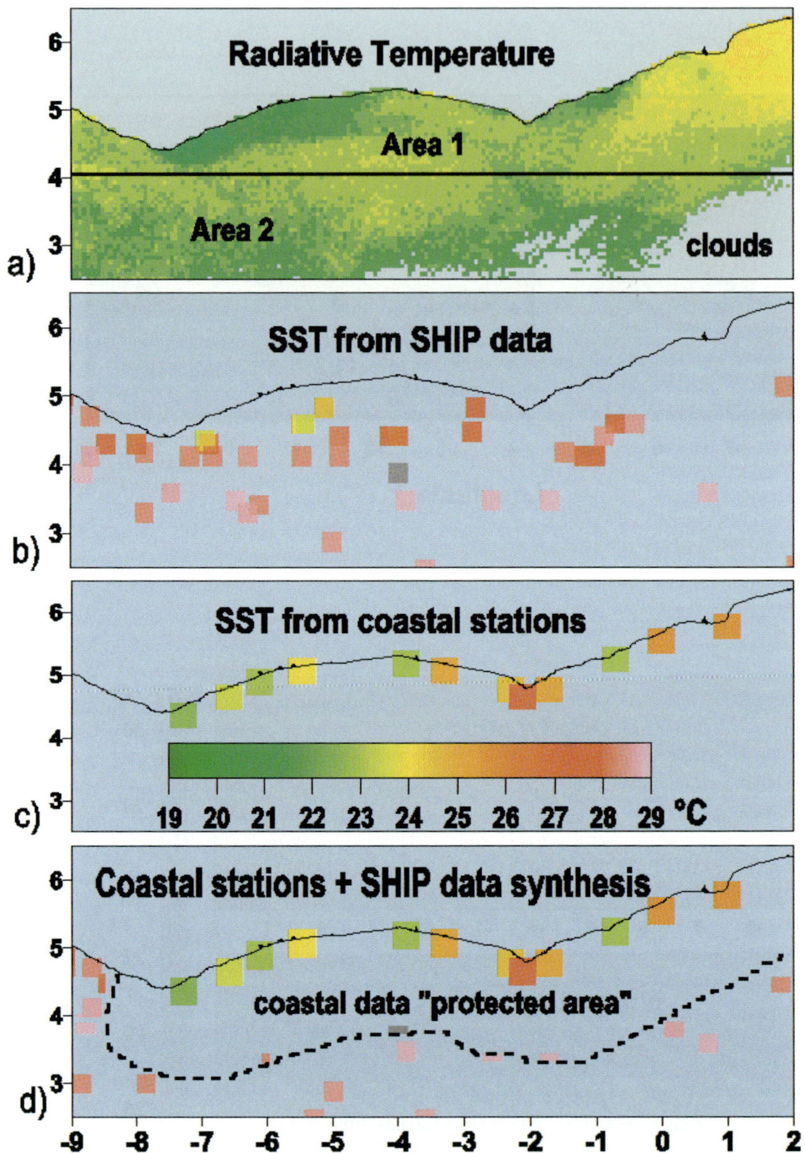

Figure 7-6 Radiative temperature field as depicted by METEOSAT on 26 January 1990 with the atmospheric subdivisions used in the processing **a** SST sampling provided during one week centred on the same day from the SHIP data set, **b** coastal SST from the coastal oceanographic stations, **c** the mixed SST selected for the processing and **d** the atmospheric absorption fields, in °C, calculated between **a** and **d**.

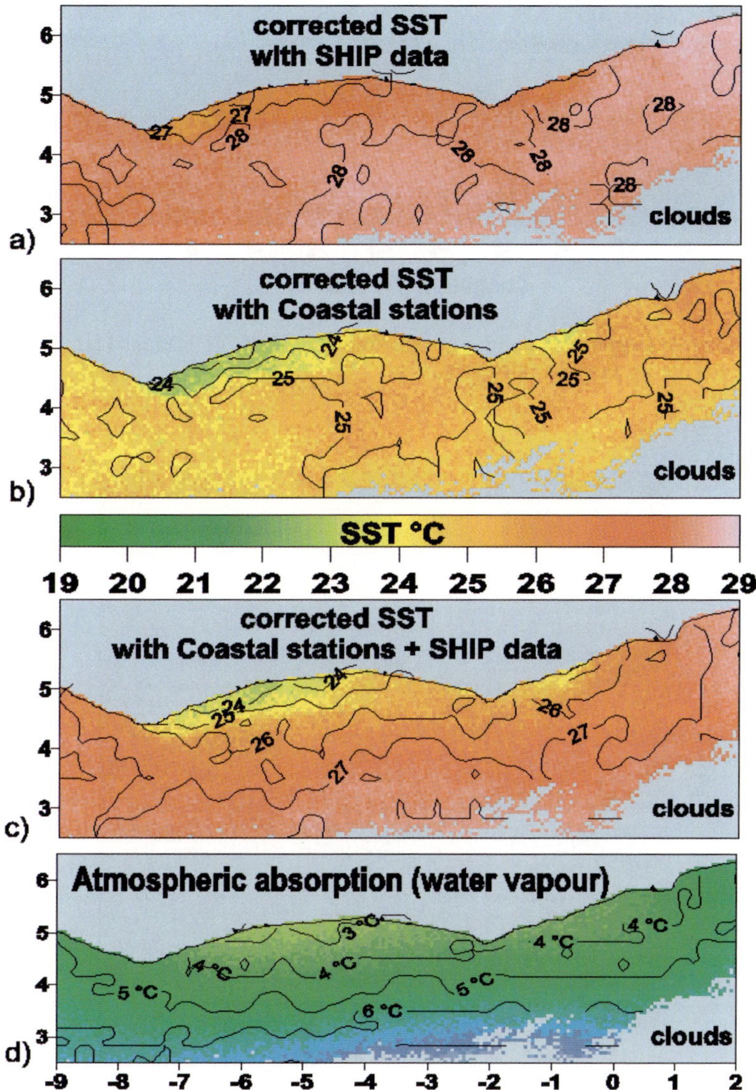

Figure 7-7 Radiative temperature field as depicted by METEOSAT on 26 January 1990, and SST processing results respectively with **a** the SHIP data set only, **b** the coastal stations only, **c** the adequate mixing of the two data sets and **d** atmospheric absorption as a measure of atmospheric water vapour.

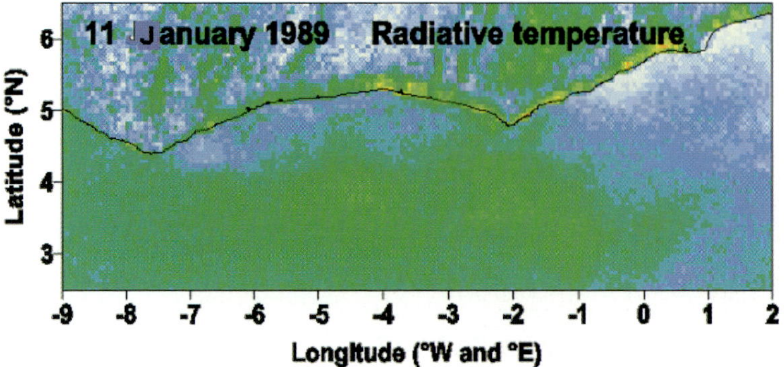

Figure 7-8 Radiative temperature daily synthesis from METEOSAT data on 11 January 1989 off Côte d'Ivoire and Ghana during an upwelling event.

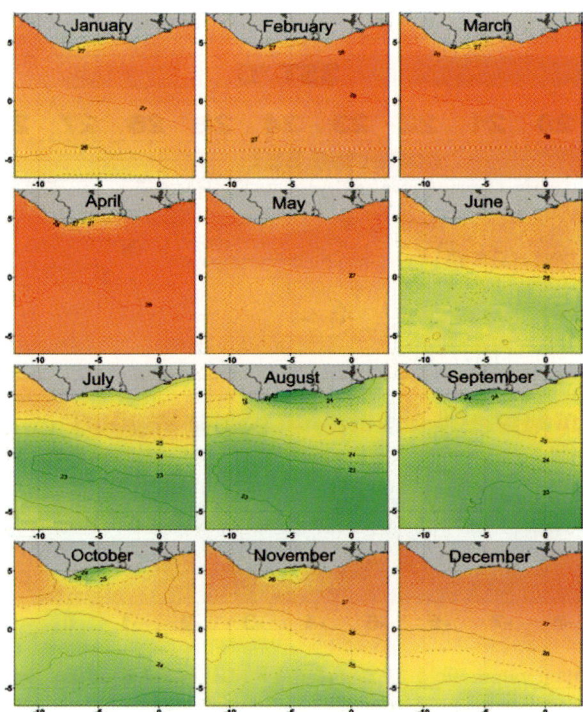

Figure 7-9 Monthly climatology of the SST off Côte d'Ivoire and Ghana as restored by the merging of METEOSAT infrared data, SHIP data and coastal oceanographic data.

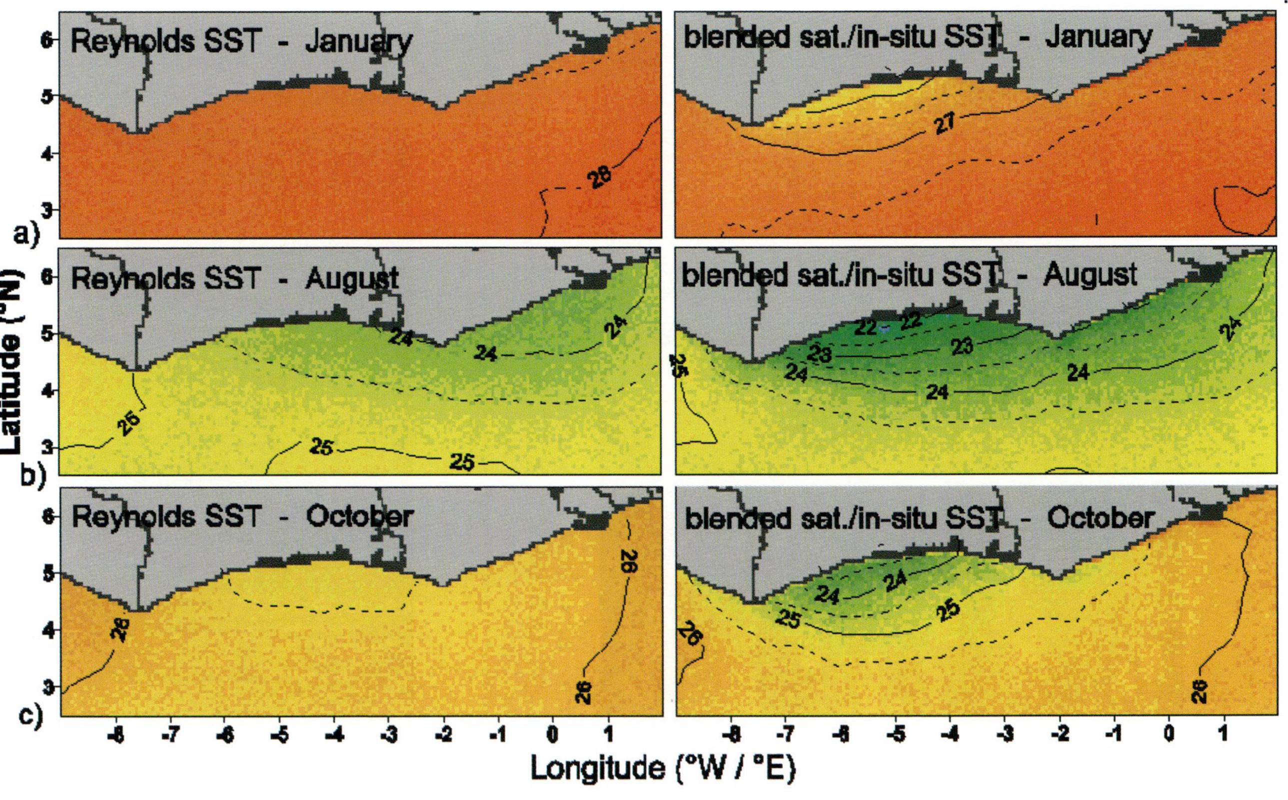

Figure 7-10 Comparison of the Ryenolds climatology of SST computed from *in situ* (ship and Buoy) and satellite (NOAA AVHRR) data (Revnolds 1995) and this study's climatology for the main upwelling seasons off Côte d'Ivoire and Ghana.

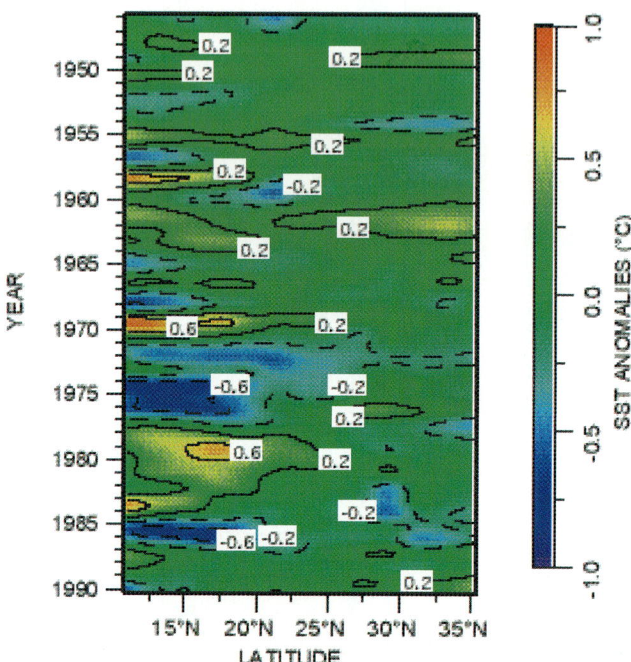

Figure 10-1 SST anomalies from 1946 to 1990 during the upwelling season, from 10°N to 36°N (source COADS dataset).

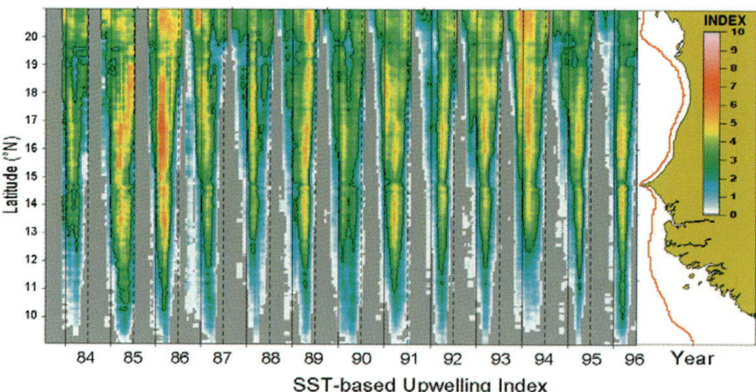

Figure 10-3 SST based upwelling index off Senegal and Mauritania processed from fortnightly means of METEOSAT/IR SST fields from 1984 to 1996.

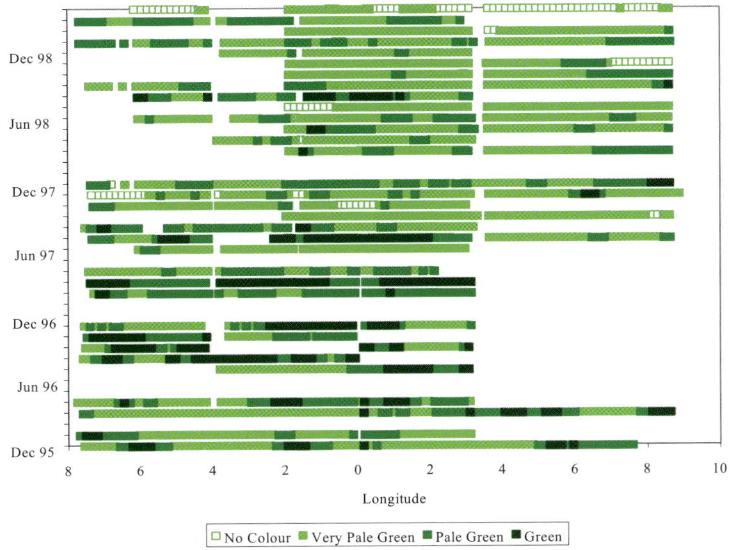

Figure 11-2 Plots of 'Phytoplankton Colour' from December 1995 to April 1999 recorded on CPR routes in the Gulf of Guinea (from 8°W to 9°E). For colour values see the Key. Phytoplankton Colour is given the following values: Very Pale Green = 1, Pale Green = 2, Green = 6.5.

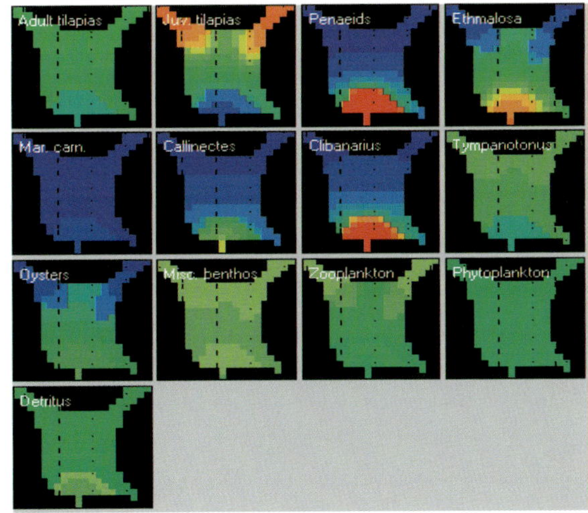

Figure 21-3 Results of a run with Ecospace, the spatial version of Ecopath (Walters *et al.* 1999) of a file representing the Sakumo Lagoon ecosystem. Note abundance (red color) of animals with marine affinities near the mouth of the lagoon, and conversely for freshwater organisms. Note that some patterns of relative abundance are not due to primary habitat assignments, but to 'cascade' effects, wherein the preys of organisms abundant in certain cells are scarce in those cells, and conversely for *their* prey (see also text).

I

Principles &
Programmes

1

The EU/INCO-DC Project: Impacts of Environmental Forcing on Marine Biodiversity & Sustainable Management of Artisanal & Industrial Fisheries in the Gulf of Guinea

J.M. McGlade, K.A. Koranteng and P. Cury

Abstract

The European Union Gulf of Guinea collaborative research project on the impacts of environmental forcing on marine biodiversity was supported by the International Co-operation with Developing Countries Programme (INCO-DC). It was a natural sequel to three earlier international research projects on environmental variability and pelagic fishery resources in West Africa (Cury and Roy 1991; Bard and Koranteng 1995; Durand *et al.* 1998). At its conclusion, the project was able to provide an assessment of the impacts of upwelling and other forms of environmental forcing on marine biodiversity, with particular reference to demersal fish, and the basis for a fisheries information and analysis system for the sustainable management of fisheries in West Africa. It also facilitated the retrieval of important fisheries and survey data that had previously been inaccessible to scientists in the region. The major achievements of the project were presented at an international workshop on "Sustainable Management of the Fish Resources in the Gulf of Guinea" held in Accra in 1998.

Introduction

In 1989, the Ministerial Conference on Fisheries Cooperation among African States Bordering the Atlantic Ocean requested help from the UN Economic Commission for Africa, the UN Office of Maritime Affairs and the Law of the Sea, and the Food and Agriculture Organisation to formulate a project to initiate the establishment of a sub-regional and regional maritime database. It was envisaged that the development of a regional database within the Ministerial Conference framework, would initiate a programme to provide member states with the strategic information necessary for optimising the exploitation of their maritime resources. The database would be used as a management tool, facilitating policy development and decision making for issues ranging from fisheries resources and their environment, to maritime transport, as well as serving as a legal base for management. The European Union agreed to seek ways of funding the establishment of this database, which became known by its French name of 'base de données régionales maritimes' (BDRM), which translates as Regional Maritime Database (Nauen 1993).

In February 1993, a workshop on the establishment of this regional maritime database was held in Dakar, Senegal (Anon. 1993). Following the meeting, it became clear that the development of BDRM would be more feasible if sub-regional nodes were first established and subsequently linked up. An example is the Fisheries Information and Analysis System (FIAS) (McGlade and Nauen 1994), operating under the auspices of the Sub-regional Fisheries Commission (CSRP in French) and covering countries in the northern CECAF (Committee for Eastern Central Atlantic Fisheries) region, i.e. Cape Verde, Mauritania, Senegal, Guinea, Guinea Bissau and the Gambia. FIAS has taken the original concept of BDRM and moved it towards a more interactive information manipulation and analysis tool to support fisheries sector planning and management.

A second, sub-regional node of BDRM was formulated for the Gulf of Guinea under the aegis of the European Union's Programme on International Co-operation in Developing Countries. The two-year project was entitled "Impacts of environmental forcing on marine biodiversity and sustainable management of artisanal and industrial fisheries in the Gulf of Guinea" (or Gulf of Guinea Biodiversity Project for short). It goals were to: assess the impacts of upwelling and other forms of environmental forcing on marine biodiversity and the dynamics of artisanal and industrial fisheries; and develop and implement an information and analysis system for the sustainable management and governance of fisheries resources in the Gulf of Guinea. More specifically, the research was aimed at:

- identifying spatial and temporal scales over which environmental forcing can be detected using remotely sensed and field data derived from a variety of sources;
- identifying biophysical habitats along the continental shelf critical to marine fisheries and biodiversity;
- generating testable hypotheses regarding causal relationships between environmental forcing factors and interannual variability in ecosystem behaviour, using output from coupled models and observational records;
- implementing BDRM and FIAS with a Gulf of Guinea focus in research institutes in Ghana and selected centres in the region;
- and establishing an information network for researchers in European member states and in the Gulf of Guinea region.

By focusing on demersal fishery resources, the project built on the *Pêche-Climat* (Cury and Roy 1991), *DUSRU* (Bard and Koranteng 1995) and *CEOS* (Durand *et al.* 1998) projects which had been concerned primarily with environmental variability and pelagic fishery resources.

Components of the Project and Linkages

Given the high level of R&D activity and interest in marine resources in the Gulf of Guinea, it was important that linkages were made with institutions outside the project. The co-ordinator of the Gulf of Guinea Biodiversity Project was Professor Jacqueline McGlade, Head of the University of Warwick's Ecosystems Analysis and Management Group (subsequently transferred to the Centre for Coastal and Marine Sciences of the Natural Environment Research Council), U.K. The other two principal researchers were Dr Kwame Koranteng from the Ministry of Food and Agriculture's Directorate of Fisheries, Ghana and Dr Philippe Cury from the Institut de recherche pour le développement (previously ORSTOM), France. Over the two years, scientists from Ghana, France, U.K. and Côte d'Ivoire worked on the project. The project had a Management Committee which was made up of representatives from the participating institutions, the European Commission and the United States' National Marine Fisheries Service.

The participation of the National Marine Fisheries Service was to ensure that there were effective links between the project and the Global Environment Facility/UN Industrial Organization (GEF/UNIDO) Gulf of Guinea Large Marine Ecosystem Project based in Abidjan, Côte d'Ivoire which is technically supported by NMFS (see Ibe and Sherman this volume). There were also links with a number of other EU-sponsored research programmes (e.g. Hogarth and McGlade 1998).

Major Achievements

The results and major achievements of the project are detailed in the following sections and chapters of this volume: Environmental Forcing & Productivity chapters 5, 6, 8 and 10; and Fish & Fisheries chapters 14, 16, 18 and 19.

Various environmental and biological data sets were constructed and used in the investigation of nearshore and climatic forcing factors on fishery resources in the Gulf of Guinea (see Demarcq and Aman; Hardman-Mountford and McGlade; Koranteng and McGlade; Roy *et al.* this volume). In particular, an assessment of the response of demersal fishery resources to environmental perturbations and a preliminary biomass budget for the Ghanaian shelf ecosystem were completed. The project brought together databases on the demersal fishery resources in the Gulf of Guinea. These included the trawl fishery data of Côte d'Ivoire, which were compiled, quality controlled and put onto CD-ROM (Ménard *et al.* this volume). The original datasheets from the Guinean Trawling Survey (Williams 1968), which were not available in West Africa and generally considered missing, were also retrieved and brought back to the region (courtesy of Professor Daniel Pauly). These data are now available on CD-ROM for the benefit of the scientific community.

The project supported five PhD and one MSc degrees at UK, French and Ivorian universities. The subjects of investigation ranged from examination of dynamics of the marine environment and the role of climate forcing, discrimination of fish stocks by molecular genetics, dynamics of fish species assemblages, exploitation and management of demersal fishery resources in the Gulf of Guinea, including the assessment of fish stocks and modelling of fishery catches.

The Dissemination Workshop

A critical part of the activities within the project was a workshop, organised in Accra, Ghana on 27-29 July 1998 (Koranteng 1998). The primary objective was to disseminate the findings of the project and relate them to research in other parts of the world. The workshop was attended by 50 scientists from Canada, Côte d'Ivoire, France, Ghana, South Africa, United States of America and the United Kingdom.

In his opening address, the Hon. Johnson Asiedu Nketiah, Deputy Minister, Ministry of Food and Agriculture, Ghana, underscored the importance of the fishing industry in the economies of countries in the Gulf of Guinea. He described the numerous management problems existing in the fishing industry and appealed to the participants to consider the issues of environmental degradation in the sub-region and the human factors leading to it, and the decline in fish catches and the impact on human populations, especially in the West African sub-region. There were also short addresses by representatives of the GEF/UNIDO Gulf of Guinea LME project, and representatives of the core partners, who underlined the importance of the findings of the project and the opportunities for building international and national awareness of the problems facing the region afforded by the workshop.

Closing the workshop, Mr Mike Akyeampong, Deputy Minister, Ministry of Food and Agriculture noted that as Deputy Minister responsible for fisheries, he had been involved in the work of the FAO Fishery Committee for the Eastern Central Atlantic (CECAF) and in the activities of the Ministerial Conference of Fisheries Cooperation Between African States Bordering the Atlantic, and as such had followed the development of the Gulf of Guinea Biodiversity Project, and was very pleased to see the workshop and project come to fruition.

The research project that culminated in this and other workshops exemplified the scientific collaboration within the EU Fishery Research Initiative under the European Union's Research and Technology for Development programme in the general framework of the Lomé IV Convention.

Acknowledgements

As scientific co-ordinator and principal researchers of the project, we would like to express our sincere gratitude to the European Commission (DG VIII and DGXII) for supporting the project and the workshops. We are also grateful to all personalities who, in many diverse ways, offered their support and advice in the formulation and implementation of the project and also for the workshops. J.McGlade would also like to acknowledge the support provided by NERC Grant GT5/00/MS/1.

References

Anon. 1993 J.M.McGlade (ed.) Atelier sur la mise en oeuvre de la base de données régionale maritime (BDRM), Dakar, Senegal, 15-19 février 1993. University of Warwick, UK.

Bard, F.X. and K.A. Koranteng (eds) 1995 Dynamics and Use of Sardinella Resources from Upwelling off Ghana and Ivory Coast. ORSTOM Editions, Paris 439p.

Cury, P. and C. Roy (eds) 1991 Pêcheries ouest-Africaines, variabilité, instabilité et changement. Editions ORSTOM, Paris. 525p.

Demarcq, H. and A. Aman (this volume, chapter 7) A multi-data approach for assessing the spatio-temporal variability of the Ivorian-Ghanaian coastal upwelling: interest for understanding pelagic fish stock dynamics.

Durand, M.-H, P. Cury, R. Mendelssohn, C. Roy, A. Bakun and D. Pauly 1998 Global versus local changes in upwelling systems. ORSTOM Editions, Paris. 594p.

Hardman-Mountford, N.J. and J.M. McGlade (this volume, chapter 5) Variability of physical processes in the Gulf of Guinea and implications for fisheries recruitment. An investigation using remotely sensed sea surface temperature.

Ibe, C. and K. Sherman (this volume, chapter 3) The Gulf of Guinea Large Marine Ecosystem project: turning challenges into achievements.

Koranteng, K.A. 1998 Two interdisciplinary workshops in the Gulf of Guinea. EC Fish.Coop.Bull./Bull.Coop.Pêche CE 11(3-4): 53-54.

Koranteng, K.A. and J. McGlade (this volume, chapter 8) Physico-chemical changes in continental shelf waters of the Gulf of Guinea and possible impacts on resource variability.

Hogarth, A.N. and J. McGlade 1998 SimCoast™: Integrating information about the coastal zone. EC Fish.Coop.Bull./Bull.Coop. Pêche CE 11(3-4):29-30.

McGlade, J.M. and C. Nauen 1994 A fisheries information and analysis system for West Africa. EC Fish.Coop.Bull./Bull.Coop. Pêche CE 7(4):6-7.

Ménard, F., V. Nordström, J. Hoepffner and J. Konan (this volume, chapter 20) A database for the trawling fisheries of Côte d'Ivoire: structure and use.

Nauen, C. 1993 Regional Maritime Database (BDRM) for coastal West Africa. EC Fish.Coop.Bull./Bull.Coop. Pêche CE 6(1):17-18.

Roy, C., P. Cury, P. Fréon and H. Demarcq (this volume, chapter 10) Environmental and resource variability off Northwest Africa.

Williams, F. 1968 Report on the Guinean Trawling Survey. Org. Afr. Unity Sci. Tech. Res. Comm. 99.

The Gulf of Guinea Large Marine Ecosystem
J.M. McGlade, P. Cury, K.A. Koranteng and N.J. Hardman-Mountford (Editors)

2

A Modular Approach to Monitoring, Assessing and Managing Large Marine Ecosystems

K. Sherman and E.D. Anderson

Abstract

In recent years, there has been a change in approach to the management of the world's natural resources from that of compartmentalised efforts focusing on maximising short-term yields and economic gain towards ecosystem management and long-term sustainability. A significant milestone was achieved in June 1992 when a majority of coastal nations adopted follow-on actions to the United Nations Conference on Environment and Development (UNCED) to: 1) *prevent, reduce, and control degradation of the marine environment so as to maintain and improve its life-support and productive capacities; 2) develop and increase the potential of marine living resources to meet human nutritional needs, as well as social, economic, and development goals;* and 3) *promote the integrated management and sustainable development of coastal areas and the marine environment.*

Regrettably, the long-term sustainability of coastal ecosystems as a resource to support healthy economies of coastal nations appears to be diminishing. A growing awareness that the quality of these ecosystems is being adversely impacted by multiple sources of stress has accelerated efforts by scientists and programme managers to monitor, assess, and mitigate such stressors from an ecosystem perspective. The linkage between science and improved global stewardship of natural resources needs to be strengthened. An ecological framework that can serve as the basis for achieving the UNCED objectives is the large marine ecosystem (LME) concept. A description is given of an ecosystems approach being developed by United States National Marine Fisheries Service for strengthening science-based resource management using five linked modules for improving ecosystem sustainability: 1) productivity, 2) fish and fisheries, 3) pollution and ecosystem health, 4) socio-economic conditions, and 5) pertinent governance regimes. Examples are presented where the LME concept and the modular approach either are in use or about to be implemented.

Introduction

A significant milestone was achieved in June 1992 with the adoption of follow-on actions to the declarations on the ocean of the UN Conference on Environment and Development (UNCED) which recommended that nations: 1) *"prevent, reduce, and control degradation of the marine environment so as to maintain and improve its life-support and productive capacities"*; 2) *"develop and increase the potential of marine living resources to meet human nutritional needs, as well as social, economic, and development goals"*; and 3) *"promote the integrated management and sustainable development of coastal areas and the marine environment"*. This followed a growing awareness world-wide that the quality of coastal marine ecosystems, and consequently their capability for long-term sustainability of natural resource use, were being adversely impacted and reduced. This, in turn, has stimulated efforts by scientists and resource managers to shift their focus from the assessment and short-term management of individual ecosystem components to that of entire ecosystems on a long-term basis. For example, the Intergovernmental Oceanographic Commission (IOC) of UNESCO is now encouraging coastal nations to establish national programs for assessing and monitoring coastal ecosystems so as to enhance the ability of national and regional management organisations to develop and implement effective remedial programs for improving the quality of degraded ecosystems.

Coastal Ecosystem Stress

Since UNCED, concern has been expressed over the deteriorating condition of the world's coastal ecosystems that produce most of the global living marine resources. Within the near-shore areas and extending seaward around the margins of global land masses, coastal ecosystems are being subjected to increased stress from toxic effluents, habitat degradation, excessive nutrient loadings, harmful algal blooms, emergent diseases, fallout from aerosol contaminants, and episodic losses of living marine resources from pollution and over-exploitation. Coastal pollution, changes in biodiversity, the degraded states of fish stocks, and the loss of coastal habitat are limiting the achievement of the full economic potential of coastal ecosystems.

Present efforts to address these problems by local, regional, national, and international institutions responsible for resource stewardship have been less than successful. Informed decisions for ensuring the long-term development and sustainability of coastal marine resources can best be made when based on sound scientifically-derived options. But for most coastal ecosystems, existing environmental data pertinent to studies of perturbations to habitats and populations at the species, population, community, and ecosystem level are difficult to synthesise because of their spatially and temporally fragmented character, lack of comparability, and inaccessibility. To overcome these shortcomings, a more coherent and integrative assessment of the changing states of coastal ecosystems, from drainage basins to the adjacent marine ecosystems and directly linked to institutions responsible for the governance of the ecosystems, is needed (McGlade 2001).

Large Marine Ecosystem Concept

In recent years, there has been a change in approach to the management of the world's natural resources from that of compartmentalised efforts focusing on maximising short-term yields and economic gain towards ecosystem management and long-term sustainability. Franklin (1994) indicated that it represents a paradigm shift from individual species to ecosystems, from small spatial scales to multiple scales, from a short-term perspective to long-term perspective, from humans being independent of ecosystems to humans being an integral part of them, from management divorced from research to adaptive management, and from managing commodities to sustaining production potential for goods and services.

An essential component of an ecosystem management regime is the inclusion of a scientifically-based strategy to monitor and assess the changing states and health of the ecosystem by monitoring changes in key biological and environmental parameters. The assessment and monitoring activity serves as a component of a management system that includes: 1) regulatory, 2) institutional, and 3) decision-making actions for managing marine ecosystems.

An ecological framework that can serve as a basis for achieving the UNCED objectives is the large marine ecosystem (LME) concept (AAAS 1986; Sherman 1986; Alexander 1986). Large marine ecosystems are relatively large regions about 200,000 km^2 or larger, characterised by distinct bathymetry, hydrography, productivity, and trophically-dependent populations. They are regions of the ocean encompassing coastal areas from river basins and estuaries to the outer boundary of continental shelves and the seaward margins of coastal current systems. LMEs are increasingly being subjected to stress from growing exploitation of fish and other renewable resources, coastal zone damage, habitat losses, river basin runoff, dumping of urban wastes, and fallout from aerosol contaminants. The theory, measurement, and modeling relevant to monitoring the changing states of LMEs are imbedded in a series of reports on ecosystems with multiple steady states and on the pattern formation and spatial diffusion within ecosystems (Holling 1973, 1986, 1993; Pimm 1984; AAAS 1986, 1989, 1990, 1991, 1993; Beddington 1986; Levin 1993; Sherman 1994).

Based on these criteria, 49 distinct LMEs have been described around the margins of the Atlantic, Pacific, and Indian Oceans (Map 1). These 49 LMEs produce 95% of the annual global fisheries biomass yields. Sherman (1994) reported on the changing states of biomass yields and health for 29 of these LMEs.

The assessments of the changing states of LMEs are based on information obtained from five operational modules that link science-based information to socio-economic benefits for countries bordering on the LMEs (Sherman 1995). The five modules are focused on ecosystem 1) productivity, 2) fish and fisheries, 3) pollution and health, 4) socio-economic conditions, and 5) governance protocols. This approach is currently being implemented in a collaborative effort among the US National Oceanographic and Atmospheric Administration (NOAA), the UN Industrial Development Organization (UNIDO) and the six coastal countries in the Gulf of Guinea.

Productivity Module

Productivity can be related to the carrying capacity of an ecosystem for supporting fish resources (Pauly and Christensen 1995). The maximum global level of primary productivity for supporting the average annual world fisheries catch has reportedly been reached. Further large-scale unmanaged increases in marine fisheries yields are likely to be at marine food chain trophic levels below fish (Beddington 1995). In some LMEs, excessive nutrient loadings of coastal waters have been related to algal blooms that have been implicated in mass mortalities of living resources, emergence of pathogens, and explosive growth of non-indigenous species (Epstein 1993).

Parameters measured in the productivity module are zooplankton species composition, biomass, diversity and size; phytoplankton species composition biomass as chlorophyll-*a*; salinity; temperature; nutrients; hydrocarbons; light; and oxygen. Plankton can be measured by deploying Continuous Plankton Recorder (CPR) systems from commercial vessels of opportunity (Glover 1967; John *et al.* this volume). Advanced plankton recorders can be fitted with sensors to measure numerous other parameters such as temperature, salinity, chlorophyll, nitrate/nitrite, petroleum, hydrocarbons, light, bioluminescence, and primary productivity (Aiken 1981; Aiken and Bellan 1990; Williams and Aiken 1990; UNESCO 1992; Williams 1993) and provide the means to monitor changes in phytoplankton, zooplankton, primary productivity, species composition and dominance, and long-term changes in the physical and nutrient characteristics of the LME, as well as longer-term changes relating to the biofeedback of the plankton to the stress of environmental change (Colebrook 1986; Dickson *et al.* 1988; Colebrook *et al.* 1991; Williams 1993).

Fish and Fisheries Module

Excessive exploitation (Sissenwine and Cohen 1991), naturally-occurring environmental shifts in climate regime (Bakun 1993, 1995; Cole and McGlade 1998), and coastal pollution (Mee 1992; Bombace 1993) are three sources of variability in fisheries yield in most LMEs. Numerous examples exist (Kullenberg 1986; MacCall 1986; Sissenwine 1986; Crawford *et al.* 1989; Mee 1992; Alheit and Bernal 1993; Bakun 1993, 1995; Bombace 1993; Tang 1993) where one or other of these sources has been linked to changes in fisheries yield, some as the primary driving force and others as the secondary or tertiary causes.

The fish and fisheries module includes fishery-independent bottom trawl surveys and acoustic surveys for pelagic species to obtain time-series information on changes in biodiversity and abundance levels of the fish community. Standardised sampling procedures, using small calibrated trawlers, can provide important information on diverse changes in fish species. The fish catch provides biological samples for stock assessments, stomach analyses, age, growth, fecundity, and size comparisons (ICES 1991), data for clarifying and quantifying multispecies trophic relationships, and the collection of samples to monitor coastal pollution. Samples of trawl-caught fish can be used to monitor pathological conditions that may be associated with coastal pollution. The trawlers can also be used as platforms for obtaining water, sediment, and benthic samples for monitoring harmful algal blooms, virus vectors of disease, eutrophication, anoxia, and changes in benthic communities.

Pollution and Ecosystem Health Module

In several LMEs, pollution has been a principal driving force in changes of biomass yields. Assessing the changing status of pollution and health of the entire LME is scientifically challenging. Ecosystem health is a concept of wide interest for which a single precise scientific definition is problematical. Methods to assess the health of LMEs are being developed from modifications to a series of indicators and indices described by various investigators (Costanza 1992; Karr 1992; Norton and Ulanowicz 1992; Rapport 1992). The over-riding objective is to monitor changes in health from an ecosystem pespective as a measure of the overall performance of a complex system (Costanza 1992). The health para-digm is based on the multiple-state comparisons of ecosystem resilience and stability (Pimm 1984, 1999; Holling 1986; Costanza 1992; McGlade 2002) and is an evolving concept.

For an ecosystem to be healthy and sustainable (Costanza 1992), an ecosystem must maintain its metabolic activity level, its internal structure and organisation, and must be resistant to external stress over time and space scales relevant to the ecosystem. Five indices

ECOSYSTEM HEALTH INDICES

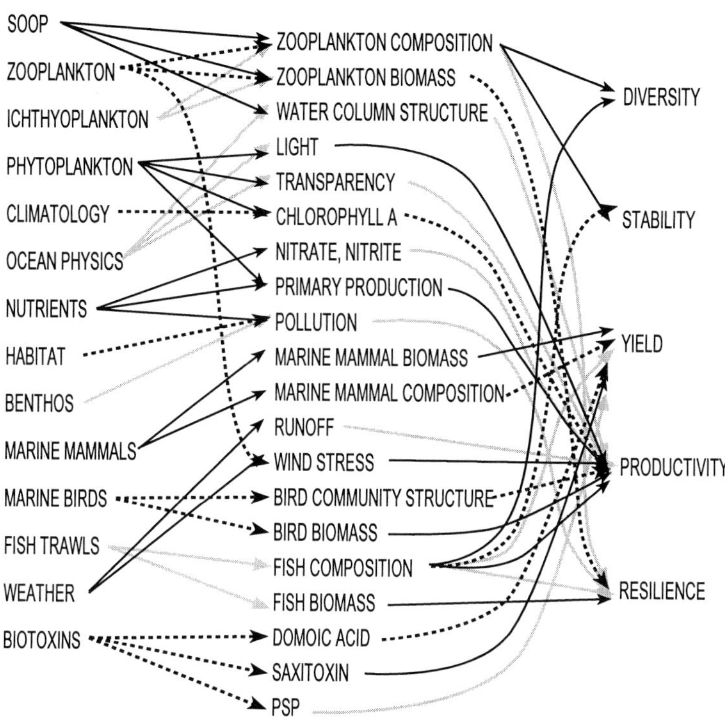

Figure 2-1 A schematic representation of the data bases and experimental parameters for indexing the changing states of large marine ecosystems. The data base (left column) represents time-series measurements of key ecosystem components (middle column) from the US Northeast Continental Shelf ecosystem. Indices (right column) will be based on changes compared with the ecosystem state in 1960.

considered as experimental measures of changing ecosystem states and health (NOAA 1993) are: 1) diversity, 2) stability, 3) yields, 4) productivity, and 5) resilience. The data from which to derive the experimental indices are obtained from time-series monitoring of key ecosystem parameters. An effort to validate the utility of these indices is currently under development by NOAA at the Northeast Fisheries Science Center. The ecosystem sampling strategy is focused on parameters relating to the resources at risk from over-exploitation, species protected by legislative authority (e.g., marine mammals), and other key biological and physical components at the lower end of the food chain (i.e., plankton, nutrients, hydrography).

The parameters of interest depicted in Figure 2-1 include zooplankton composition, zooplankton biomass, water column structure, photosynthetically active radiation, transparency, chlorophyll-*a*, NO$_2$, NO$_3$, primary production, pollution, marine mammal biomass, marine mammal composition, runoff, wind stress, seabird community structure, seabird counts, finfish composition, finfish biomass, domoic acid, saxitoxin, and paralytic shellfish poisoning. The experimental parameters selected incorporate the behavior of individuals, the resultant responses of populations and communities, as well as their interactions with the physical and chemical environment. The selected parameters provide a basis for comparing changing health status among ecosystems. The interrelationship between the data sets and the selected parameters are indicated by the arrows leading from column 1 to column 2.

The measured ecosystem components are shown in relation to ecosystem structure in a diagrammatic conceptualisation of patterns and activities within the LME at different levels of complexity (Likens 1992): the individual, the population, and the community (Figure 2-2). Each sphere represents an individual abiotic or biotic entity. Broad, double-headed arrows indicate feedback between entities and the energy matrix for the system, and thin arrows represent direct interactions between individual entities.

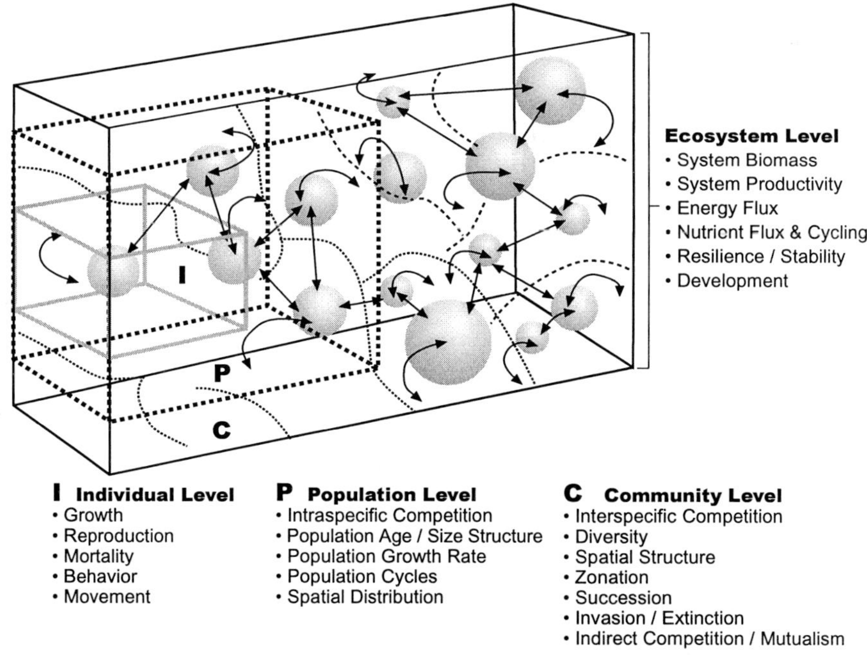

Figure 2-2 Diagrammatic conceptualisation of patterns and activities at different levels of ecosystem complexity. Each sphere represents an individual abiotic or biotic entity. Abiotic is defined as nonliving matter. Double-headed arrows either indicate feedback between entities and the energy matrix for the system, or represent direct interactions between individual entities. Much of ecology is devoted to studying interactions between biotic and abiotic entities with a focus on the effects of such interactions on individuals (I), populations (P), or communities (C) of organisms. Ecosystem ecology studies these interactions from the viewpoint of their effect on both the biotic and abiotic entities and within the context of the system. The boundaries of the system must be established to conduct quantitative studies of flux (from Likens 1992).

Fish, benthic invertebrates, and other biological indicator species are used in the pollution and ecosystem health module to measure pollution effects on the ecosystem including the bivalve monitoring strategy of "Mussel-Watch", the pathobiological examination of fish (Goldberg 1976; Farrington *et al.* 1983; ICES 1988; O'Connor and Ehler 1991), and the estuarine and nearshore monitoring of contaminants in the water column, substrate, and selected groups of organisms. An important component of the associated research to support the assessment is the definition of routes of exposure to toxic contaminants of selected finfish and shellfish and the assessment of exposure to toxic chemicals by several life history stages. The routes of bioaccumulation and trophic transfer of contaminants is assessed, and critical life history stages and selected food-chain organisms are examined for a variety of parameters that indicate exposure to, and effects of, contaminants. Contaminant-related effects measured include diseases, impaired reproductive capacity, and impaired growth. Many of these effects can be caused by direct exposure to contaminants, or by indirect effects such as those resulting from alterations in prey organisms.

The assessment of chemical contaminant exposure and effects on fisheries resources and food-chain organisms consists of a suite of parameters including biochemical responses that are clearly linked to contaminant exposure coupled with measurements of organ disease and reproductive status that have been used in previous studies to establish links between exposure and effects. The specific suite of parameters measured will cover the same general responses and thus allow for comparable assessment of the physiological status of each species sampled as it relates to chemical contaminant exposure and effects at the individual, species, and population levels. The implementation of protocols for assessing the frequency and effect of harmful algal blooms (Smayda 1991) and emergent diseases (Epstein 1993) are also included in the pollution module.

Socioeconomic Module

The socioeconomic module is characterised by its emphasis on practical applications of its scientific findings in managing the LME and on the explicit integration of economic analysis with the scientific research to assure that prospective management measures are cost-effective. Economists and policy analysts will need to work closely with ecologists and other scientists to identify and evaluate management options that are both scientifically credible and economically practical.

The economic and management research is closely integrated with the science throughout, and is designed intentionally to respond adaptively to enhanced scientific information. This component of the LME approach to marine resources management was developed by the late James Broadus, former Director of the Marine Policy Center, Woods Hole Oceanographic Institution, Woods Hole, MA, and contains six interrelated elements:

• Human forcing functions

The natural starting point is a generalised characterisation of the ways in which human activities affect the natural marine system and the expected "sensitivity" of these forcing functions to various types and levels of human activity. Population dynamics, coastal development, and land-use practices in the system's drainage basis are clear examples. Work integrating the efforts of natural and social scientists should concentrate further on resolving apparent effects (such as eutrophication-associated red tide events or changing fish population structures) that are confounded by cycles or complex dynamics in the natural system itself.

Progress is possible, too, in achieving better characterisations of the way in which human forcing is mediated by alternative management options. Emphasis should be on isolating and quantifying those forcing activities (e.g., sewage discharge, agricultural runoff, fishing effort) likely to be expressed most prominently in effects on the natural system.

- Feedback to environmental impacts and stresses

Collaborative effort should also be devoted to identifying and estimating the feedbacks of economic impacts into the human forcing function. Extensive coastal eutrophication, for example, associated with coastal development and runoff, might reduce the suitability of coastal areas for aquaculture production and increase exposure to red tide damage, thereby putting a premium on capture fisheries and increasing pressure on wild stocks. Similar feedbacks, both negative and positive, should be addressed and expressed in economic terms for all the major sectors.

- Assessing economic impacts

Another element is the estimation and possibly prediction of the economic impacts of unmanaged degradation in the natural system and, obversely, the expected benefits of management measures. Such an assessment is a form of standard benefit-cost analysis, but it requires scientific information to describe the effects of human forcing so as to quantify them in economic terms. Initial analysis should focus on the social and economic sectors likely to experience the largest impacts (i.e., fishing, aquaculture, public health, recreation, and tourism).

- Ecosystem service versus value of biodiversity

Special consideration should be given to improving knowledge of how the natural system generates economic values. Many valuable services provided by natural systems are not traded in markets or included in planning evaluations, so extra care must be made to assure that they are not sacrificed through ignorance. The services provided by coastal wetlands as nurseries for fisheries, natural pollution filters, and storm buffers are well-known examples that have particular relevance to coastal reclamation activities. Other examples are more subtle, including the importance of predator-prey relationships and the possibility of losing unrecognised "keystone" species in a valuable ecosystem. Experience suggests that growing economic values on aesthetic and recreational/tourism amenities may be expected in the LME setting as well. A variety of sources of economic value arising from the natural diversity of the LME should be identified and assessed in regard to existing uses and potential management innovations.

- Environmental economics (value setting)

Many of the elements described here comprise topics in the field of environmental economics. Specialists in that field attempt to estimate the economic values (both use and non-use) associated with environmental resources and to identify the conditions associated with their optimal management (to derive the greatest net benefits for society) (Milner-Gulland 1999). An important element is the collaboration between scholars from developing nations and those from the developed countries to transfer and adapt to the needs and techniques of environmental economics.

- Integrating assessments (decision-support network)

The ultimate objective is the integration of all the results achieved above, with scientific characterisations of the LME, into a comprehensive analytic framework (decision-support environment) that will permit integrated assessment of human practices, effects, and management options in the region. Such work is at the forefront of recent research on the human dimension of global environmental change as well as research on human interactions with natural marine systems.

Governance Module

The governance module is a critical element for achieving sustainable management in the coastal zone (McGlade 2001). Methods now evolving are based on case studies underway among ecosystems to be managed from a more holistic perspective in projects supported by the Global Environmental Facility (GEF), including the Yellow Sea Ecosystem, where the principal effort is underway by the People's Republic of China (Tang 1989) and South Korea (Zhang and Kim 1999), for the Gulf of Guinea LME by Côte d'Ivoire, Ghana, Nigeria, Benin, Cameroon, and Togo (UNIDO/UNDP/NOAA/UNEP 1997), and by the governments of South Africa, Namibia, and Angola to conduct joint assessments of the resources of the Benguela LME to guide the development of the shared resources of the ecosystem to ensure their long-term sustainability, particularly with regard to food security. The Great Barrier Reef Ecosystem is also being managed from an holistic ecosystem perspective (Bradbury and Mundy 1989; Kelleher 1993), along with the Northern Australian Continental Shelf Ecosystem (Sainsbury 1988), by the state and federal governments of Australia. The Antarctic Marine Ecosystem is being managed under the Commission for the Conservation of Antarctic Marine Living Resources (CCAMLR) and its 21-nation membership (Scully *et al.* 1986; Sherman and Ryan 1988). Movement towards ecosystem management is emerging for the North Sea (NSQSR 1993), Barents Sea (Eikland 1992), Black Sea (Hey and Mee 1993), and Baltic Sea (IOC 1998).

Post-UNCED Support

The Global Environment Facility (GEF), a unique international entity located within the World Bank in Washington, DC, was organised to implement the mandates of the 1992 UNCED by providing financial grants and low interest loans to developing nations and economies in transition. The GEF collaborates with the World Bank, the United Nations Development Programme (UNDP) and the United Nations Environment Programme (UNEP) to fund national, regional, and global development projects that benefit the world's environment by addressing issues related to climate change, biological diversity, and international waters. In international waters activities, the GEF addresses priority transboundary concerns consistent with the UNCED declaration. Various donor nations (Germany, France, Sweden, United Kingdom, Denmark, Norway, Canada, Japan, and the United States) have contributed a total of US$2 billion to the first round of GEF projects and about US$2.7 billion to the second round. GEF, in collaboration with other UN agencies, and with NOAA, IUCN (International Union for the Conservation of Nature and Natural Resources), and various other national and international organisations and agencies, and with the aid of many scientists from national marine resource agencies from a number of developed countries, is assisting developing nations in implementing coastal ecosystem assessment, monitoring, and management projects aimed at improving the prospects for the long-term development and sustainability of multi-sectoral marine resources.

Implementation of Regional LME Projects

Scientific and managerial projects are currently underway or in the planning stages in a number of large marine ecosystems under the auspices or support of either GEF, in co-operation with UN, regional, and national agencies or national governments or intergovernmental organisations. Such activities in some of the LMEs are reviewed below (IOC 1998).

Yellow Sea LME

A World Bank project for South Korea, including productivity, fisheries, and pollution modules, was implemented in March 1998. A large bilateral Korea-China project is in the planning stages with support from the GEF.

Gulf of Guinea LME

Some of the major milestones and developments include a positive mid-point review, a workshop in October 1997 to present and discuss results from the first two years, Togo joining Côte d'Ivoire, Ghana, Nigeria, Benin, and Cameroon in the project, the financial investment by petroleum-producing corporations towards environmental issues in the Niger Delta region, and a first Regional Symposium held in January 1998 in Abidjan, Côte d'Ivoire. The Gulf of Guinea LME provides resources currently representing annually an estimated US$3.8 billion and US$9 billion potentially to the economies of the adjoining countries.

The Gulf of Guinea LME project has been ongoing for two years and has documented a number of accomplishments including reaching consensus on methods for restoration of damaged mangroves and other habitats, initiation of community-based mangrove restoration projects, and agreement by several of the participating countries to initiate non-hazardous waste-stock exchange programmes with important gas, oil, mining, steel, agricultural, and food production enterprises to control pollution and apply new technologies for the profitable recycling of wastes as a principal means for protecting fisheries and other living resources.

This is the largest single project presently underway in Africa for increasing the socio-economic benefits of the natural productivity of a large marine ecosystem. In addition to GEF support, assistance from within Africa is being provided by the Organization of African Unity and the African Development Bank. A task force of about 350 specialists from government and non-governmental organisations, community-based organisations, the private sector, and academic institutions from the six participating countries is involved in the project.

Bay of Bengal LME

The coastal portion of the countries bordering this LME (Bangladesh, India, Indonesia, Malaysia, Maldives, Sri Lanka, and Thailand) has a combined population of about 400 million people. Problems requiring resolution include land- and sea-based pollution, terrestrial degradation, inadequate management of living marine resources, and natural disasters. The project's original objective, to improve the socio-economic conditions of small-scale fishing communities by the development and promotion of new and innovative technologies, has been expanded to include a greater emphasis on the development of community-based fisheries management policies and stakeholder participation. Objectives of the project's strategic action programme (SAP) include prioritising transboundary water-related environmental concerns and root causes of problems and threats, agreeing on actions and priorities required to resolve transboundary problems, identifying responsible parties, setting realistic targets, performing cost-benefit analyses, determining associated incremental costs, and generating potential investment, technical assistance, and capacity-building interventions.

Benguela Current LME

This LME, which encompasses one of the four eastern boundary upwelling systems in the world, has a high primary productivity level which supports an important reservoir of biodiversity and a large biomass of fish, crustaceans, sea birds, and marine mammals. The near-shore and shelf sediments hold rich deposits of diamonds, phosphorite, diatomite and oil and natural gas reserves. It is subject to considerable environmental variability which has severely impacted the ecosystem and led to a marked decline in fish abundance and availability. Increased fishing pressure coupled with widespread toxic algal blooms and a high pollution risk associated with ongoing seabed mining and petroleum exploration and production has stimulated the need for an integrated and co-ordinated approach to ecosystem management within the three bordering countries (South Africa, Namibia, and Angola). The project's proposal of enhancing national and regional efforts for sustainable integrated management, which has been approved by GEF with full funding from the end of 1998, will be accomplished by establishing a regional cooperative mechanism, undertaking a review of existing knowledge of the status and threats to the LME, and developing an SAP to address the threats and the gaps in current knowledge.

Caribbean Sea LME

This LME has been seriously damaged from the loss of critical coral reef and mangrove habitat, diminishing food security from over-exploitation of coastal fisheries, and growing eutrophication and degradation of coastal waters from excessive nutrient loading and chronic oil spills. Project objectives include enhancing national and regional efforts toward collaborative assessment, monitoring, and management by the Bahamas, Barbados, Belize, Brazil, Colombia, Costa Rica, Cuba, Jamaica, Mexico, Panama, St. Lucia, Trinidad and Tobago, and Venezuela. Initial efforts will include the establishment of a mechanism for regional co-operation, review of the existing knowledge of the threat to the system, and development of an SAP to redress the damage to the ecosystem and overcome the gaps in knowledge. The project will also focus on strengthening co-ordination for the assessment and monitoring of ecosystem changes, the development of institutional capacity, direct support of priority activities identified in the SAP, and support for directing national and international activities toward long-term sustainability and economic development of marine resources.

Humboldt Current LME

Work is underway by Chile, Peru, and Ecuador in preparation for the submission of project proposals. The first steps will be the establishment of a mechanism for regional co-operation, a review of the existing knowledge of the status and threats to the LME, and the development of a SAP. This is the most productive LME in the world and supports an important reservoir of biodiversity and a large biomass of phytoplankton, zooplankton, fish, sea birds, and marine mammals. Global markets for fish protein are very dependent on the annual yield from this LME. The consequences of El Niño conditions and the resulting frequency, extent, and intensity of coastal upwelling of the LME have a major impact on the fishing, agriculture, and tourism sectors, and hence the overall economies, of the three coastal states.

The principal issues associated with managing this LME include: habitat loss and pollution of the fragile and relatively pristine nature of the coast of the Humboldt Current region due to uncontrolled tourism development and ongoing expansion of urban areas; serious degradation of coastal areas adjacent to urban centers as a result of pollution, habitat loss, and extreme fluctuation in marine and coastal natural resources; significant and widespread increase in the frequency of marked environmental changes in the ecosystem induced by El Niño and resulting in very large losses of biomass of fish, sea birds, and marine mammals; and an apparent opportunity for important climate change assessment, monitoring, and prediction since the Humboldt Current is both a source and sink of carbon dioxide and a known predictor of climatic variations in the region.

Baltic Sea LME

Discussion has taken place among various Baltic countries, and the GEF has indicated an interest in supporting a joint ICES-HELCOM initiative for the Baltic Sea LME. ICES is convening a coordination meeting to submit a project proposal from the Baltic countries for submission to GEF. The Baltic Sea is threatened by pollution and habitat loss from extensive economic development in the coastal zones of the nine bordering countries (Poland, Estonia, Latvia, Lithuania, Russia, Finland, Sweden, Germany, and Denmark) and by unsustainable exploitation of natural resources including both fisheries and significant hydrocarbon reserves. The project objective is to prepare an SAP that will complement previous accomplishments of HELCOM.

Canary Current LME

A UNEP proposal to GEF has been funded to support a transboundary diagnostic analysis to be conducted jointly by NOAA, UNEP, and IUCN. The coastal zones of the seven countries bordering this LME (Guinea, Guinea Bissau, Gambia, Cape Verde, Mauritania, Morocco, and Senegal) have been major areas for industrial development, with the consequent migration of workers from rural inland areas to coastal areas and the associated overloading of the coastal capacity to provide infrastructure and services.

Somali Current LME

Somalia, Kenya, and Tanzania have applied for a grant to address the reduction of stress and degradation of the coastal habitats of the countries bordering the LME, and the provision of initial assessments of the size and extent of renewable resources of the LME in relation to the potential for systematic economic expansion of the fisheries to reduce malnutrition in significant segments of the human populations of the coastal states of the region.

South China Sea LME

Countries involved include Sumatra, Malaysia, Cambodia, Thailand, Vietnam, China, Philippines, and Borneo. Batangas Bay in the Philippines was developed as a demonstration project in Integrated Coastal Management (ICM). The approach provides a framework for the management of the coast of the Batangas Bay region of the Province of Batangas in the Philippines including integrated waste management, water pollution abatement, conservation of stressed mangrove and coral reef areas, coastal tourism development, and improvements in municipal fisheries.

References

AAAS (American Association for the Advancement of Science) 1986 K. Sherman and L.M. Alexander (eds) *Variability and management of large marine ecosystems.* AAAS Selected Symposium 99. Westview Press, Inc. Boulder, CO.

AAAS (American Association for the Advancement of Science) 1989 K. Sherman and L.M. Alexander (eds) *Biomass yields and geography of large marine ecosystems.* AAAS Selected Symposium 111. Westview Press, Inc. Boulder, CO.

AAAS (American Association for the Advancement of Science) 1990 K. Sherman and L.M. Alexander (eds) *Large marine ecosystems: patterns, processes and yields.* AAAS Press, Washington, DC.

AAAS (American Association for the Advancement of Science) 1991 K. Sherman and L.M. Alexander (eds) *Food chains, yields, models, and management of large marine ecosystems.* Westview Press, Inc. Boulder, CO.

AAAS (American Association for the Advancement of Science) 1993 K. Sherman, L.M. Alexander, and B.D. Gold (eds) *Large marine ecosystems: stress, mitigation, and sustainability.* AAAS Press, Washington, DC.

Aiken, J. 1981 The undulating oceanographic recorder mark 2. J. Plank. Res. 3: 551-560.

Aiken, J. and I. Bellan 1990 Optical oceanography: an assessment of a towed method. pp39-57. *In*: P.J. Herring, A.K. Campbell, M. Whitfield, and L. Maddock (eds). *Light and life in the sea.* Cambridge Univ. Press, London.

Alexander, L.M. 1986 Large marine ecosystems as regional phenomena. pp239-240. *In*: K. Sherman and L.M. Alexander (eds). *Variability and management of large marine ecosystems.* AAAS Selected Symposium 99, Westview Press, Boulder, CO.

Alheit, J. and P. Bernal 1993 Effects of physical and biological changes on the biomass yield of the Humboldt Current ecosystem. pp53-68. *In*: K. Sherman, L.M. Alexander, and B.D. Gold (eds). *Large marine ecosystems: stress, mitigation, and sustainability.* AAAS Press, Washington, DC.

Bakun, A. 1993 The California Current, Benguela Current and Southwestern Atlantic Shelf ecosystems: a comparative approach to identifying factors regulating biomass yields. pp199-214. *In*: K. Sherman, L.M. Alexander, and B.D. Gold (eds). *Large marine ecosystems: stress, mitigation and sustainability.* AAAS Press, Washington, DC.

Bakun, A. 1995 A dynamical scenario for simultaneous 'regime-scale' marine population shifts in widely separated LMEs of the Pacific. pp47-72. *In*: Q. Tang and K. Sherman (eds.). *The large marine ecosystems of the Pacific Rim: a report of a Symposium held in Qingdao*, People's Republic of China, 8-11 October 1994. IUCN Marine Conservation and Development. Gland, Switzerland.

Beddington, J.R. 1986 Shifts in resource populations in large marine ecosystems. pp9-18. *In*: K. Sherman and L.M. Alexander (eds). *Variability and management of large marine ecosystems*. AAAS Selected Symposium 99. Westview Press, Inc. Boulder, CO.

Beddington, J.R. 1995 The primary requirements. Nature 374: 213-214.

Bombace, G. 1993. Ecological and fishing features of the Adriatic Sea. pp119-136. *In*: K. Sherman, L.M. Alexander, and B.D. Gold (eds). *Large marine ecosystems: stress, mitigation and sustainability*. AAAS Press, Washington, DC.

Bradbury, R.H. and C.N. Mundy 1989 Large-scale shifts in biomass of the Great Barrier Reef ecosystem. pp143-167. *In*: K. Sherman and L.M. Alexander (eds). *Biomass yields and geography of large marine ecosystems*. AAAS Selected Symposium 111. Westview Press, Inc. Boulder, CO.

Cole, J. and J. McGlade 1998 Clupeoid population variability, the environment and satellite imagery in coastal upwelling systems. Reviews in Fish Biology and Fisheries 8: 445-471.

Colebrook, J.M. 1986 Environmental influences on long-term variability in marine plankton. Hydrobiologia 142: 309-325.

Colebrook, J.M., A.J. Warner, C.A. Proctor, H.G. Hunt, P. Pritchard, A.W.G. John, D. Joyce, and R. Barnard 1991 60 years of the Continuous Plankton Recorder survey: a celebration. The Sir Alister Hardy Foundation for Ocean Science, Plymouth, England.

Costanza, R. 1992 Toward an operational definition of ecosystem health. pp239-256. *In*: R. Costanza, B.G. Norton, and B.D. Haskell (eds). *Ecosystem health: new goals for environmental management*. Island Press, Washington, DC.

Crawford, R.J.M., L.V. Shannon, and P.A. Shelton 1989 Characteristics and management of the Benguela as a large marine ecosystem. pp169-219. *In*: K. Sherman and L.M. Alexander (eds). *Biomass yields and geography of large marine ecosystems*. AAAS Selected Symposium 111. Westview Press, Inc. Boulder, CO.

Dickson, R.R., P.M. Kelly, J.M. Colebrook, W.S. Wooster, and D.H. Cushing 1988 North winds and production in the eastern North Atlantic. J. Plank. Res. 10: 151-169.

Eikland, P.O. 1992 Multispecies management of the Barents Sea Large Marine Ecosystem: a framework for discussing future challenges. The Fridtjof Nansen Institute, Polhogda, Postboks 326, Fridtjof Nansensvei 17, N-1324 Lysaker, Norway.

Epstein, P.R. 1993 Algal blooms and public health. World Resource Review 5(2): 190-206.

Farrington, J.W., E.D. Goldberg, R.W. Risebrough, J.H. Martin, and V.T. Bowen 1983 U.S. 'Musselwatch' 1976-1978: an overview of the trace metal, DDE, PCB, hydrocarbon and artificial radionuclide data. Environ. Sci. Tech. 17: 490-496.

Franklin, J.F. 1994 Ecosystem management: an overview. *In*: M.S. Boyce and A.W. Haney (eds). *Ecosystem management: concepts and methods*. Yale Univ. Press, New Haven, CT.

Glover, R.S. 1967 The continuous plankton recorder survey of the North Atlantic. Symp. Zool. Soc. Lond. 19: 189-210.

Goldberg, E.D. 1976 The health of the oceans. UNESCO Press, Paris.

Hey, E. and L.D. Mee 1993. Black Sea. The ministerial declaration: an important step. Environ. Pollut. Law 2315: 215-217, 235-236.

Holling, C.S. 1973 *Resilience and stability of ecological systems*. Institute of Resource

Ecology, Univ. British Columbia, Vancouver, BC, Canada.

Holling, C.S. 1986 The resilience of terrestrial ecosystems: local surprise and global change. pp292-317. *In*: W.C. Clark and R.E. Munn (eds). *Sustainable development of the biosphere.* Cambridge Univ. Press, London.

Holling, C.S. 1993 Investing in research for sustainability. Ecol. Appl. 3: 552-555.

ICES (International Council for the Exploration of the Sea) 1988 Results of the 1985 baseline study of contaminants in fish and shellfish. ICES Coop. Res. Rept. 151.

ICES (International Council for the Exploration of the Sea) 1991 Report of the Multispecies Assessment Working Group. ICES C.M. 1991/Assess:7.

IOC (Intergovernmental Oceanographic Commission) 1998 Meeting report of the IOC Consultative Meeting on Large Marine Ecosystems. 15-16 March 1998, Paris. Draft IOC rept, 21 p.

John, A.W.G., P.C. Reid, S.D. Batten and E.R. Anang (this volume, chapter 11) Monitoring levels of "Phytoplankton Colour" in the Gulf of Guinea using ships of opportunity.

Karr, J. 1992 Ecological integrity; protecting earth's life support systems. pp223-238. *In*: R. Costanza, B.G. Norton, and B.D. Haskell (eds). *Ecosystem health: new goals for environmental management.* Island Press, Washington, DC.

Kelleher, G. 1993 Sustainable development of the Great Barrier Reef as a large marine ecosystem. pp272-279. *In*: K. Sherman, L.M. Alexander, and B.D. Gold (eds.). *Large marine ecosystems: stress, mitigation, and sustainability.* AAAS Press, Washington, DC.

Kullenberg, G. 1986 Long-term changes in the Baltic ecosystem. pp19-32. *In*: K. Sherman and L.M. Alexander (eds). *Variability and management of large marine ecosystems.* AAAS Selected Symposium 99. Westview Press, Inc., Boulder, CO.

Levin, S.A. 1993 Approaches to forecasting biomass yields in large marine ecosystem. pp36-39. *In*: K. Sherman, L.M. Alexander, and B.D. Gold (eds). *Large marine ecosystems: stress, mitigation, and sustainability.* AAAS Press, Washington, DC.

Likens, G.E. 1992 The ecosystem approach: its use and abuse. *In* O. Kinne (ed.). *Excellence in ecology*, Vol. 3. Ecology Institute, Oldendorf/Luhe, Germany.

MacCall, A.D. 1986 Changes in the biomass of the California Current system. pp 33-54. *In*: K. Sherman and L.M. Alexander (eds). *Variability and management of large marine ecosystems.* AAAS Selected Symposium 99. Westview Press, Inc. Boulder, CO.

McGlade, J.M. 2001 Governance and sustainable fisheries. pp307-326. *In*: B.Bodungen and K. Turner. (eds). *Science & Integrated Coastal Zone Management*, Dahlem University Press, Berlin.

McGlade, J.M. 2002 Governance of transboundary pollution in the Danube River. Aquatic Ecosystem Health and Management (in press).

Mee, L. 1992 The Black Sea in crisis: a need for concerted international action. Ambio 21(4): 1278-1286.

Milner-Gulland, E.-J. 1999 Ecological economics. pp282 – 308. *In*: J.M. McGlade (ed.) *Advanced Ecological Theory.* Blackwell Science, Oxford

NOAA (National Oceanic and Atmospheric Administration) 1993 Emerging theoretical basis for monitoring the changing states (health) of large marine ecosystems: summary report of two workshops: 23 April 1992, National Marine Fisheries Service, Narragansett, RI and 11-12 July 1992, Cornell University, Ithaca, NY. U.S. Dept. Comm., NOAA Tech. Memo. NMFS-F/NEC-100. Woods Hole, MA.

Norton, B.G. and R.E. Ulanowicz 1992 Scale and biodiversity policy: a hierarchical approach. Ambio 21(3): 244-249.

NSQSR (North Sea Quality Status Report) 1993 Oslo and Paris Commissions, London. Olsen and Olsen, Fredensborg, Denmark. 132 p.

O'Connor, T.P. and C.N. Ehler 1991 Results from the NOAA National Status and Trends Program on distribution and effects of chemical contamination in the coastal and estuarine United States. Environ. Monitor. Assess. 17: 33-49.

Pauly, D. and V. Christensen 1995 Primary production required to sustain global fisheries. Nature 374: 255-257.

Pimm, S.L. 1984 The complexity and stability of ecosystems. Nature 307: 321-326.

Pimm, S.L. 1999 The dynamics of the flows of matter and energy pp172–193. *In*: J. McGlade (ed.). *Advanced Ecological Theory*. Blackwell Science, Oxford

Rapport, D.J. 1992 What is clinical ecology? pp223-228. *In*: R. Costanza, B.G. Norton, and B.D. Haskell (eds). Ecosystem health: new goals for environmental management. Island Press, Washington, DC.

Sainsbury, K.J. 1988 The ecological basis of multispecies fisheries, and management of a demersal fishery in tropical Australia. pp349-382. *In*: J.A. Gulland (ed.). *Fish population dynamics*, 2nd edition. John Wiley and Sons, New York.

Scully, R.T., W.Y. Brown, and B.S. Manheim 1986 The Convention for the Conservation of Antarctic Marine Living Resources: a model for large marine ecosystem management. pp281-286. *In*: K. Sherman and L.M. Alexander (eds). *Variability and management of large marine ecosystems*. AAAS Selected Symposium 99. Westview Press, Inc., Boulder CO.

Sherman, K. 1986 Introduction to Parts One and Two: large marine ecosystems as tractable entities for measurement and management. pp3-7. *In*: K. Sherman and L.M. Alexander (eds). *Variability and management of large marine ecosystems*. AAAS Selected Symposium 99, Westview Press, Boulder, CO.

Sherman, K. 1994 Sustainability, biomass yields, and health of coastal ecosystems: an ecological perspective. Mar. Ecol. Prog. Ser. 112: 277-301.

Sherman, K. 1995 Achieving regional co-operation in the management of marine ecosystems: the use of the large marine ecosystem approach. Ocean Coast. Mgmt, 29 (1-3): 165-185.

Sherman, K. and A.F. Ryan 1988 Antarctic marine living resources. Oceanus 31(2): 59-63.

Sissenwine, M.P. 1986 Perturbation of a predator-controlled continental shelf ecosystem. pp55-85. *In*: K. Sherman and L.M. Alexander (eds). *Variability and management of large marine ecosystems*. AAAS Selected Symposium 99. Westview Press, Inc., Boulder CO.

Sissenwine, M.P. and E.B. Cohen 1991 Resource productivity and fisheries management of the northeast shelf ecosystem. pp1077-1123. *In*: K. Sherman, L.M. Alexander, and B.D. Gold (eds.). *Food chains, yields, models and management of large marine ecosystems*. Westview Press, Inc., Boulder, CO.

Smayda, T. 1991 Global epidemic of noxious phytoplankton blooms and food chain consequences in large ecosystems. pp275-308. *In*: K. Sherman, L.M. Alexander, and B.D. Gold (eds). *Food chains, yields, models and management of large marine ecosystems*. Westview Press, Inc., Boulder, CO.

Tang, Q. 1989 Changes in the biomass of the Yellow Sea ecosystems. pp7-35. *In*: K.

Sherman and L.M. Alexander (eds). *Biomass yields and geography of large marine ecosystems*. AAAS Selected Symposium 111. Westview Press, Inc. Boulder, CO.

Tang, Q. 1993 Effects of long-term physical and biological perturbations on the contemporary biomass yields of the Yellow Sea ecosystem. pp79-93. *In*: K. Sherman, L.M. Alexander, and B.D. Gold (eds.). *Large marine ecosystems: stress, mitigation, and sustainability*. AAAS Press, Washington, DC.

UNESCO (United Nations Educational, Scientific, and Cultural Organization) 1992 Monitoring the health of the oceans: defining the role of the Continuous Plankton Recorder in global ecosystems studies. The Intergovernmental Oceanographic Commission and the Sir Alister Hardy Foundation for Ocean Science. IOC/INF-869, SC-92/WS-8.

UNIDO/UNDP/NOAA/UNEP (United Nations Industrial Development Organization, United Nations Development Programme, National Oceanic and Atmospheric Administration, United Nations Environment Programme) 1997 Gulf of Guinea Large Marine Ecosystem Project, Second Meeting of the Steering Committee, Cotonou, Bénin, 11-12 March 1997. Report GEF/GOG-LME/Sc-2.

Williams, R. 1993 Evaluation of new techniques for monitoring and assessing the health of large marine ecosystems. pp257-272. *In*: D. Rapport (ed.). *NATO Advanced Research Workshop Evaluating and Monitoring the Health of Large-scale Ecosystems*. Springer-Verlag, Berlin.

Williams, R. and J. Aiken 1990 Optical measurements from underwater towed vehicles deployed from ships-of-opportunity in the North Sea. pp186-194. *In*: H.O. Nielsen (ed.). *Environment and pollution measurement, sensor and systems*. Proc. SPIE 1269.

Zhang, C.I. and S. Kim 1999 Status and perspectives of living marine resources of the Yellow Sea ecosystem in Korean waters. pp163-178. *In* K. Sherman and Q. Tang (eds). *The large marine ecosystems (LMEs) of the Pacific Rim. Assessment, sustainability, and management*. Blackwell Science, Inc. Malden, Massachusetts USA.

3

The Gulf of Guinea Large Marine Ecosystem Project: Turning Challenges into Achievements

C. Ibe and K. Sherman

Abstract

Between 1995 and 1999, six countries in the Gulf of Guinea, a subset of the Guinea Current Large Marine Ecosystem which covers 16 countries between Guinea Bissau and Angola, co-operated in a Global Environment Facility funded pilot project. The project focussed on reversing the degradation of the coastal and marine environment and ensuring long term sustainable utilisation of the region's ample but shared living resources. It was based on an ecosystem approach to management, implicit in the innovative Large Marine Ecosystem programmes described by Sherman and Anderson (this volume). Besides those actions focussed on reducing pollution from land based activities especially from industries, achievements included a sustained survey of the productivity of the ecosystem using ships of opportunity, a successful joint region-wide fish trawl survey, community based mangrove restoration efforts and the formulation of National Integrated Coastal Areas Management Plans based on common regional policies and strategies. The capacity to intervene meaningfully in the protection of the environment and more so in taking preventive measures was enhanced, and definitive steps were taken towards the institutionalisation of GIS based decision making support systems. A Ministerial level "Accra Declaration" was adopted in 1998 by Environment Ministers from the six countries embodying political, intellectual and material commitments to environmentally sustainable development of the Guinea Current Large Marine Ecosystem. The ten additional countries implicated have since given Ministerial level endorsements to the provisions of the Accra Declaration.

Introduction

The concerns shared by countries in the Gulf of Guinea about the deterioration of their marine environment, coupled with experiences gained from their participation in existing regional and global conventions and protocols in environmental protection, make them eager and ready to move forward collectively and boldly with new, sound initiatives for the protection of their shared waters and natural resources. The pilot project, "Water Pollution Control and Biodiversity Conservation in the Gulf of Guinea Large Marine Ecosystem (LME)", funded by the Global Environment Facility, is one such initiative being undertaken by Benin, Cameroon, Côte d'Ivoire, Ghana Nigeria and Togo, and which are making large in-kind and cash contributions to the project. It is anchored in the concept of Large Marine Ecosystems as geographic units for improving the assessment and management of marine resources (Sherman 1993; 1994, 1996; Ibe and Brown 1997; Ibe and Csizer 1998; Scheren and Ibe this

27

	Benin	Cameroon	Côte d'Ivoire	Ghana	Nigeria	Togo
Total population (Million)	5,8	14,1	14,7	18,0	117.9	4.3
% Urban population out of total population	40	46	51	37	41	32
% Urban population in large coastal cities	100	54	84	72	20	100
Population growth (%)	2,9	2,9	2,9	2,7	2.9	3.0
Major coastal cities	Cotonou Porto Novo	Douala Limbé	Abidjan Sassandra San Pedro	Accra Takoradi Cape Coast	Lagos Warri Sapele Port harcourt Calabar	Lomé Aného
Major ports	Cotonou	Douala Limbé Kribi	Abidjan San Pedro	Tema Takoradi	Lagos Warri Port Harcourt Calabar	Lomé
Major Lagoon systems	Nokoué Porto-Novo	No major lagoons	Ebrié Aby-Tendo-Ehy Grand Lahou	Keta Sakumo Songaw Korle	Lagos Lekki	Togo
Length of coastline (km)	111	402	515	539	853	50
Marine area (km^2)	7900	4500	30500	63600	61500	600
Area occupied by mangroves (km^2)	30	4860	640	630	12200	10
Surface area continental shelf (km^2)	2800	12900	10200	23700	11470	1306
Width of the continental shelf (km)	30	30 - 50	20 - 35	20 - 90	15 – 85	21 – 32
Coastal erosion rate (m/y)	6,7 (30-year average)	-	1,5	3	15-30	20
Tidal range (m)	1,2	0,5 - 2,7 (location dependent)	1,2	1,3	1,5	1,5
Currents (m/s)	0,5 – 1,5	0,5 – 3	0,5 – 1,5	0,5 – 1,5	0,5 – 1,5	0,5 – 1,5

Table 3-1 Physical and demographic characteristics of the six countries in the Gulf of Guinea Large Marine Ecosystem project.

volume ; Sherman and Anderson this volume). Table 3-1 provides an overview of the physical and demographic characteristics of the Gulf of Guinea.

The long-term objective of the Global Environment Facility project is to restore and sustain the health of the Gulf of Guinea Large Marine Ecosystem and its living resources, particularly with regard to biological diversity and the control of water pollution. To achieve this overall goal, the following strategic objectives were established:

- Strengthening regional institutional capacities to prevent and remedy pollution of the Gulf of Guinea LME and associated degradation of critical habitats;
- Developing an integrated information management and decision making system for ecosystem management;
- Establishing a comprehensive programme for monitoring and assessing the living marine resources, health, and productivity of the Gulf of Guinea LME;
- Preventing and controlling land based sources of industrial and urban pollution;
- Developing national and regional strategies and policies for the long-term management and protection of the Gulf of Guinea LME.

Achieving the project objectives has involved networks of national environmental protection agencies/departments, public health administrations, sewage work authorities, industries, and universities/research institutions in the participating countries. Non-governmental organisations (NGOs) and community based organisations (CBOs), have also been very active particularly in areas relating to public awareness and environmental education aspects. In order to provide the necessary focus, National Focal Point Agencies and National Focal Point Institutions were designated.

At the international level, the project has the UN Development Programme (UNDP) as the Implementing Agency, UN Industrial Development Organization (UNIDO) as the Executing Agency and UN Environment Programme (UNEP) as the Co-operating Organization. The United States Department of Commerce through its National Oceanic and Atmospheric Administration (NOAA) is providing technical support particularly in capacity building initiatives in addition to in kind contribution to the funding of the project. Other United Nations and non-United Nations Agencies such as the Intergovernmental Oceanographic Commission (IOC) of UNESCO, the International Maritime Organization (IMO), the Food and Agriculture Organisation (FAO) and the World Conservation Union (IUCN), have provided guidance at specific stages in project implementation. Table 3-2 depicts international collaboration under the project.

Based on a recognition that pollutants and living resources in the coastal and marine environment respect no political boundaries and few geographical ones, the project has put considerable emphasis on 'uniting' neighbouring countries which share international waters and natural resources in the Gulf of Guinea. Actions include joint identification of major trans-border environmental and living resources management issues and problems, and adoption of common regional approaches, in terms of strategies and policies for addressing these priorities in the national planning process at all levels of administration, including local governments.

This paper presents the highlights of project implementation since 1995. For brevity and clarity, this is done along the lines of the strategic objectives established, and the modular approach outlined in Sherman and Anderson (this volume).

Strengthening Regional Institutional Capacities

The project has established intra- and international networks of scientific institutions and non-governmental organisations, with a total of more than 500 scientists, policy makers and other participants (making it the African continent's single largest network for marine and coastal area management), to undertake studies on ecosystem degradation, to assess living resources availability and biodiversity, and to measure socio-economic impacts of actions and non-actions. The capacity of the networks has been reinforced through the supply of appropriate equipment and by a series of group training workshops aimed at standardising methodological approaches around five project modules.

More than 40 region-wide workshops, involving a total of nearly 900 participants, have been held on issues varying from pollution monitoring, ecosystem productivity studies, natural resources management and planning, development of institutional capacities (including administrative and legal structures) and data and information management and exchange. Intercalibration exercises with the International Atomic Energy Agency (IAEA) have been promoted as a further means of ensuring comparability of results from different participating laboratories and countries. On a national level, numerous workshops and

Organisation	Areas of collaboration
1. UN Environment Programme	LBS; Effluent Standards; CAM, etc. (GRID, GFMS, etc.)
2. National Oceanographic and Atmospheric Administration	LME Monitoring/Assessment (Fish Trawl; Productivity, etc.)
3. UN Development Programme	ICAM; Training; Round Table (Investments) Sustainable Development Network, etc.
4. International Oceanographic) Commission (UNESCO)	GIS; Marine Debris; ICAM; Pollution
5. Food and Agriculture Organization/CECAF	Fisheries; Mangrove ICAM; Pollution; Data Management (FAO/GIS-OF/WA)
6. International Maritime Organization	Oil Spill Contingency Planning; ESI Mapping; Waste Management; Marine Debris
7. IOI	ICAM and EEZ Management
8. World Conservation Union (IUCN)/WWF	Mangrove; Fisheries; Biodiversity, etc.; Natural Resource Management.
9. Institut de recherche pour le développement (IRD)	Fisheries; Oceanographic cruises, etc.
10. UN Education, Scientific and Cultural Organisation	ICAM; Environmental Education, etc.
11. European Union Biodiversity project	Regional Marine Data Base Project (BDRM); Gulf of Guinea
12. World Health Organization	Pollution; Public Heath
13. International Atomic Energy Agency	Inter-calibration and Inter-comparison; Reference Materials and Standards
14. Marine Institutions	(in Europe, Americas, Asia, etc.) Routine collaboration.
15. African Development Bank	Investment Linkages and Mechanisms
16. World Bank	Investment Linkages and Mechanisms
17. Office of African Unity	Joint Programming

Table 3-2 International collaborations.

training activities have been organised under the project, while individuals benefited from fellowships and study tours abroad.

The project's emerging results have provided the technical underpinnings for management decisions at regional and national levels. For example, under the Productivity module of the project (with the Sir Alister Hardy Foundation for Ocean Science, UK), the monthly region-wide, continuous plankton recorder tows are providing indications on the health and carrying capacity of the Gulf of Guinea Large Marine Ecosystem (John *et al.* this volume). The information from the collection of plankton, a group of organisms which represent the base of the marine food chain, enables projections to be made on future food security for the burgeoning populations of the region. The fish trawl surveys (with IRD and FAO) are providing insights into exploitable yields and into undesirable shifts in biological diversity. Capacities for anticipatory actions such as oil spill contingency planning (with IMO), environmental impact assessment (with IOC) and environmental education (with UNESCO) have also been strengthened. At the same time, countries have been encouraged to ratify and adhere to pertinent international treaties, conventions and protocols.

Development of an Integrated Information Management and Decision Making Support System

In developing countries in particular, effective management at all levels and for all purposes is often vitiated by a lack of data and information upon which decisions and management choices can be based. An important target of the project has been to create a basic data and information management system in the participating countries, for the collection and systematisation of data and information in user-friendly formats, such as a geographic information system (GIS).

State of the Marine Environment Country Reports have been produced and issued as 'initial assessments', encompassing published and unpublished data and including policy options and past interventions. These assessments constitute the preliminary 'inputs', with retroactive georeferencing where necessary, for national and regional data and information banks. Standard formats have been adopted and available equipment has been upgraded for receiving and systematizing data and information from ongoing and future activities. Thematic GIS maps have been produced both at the regional and national levels. Effective e-mail networks with Internet connections have been set up for networking and data and information exchange.

These mechanisms provide decision-makers with readily available tools, literally 'at the touch of a button', that can contribute to the effective planning and management of the coastal environment and associated natural resources. For example, the regional mangrove distribution maps prepared under the project - when combined with data from field studies of mangrove ecosystems - have given a reliable picture of the extent of anthropogenic impact (principally from overcutting and/or pollution). They have provided a basis for preliminary decisions in project countries on appropriate reforestation sites for this favoured 'spawning and nursery ground' of many living resources that occur within the coastal waters of the Gulf of Guinea LME.

Monitoring and Assessment of the Health and Productivity of the Gulf of Guinea Large Marine Ecosystem

The aim of the Large Marine Ecosystem monitoring and assessment program is to address management needs which are often regional in scope but for which responsibility is divided among a wide array of jurisdictions. In this respect, the programme has provided crucial data and information that enhance the ability of national and regional authorities to develop and implement effective management and remedial programs for the coastal and marine environment.

A comprehensive program for monitoring and assessment of the living marine resources, the health, and productivity of the Large Marine Ecosystem was carried out, consisting of six main components:

- Living marine resources survey
- Plankton survey
- Mangrove survey
- Pollution monitoring in coastal lagoons
- Pollution monitoring of nearshore waters, sediments and fish
- Nutrients survey

National and regional experts were designated for each of the six components of the monitoring and assessment module, to execute the jointly agreed monitoring programmes at the national and regional level. The capacity of national institutes and experts was reinforced through the supply of appropriate equipment and by a series of workshops aimed at standardising methodological approaches in the afore mentioned components. Activity groups on specific topics (i.e. pollution monitoring, nutrients, mangroves) were convened regularly to discuss the progress made and problems encountered, and to undertake joint assessments. Emphasis was planed on inter-calibration and inter-comparison exercises (with assistance of IAEA, Monaco) to ensure that results obtained were intercomparable through meeting established Quality Assurance/Quality control Criteria (Topping 1992; Ibe and Kullemberg 1995)

One sub-regional fish trawl (western Gulf of Guinea), one regional fish trawl and marine pollution surveys were carried out, in August 1996 and in February/March, 1999 respectively, bringing together on board scientists from the six project countries for the first time since the Guinean Trawling Survey (GTS) in 1963/64. The results of the surveys provide an insight in the distribution and evolution of fish resources throughout the Gulf of Guinea continental shelf, and particularly the exploitable yields and undesirable shifts in biological diversity (Konan *et al.* 1999). The surveys furthermore included sampling of water, sediments and fish for study of their state of pollution, resulting in information on the state of the health of the Gulf of Guinea. Partly as a result of this effort, a manual on standardized methods of pollution monitoring was developed (Mbome *et al.* in press).

Monthly continuous plankton recorder surveys of the Gulf, carried out in co-operation with the Sir Alister Hardy Foundation for Ocean Science (SAHFOS) in the United Kingdom, has provided invaluable information on the Gulf's productivity and marine life support systems. It reveals at least 3 areas along the routes with high phytoplankton levels which may represent upwelling (SAHFOS 1997; John *et al.* this volume). The information from the plankton communities, representing the base of the marine food chain gives an indication of the carrying capacity of the ecosystem, which will enable projections to be made on future food security for the burgeoning populations of the region.

Overcutting and pollution of mangroves which are important as spawning and nursery grounds for many fish species, as habitat for many species of plants and animals, as well as serving as a natural filter for pollutants and silt from land, has led to a decrease in the reproduction of fish and declining fish catches as a result. A large scale inventory of mangrove forests in the region, including an evaluation of deforestation and the effects of pollution, was executed under the project. Both on-the-ground surveys, using standardised questionnaires, as well as advanced remote sensing technologies were applied, resulting in the production of mangrove disturbance and distribution maps (Saenger *et al.* 1996; Sankare *et al.* 1999; Isebor 1999). Pilot community based mangrove re-plantation projects and intensive public awareness campaigns on sustainable utilization and restoration of mangrove ecosystems are now being carried out in all of the six project countries.

Controlling Land-based Sources of Pollution

Recognising that, by far, the greater part of the pollution that causes degradation of the marine environment is from land-based activities, the project has put a premium on the assessment, prevention and control of such pollution (Scheren and Ibe this volume).

Industrial pollution

To put the problem into perspective, a rapid semi-quantitative assessment of land-based sources of pollution in the region was undertaken (with UNEP) on a country-by country basis (e.g. Métongo 1997; Ajao and Anurigwo 1998). As part of a more detailed survey on industrial pollution, manufacturing industries in the various countries have co-operated in completing a series of questionnaires and in allowing on-site investigations aimed at obtaining an in-depth assessment of the nature of the manufacturing process, the types and quantities of wastes generated, and the nature of waste treatment and discharge practices (Ehn 1999).

The significance of nutrients in the Gulf of Guinea Region lies in the high levels of productivity in the coastal ecosystems of the region. The loading of nutrients leads to eutrophication and concomitant harmful algal blooms generating much concern among the participating countries. The project participants paid special attention to the issue of nutrients, with the establishment of a Nutrient Activity Group as a first pragmatic step. This led to the development of a common reference document for sampling, monitoring and laboratory practice adapted to the regional context (Anurigwo 1999).

In addition to the specific monitoring and assessment modules defined, an extensive study of the physical oceanographic processes in the Gulf of Guinea, in conjunction with other programmes in the region, has shed light on the causal and influencing factors for the occurrence and variability of marine living resources, the levels, distribution and fate of pollutants, and the problem of extensive coastal erosion. The latter was the subject of a special Working Group that incorporated entrepreneurs in this field of endeavour, and which resulted in the recommendation of novel low technology and low cost options for combating this significant menace, identified by the Committee of Ministers as needing the highest and most urgent attention.

The assessments provide a basis upon which to develop suggestions concerning environmental controls, including alterations in production process technology. This is linked to UNIDO's ongoing drive for the adoption by countries of cleaner production technologies and the concomitant programme at national and regional levels to establish cleaner production centres. The results from these surveys have been invaluable in fixing national and regional guideline effluent standards, and in formulating recommendations for Governments to control industrial pollution.

In addition, a parallel attempt has been made to develop strategies and policies that will encourage reduction, recycling, recovery and reuse of industrial wastes. One such initiative, now at the pilot stage in Ghana, is the establishment of Waste Stock Exchange Management System. This concept, which has been enthusiastically embraced by manufacturing industries in Ghana, has as a slogan 'one person's waste, another person's raw material' and holds considerable promise as a cost effective waste management tool in Ghana and beyond (MAMSCO 1998).

Urban waste / sewage

A detailed study of the nature and quantities of urban wastes and sewage and the present status of their management has also been executed. With due recognition of the possible reinforcement of present government efforts, to evolve and implement masterplans for urban waste and sewage management, the project has focused with municipal authorities on novel and low technology options, such as the use of settling pits in Ghana for sewage treatment in small communities and the sorting of domestic waste prior to disposal as a means of encouraging recycling and reuse.

As part of the drive for low-cost, low-technology options, a pilot project on the foreshores of the Ebrié Lagoon in the Bay of Cocody area of Abidjan is based on the use of mangroves as 'purifiers' of urban wastes and sewage that presently run into the lagoon, provoking offensive odours and biological oxygen demand (BOD) problems. Preliminary proposals for effective urban waste and sewage control that are now being directed to the governments are a mix of conventional and innovative applications.

Marine debris

Analyses of results from four years of marine debris (beach litter) monitoring (with IOC-UNESCO) have resulted in advice and recommendations comprising preventive and control actions to municipal and local authorities on solid waste management. At the last review meeting on Marine Debris (Abidjan, 19-21 April 1999) recommendations were made for the establishment of an effective region-wide program of control of marine debris. A decision was made to formulate a protocol to the Abidjan Convention in this regard. The text of such protocol is currently under preparation (Folack *et al.* 1999).

Formulating National and Regional Strategies and Policies

Perhaps more than in any other region of the world, the coastal zone in the Gulf of Guinea region is both its heart and soul. Deriving from the history of the region's early European contact and the richness of natural resources in this zone, all major cities, harbours, airports, industries, plantations, and a host of other socio-economic infrastructures are located on or near the coast.

The multitude of competing interests and stakeholders in the coastal zone dictates the preparation of management plans that are multidimensional and cross-sectoral. At a World Bank sponsored meeting on Environment and Sustainable Development for the countries of West and Central Africa, held in Yamoussoukro (Côte d'Ivoire) in July 1996, the GOG/LME project was adopted as an appropriate mechanism for joint initiatives on the coastal zone and the adjacent Marine Catchment Basins (MCBs) of the region. Furthermore, the project was called upon to develop a clearing house for coastal zone data and to prepare national action plans for coastal zone management.

Coastal Profiles and draft national Integrated Coastal Areas Management (ICAM) plans have been prepared, following common guidelines established at an earlier regional consultation on Integrated Coastal Areas Management held in April 1996 in Accra (Ghana). They include: (Rep. of Benin/UNIDO 1998; Rep. of Benin 1998; Rep. of Cameroon/UNIDO 1998; Rep. of Cameroon 1998; Rep. of Côte d'Ivoire/UNIDO 1998; Rep. of Côte d'Ivoire 1998; Rep. of Ghana/UNIDO 1997; Rep. of Ghana 1998; Rep. of Nigeria/UNIDO 1998; Rep. of Nigeria 1998; Rep. of Togo/UNIDO 1998; Rep. of Togo 1998). The plans were developed as elaboration of existing National Environmental Action Plans (NEAPS) for the littoral zone. The intention is that implementation of these plans will reduce, if not eliminate, the inevitable conflicts resulting from multiple uses of the coast, and will contribute to a more orderly development process that protects the environment and promotes the sustainable use of natural resources. National Steering Committees have been set up by decrees to promote the cross-sectoral management implied in the plans, which are currently in the process of adoption in the various participating countries.

In collaboration with the International Maritime Organisation (IMO) a specific focus was given to Oil Spill Contingency Planning, including a discussion on areas of international liability, and the development of regional oil spill contingency plans.

Through a regional consultative process, involving experts from the six project countries, the project furthermore developed a preliminary Transboundary Diagnostic Analysis document. The latter report provides an overview of the regional issues and problems regarding the Gulf of Guinea Large Marine Ecosystem, and suggests options for solutions in their regard. The document therefore provides the basis for development of a Strategic Action Programme for the region.

At the prompting of the Committee of Ministers of Environment of participating countries, and with the active participation of the private sector, a regional group of experts on coastal erosion and flooding convened in August, 1998 and agreed on a regional strategy to tackle the problem; one that will ensure that a solution applied by one country does not provoke additional problems for its downstream neighbours.

Future Perspectives

The five Ministers of Environment from Benin, Cameroon, Côte d'Ivoire, Ghana and Togo, and the Director General/Chief Executive of the Federal Environmental Protection Agency of Nigeria, which constitute the Committee of Ministers for the Gulf of Guinea Large marine Ecosystem Project convened their first meeting in Accra, Ghana, on 9 and 10 July 1998. Taking note of the extensive achievements of the project in fostering effective consultation, coordination and monitoring mechanisms between participating countries, and in executing joint actions in environmental and living resources management, the Committee of Ministers recognized the project as a viable mechanism for regional co-operation. The Committee adopted the Accra Declaration (Table 3-3) as an expression of common political will for the environmentally sustainable development of marine and coastal areas of the Gulf of Guinea, and called for the development of a Strategic Action Plan including a full Transboundary Diagnostic Analysis, thus leading to an expanded second phase to include all the 16 countries between Guinea Bissau in the north to Angola in the south.

The achievements made by the GOG-LME Project, culminating in the signing of the Accra Declaration, attracted a special tribute in the Ministerial Statement of the Pan African Conference on Sustainable Integrated Coastal Management (PACSICOM) held in Maputo, Mozambique, 18-24 July. Furthermore, on 3 and 4 December 1998, on the occasion of the Ministerial segment of the high level Conference on Co-operation for the Development and Protection of the Coastal and Marine Environment in Sub-Saharan Africa (ACOPS), the Ministers recognised the validity of the LME concept in providing the scientific underpinning for the effective integrated management of the environment and natural resources of the coastal zone in Africa, and applauded the GOG-LME project as a model of a successful implementation of this concept.

The GOG/LME project has issued over a dozen substantive publications (e.g. Adam 1998; Ibe 1998; Ibe *et al.* 1998a,b; Ibe and Zabi 1998; Ibe and Abé 1999) in addition to a quarterly Newsletter and over 100 technical and meetings reports that with proper availability will provide effective means of information dissemination, transfer of knowledge and making outputs available for others to implement as opportunities arise.

A multimedia CD-ROM of the Gulf of Guinea Large Marine Ecosystem Project was issued in April 1999 for distribution during the Convention on Sustainable Development 7 (New York, 19-30 April 1999). It presents briefly, but illustratively, the objectives, approaches and results of four years of project implementation. Copies of all publications can be obtained from the Project Regional Co-ordination Center in Abidjan, or from UNIDO in Vienna.

The future is indeed auspicious, and UNIDO looks forward eagerly to continuing its partnership with the 16 countries along with the Global Environmental Facility and collaborators in and outside of the UN including NGOs and CBOs as the 16 countries strive not only to attain equitable and sustainable socio-economic development, but also to contribute towards a safer global environment and a rational utilization of global resources.

The first meeting of the Ministerial Committee of the Gulf of Guinea Large Marine Ecosystem (GOG-LME) Project took place in Accra, Ghana, on 9th and 10th July, 1998. The meeting was attended by the five Ministers with responsibility for the environment in Benin, Cameroon, Côte d'Ivoire, Ghana and Togo and the Director General/Chief Executive of the Federal Environmental Protection Agency of Nigeria. Basing their deliberations on extensive and substantive preparations, the Committee of Ministers has adopted the Accra Declaration on Environmentally Sustainable Development of the GOG-LME.

PREAMBLE

We, the Ministers of Environment of Benin, Cameroon, Côte d'Ivoire, Ghana and Togo and the Director General/Chief Executive of the Federal Environmental Protection Agency of Nigeria responsible for the GOG-LME Project,

Conscious of the fundamental importance of the health of the Gulf of Guinea Large Marine Ecosystem, including its coastal areas, to the well-being of the coastal communities, the economies and food security of the coastal states and the socio-cultural life of the Gulf of Guinea Region;

Recognising the transboundary nature of the marine environmental and living resource management problems confronting the Gulf of Guinea Region;

Concerned about the severe rates of coastal erosion, the threat of flooding, the seriousness of pollution, loss of biological diversity and depletion of fishery resources;

Conscious of the necessity to adopt a standardised regional approach in a cooperative effort to their control;

Conscious of the importance of having the means to combat the problem of coastal erosion;

Convinced of the validity of the integrated and sustainable management of the Large Marine Ecosystem to the resolution of problems, including strengthening regional cooperation and development, as well as establishing proper linkages between local, national, regional and global decision-making, and which is in fact unachievable without these said linkages;

Aware of the need to strengthen project implementation and to integrate more countries bordering the Guinea Current Large Marine Ecosystem and the necessity to enlarge the partnership notably with the inclusion of the private sector and other bilateral and multilateral donors;

Believing, therefore, that regional networking is an essential component of the system of ocean and coastal governance for the next century and beyond;

Noting and fully supporting the important achievement by the UNDP-GEF funded GOG-LME Project over the past three years, in the context of project execution by the project countries assisted by UNIDO, UNEP and US-NOAA, especially in forging a regional approach to ecosystem management;

Cognisant of the coming into force of the UN Conventions on the Law of the Sea, of the Framework Convention on Climate Change, of the Biodiversity Convention and the Abidjan Convention on Cooperation for the Protection and Development of Marine and the Coastal Zones of West and Central Africa (WACAF, 1981);

Determined to prevent, control and reduce coastal and marine environmental degradation in our respective countries, with a view to improve living conditions and productivity,

DECISIONS Have agreed that:

The countries within the Gulf of Guinea should as soon as possible, establish appropriate institutional mechanisms for the planning, implementation and evaluation of Integrated Coastal Areas Management (ICAM) plans;

Management plans and strategies, which may vary from country to country, should follow general guidelines adopted at the regional level. They should balance economic development with environmental protection and living resources conservation concerns and harmonise long-term ecosystem requirements with short-term political and economic interests;

Efforts shall be made to initiate, encourage and work synergistically with current and/or programmed national and international programmes on integrated coastal zone management in the region. The national concerns of flooding, and pollution caused by hydrocarbons, toxic chemical products, fisheries productivity and over-exploitation and, above all, coastal erosion call for the special attention of donors;

Data and information networking between the GOG-LME countries should be improved. National and Regional databases on the coastal and marine environment should be established using the Geographical Information System (GIS) to support decision-making, to be available to all users;

Transfer of knowledge and experiences among the countries of the GOG-LME, through the consolidation of networks for joint monitoring, research and capacity building in the field of marine environmental and natural resource management, should be enhanced;

Adequate and timely material and financial resources should be provided by Our Governments with support from UNDP/GEF, UNIDO as well as our private sector, bilateral and multilateral partners to the GOG-LME Project to ensure its efficient implementation and harmonious development;

Implementation of programmes should be monitored and rigorous and objective evaluations should be conducted on a periodic basis to determine the effectiveness of programmes and the efficiency of the system in achieving the goals and objectives of the GOG-LME Project;

The existing networks of non-governmental organisations (NGOs) in and among countries should be consolidated and expanded to ensure efficient and effective grassroots community involvement and information dissemination;

The development of a Strategic Action Plan including a full Transboundary Diagnostic Analysis leading to the second phase of the Project to include all the countries bordering the Guinea Current Large Marine Ecosystem, should be accelerated.

Table 3-3 The Accra Declaration on the environmentally sustainable development of the GOG-LME.

References

Adam, S.K. 1998 *Towards an integrated coastal zone management in the Gulf of Guinea: a framework document.* Edition Flamboyant, Cotonou. 85p.

Ajao, E.A. and S. C. Anurigwo 1998 Assessment of land-based sources of pollution in Nigeria. UNIDO Project EG/RAF/92/G.34.

Anurigwo, S. C. 1999 Manual on analysis and water quality monitoring in the Gulf of Guinea, UNIDO Project EG/RAF/92/G.34. 83p.

Ehn, J. 1999 Assessment of representative land-based polluting industries in the coastal area. UNIDO Project EG/RAF/92/G.34.

Folack, J., J. Ofori-Mensah and J. Abé 1999 Marine Debris Survey and Solid Waste Management in the Gulf of Guinea. IOC and UNIDO Report

Ibe, A. C. (ed.) 1998 *Perspectives in Coastal Areas Management in the Gulf of Guinea.* CEDA, Cotonou. 81p.

Ibe, A. C and J. Abé (eds) 1999 Physical oceanographic processes of the Gulf of Guinea, UNIDO, Vienna.

Ibe, A. C., L. Awosika and K Aka (eds) 1998 *Nearshore dynamics and sedimentology of the Gulf of Guinea.* CEDA, Cotonou. 221p.

Ibe, A.C. and B.E. Brown 1997 Ensuring living resources availability and biodiversity conservation in the Gulf of Guinea through large scale integrated ecosystem management. Conference on Fisheries, Habitat and Pollution, Charleston, South Carolina, USA.

Ibe, A. C. and Z. Csizer, 1998 The Gulf of Guinea as a Large Marine Ecosystem. Nature and Resources 34 (3): 30-39.

Ibe, A.C. and G. Kullenberg 1995 Quality Assurance /Quality Control (QA/QC) Regime in marine pollution monitoring programmes. The GIPME Perspective. Marine Pollution Bulletin 31: 209-213.

Ibe, A. C. and S.G. Zabi, (eds) 1998 *State of the coastal and marine environment of the Gulf of Guinea.* CEDA, Cotonou. 158p.

Ibe, A. C., A.A. Oteng-Yeboah, S.G. Zabi and O.A. Afolabi (eds) 1998 *Integrated environmental and living resources management in the Gulf of Guinea: the Large Marine Ecosystem approach* (Proceedings of the first Regional Symposium on the Gulf of Guinea Large Marine Ecosystem), CSIR Press, Accra. 281p.

Isebor, C. E. 1999 Mangroves of the Gulf of Guinea. UNIDO Project EG/RAF/92/G.34.

John, A.W.G., P.C. Reid, S.D. Batten and E.R. Anang (this volume, chapter 11) Monitoring levels of 'Phytoplankton Colour' in the Gulf of Guinea using ships of opportunity.

Konan, J., C. E. Isebor, K.A. Koranteng, N. N'goran, and P. Amiengheme, 1999 Report of the second Regional Fish Trawl Survey of the Gulf of Guinea, 25 February – 25 March 1999. A Report for UNIDO Project EG/RAF/92/G.34

MAMSCO Management Ventures Ltd 1998 Feasibility study on waste stock exchange management systems (WSEMS). UNIDO Project EG/RAF/92/G.34.

Mbome, I. L., H. Soclo, R. Arfi, C. Biney, D. Afolabi, N.M. Dibi, E.A. Ajao, A. Adingra and P. Manizan (in press) Manual on standardised methodologies for pollution monitoring in the Gulf of Guinea, UNIDO.

Métongo, B.S. 1997 Rapid assessment of sources of air, water and land pollution in Côte d'Ivoire. UNIDO Project EG/RAF/92/G.34

Republic of Benin / UNIDO 1998 Coastal Profile of Benin.

Republic of Cameroon / UNIDO 1998 Coastal Profile of Cameroon.

Republic of Côte d'Ivoire / UNIDO 1997 Coastal Profile of Côte d'Ivoire.

Republic of Ghana / UNIDO 1997 Coastal Profile of Ghana.

Republic (Federal) of Nigeria / UNIDO 1998 Coastal Profile of Nigeria.

Republic of Togo / UNIDO 1998 Coastal profile Togo.

Republic of Benin 1998 Draft National ICAM Plan Benin.

Republic of Cameroon 1998 Draft National ICAM Plan Cameroon.

Republic of Côte d'Ivoire 1998 Draft National ICAM Plan Côte d'Ivoire.

Republic of Ghana 1998 Draft National ICAM Plan Ghana.

Republic (Federal) of Nigeria 1998 Draft ICAM Master Plan Nigeria.

Republic of Togo 1998 Draft National ICAM Plan Togo.

Saenger, P., Y. Sankaré and T. Perry 1996 Effects of pollution and over-cutting on mangroves. UNIDO Project EG/RAF/92/G.34.

SAHFOS (UK), 1997 Annual summary of CPR tows in the Gulf of Guinea for 01 Dec.1996 - 30Nov 1997. UNIDO Project No EG/RAF/92/G34.

Sankare, Y, J-B. L.F. Avit, W. Egnankou and P. Saenger 1999 Etude floristique des mangroves des milieux margino-littoraux de Côte d'Ivoire. Bull. Jard. Bot. Nat. Belg./Bull. Nat. Plantentuin Belg. 67: 335-360.

Scheren, P and A.C. Ibe (this volume, chapter 22) Environmental pollution in the Gulf of Guinea – a regional approach.

Sherman, K. (ed.) 1993 Emerging theoretical basis for monitoring the changing states (health) of Large Marine Ecosystems. Summary Report of Two Workshops: 23 April 1992 National Marine Fisheries Service, Narragansett Rhode Island, and 11 -12 July 1992 Cornell University, Ithaca, New York. NOM Tech. Mem. NMFS-F/NEC-100. 27p.

Sherman, K. 1994 Sustainability, biomass yields, and health of coastal ecosystems: an ecological perspective. Marine Ecology Progress Series 112:277-301.

Sherman, K. 1996 Achieving regional co-operation in the management of marine ecosystems: the use of the Large Marine Ecosystem approach. Ocean and Coastal Management 29: 165-185.

Sherman, K. and E.D. Anderson (this volume, chapter 2) A modular approach to monitoring, assessing and managing Large Marine Ecosystems.

Topping, G. 1992 The role and application of quality assurance in marine environmental protection. Marine Pollution Bulletin 25: 61-65.

4

Consilience in Oceanographic and Fishery Research: A Concept and Some Digressions

D. Pauly

Abstract

The concept of *consilience*, describing the 'jumping together' of different scientific disciplines, and recently revived in a book of the same title, authored by E.O. Wilson, is presented, along with some of its implications for work conducted by oceanographers, marine biologists and fisheries and social scientists in the Gulf of Guinea area. It is suggested that maintenance and analysis of time series data, remote sensing of marine primary production, trophic mass balance modelling, and analysis of multi-sectoral coastal transects are eminently *consilient* in that they not only invite interdisciplinary co-operation, but also impose standards and provide common currencies that makes such co-operation meaningful.

Consilience

The text below is not an attempt to summarise the contributions to the Accra workshop documented in this volume. Rather, a few major threads - we would call them *lignes de force* in French - are brought together, showing the inherent coherence of the themes covered during our deliberations. To document this, however, I shall however, use a new, or rather newly revived concept, that of *consilience* (Wilson 1998). Most advances, within the various disciplines practised by the participants of this workshop, such as biology, or the environmental sciences have been the result of synthesis of previously unconnected data or ideas (Simonton 1988). Thus, the furious pace at which data continue to be collected and ideas continue to sprout guarantees that new knowledge will continue to be generated within our disciplines, if only by integrating these data and ideas into common explanatory frameworks.

In analogy to this, if perhaps at a grander scale, major scientific breakthroughs have often been the result of convergence among traditionally distinct disciplines. Herein, these disciplines usually articulate themselves in a hierarchy, with logic/mathematics (and the usual criterion of parsimony; Pauly 1994) providing the backbone of any given breakthrough, physics and chemistry providing the basic rules constraining the changes of its material substrate, and evolutionary biology providing the framework that constrains its living organisms (if any), including humans and their culture. Wilson (1998) called *consilience* (from 'jumping together') the explicit search for scientific explanations within the context of this hierarchy, and provided examples of research issues whose resolution, he felt, would occur faster if consilience were used as an explicit criterion - in addition to parsimony - for structuring the relevant research programmes.

41

My favourite examples (not Wilson's) of consilience are the Cretaceous/Tertiary (K/T) extinction of 65 million years ago, of dinosaur fame, and the origins of Homo sapiens, both of which made mutually compatible data and concepts from an enormous range of disciplines, previously not interacting with each other. A nuclear scientist, his geologist son, and two chemists colleagues proposed that the K/T extinction was caused by the impact of a large meteorite (Alvarez *et al.* 1980). This hypothesis, then based mainly on evidence from an Italian dig, was subsequently corroborated by petroleum geologists, who had previously identified, then ignored the Chicxulub impact crater, in Yucatan, Mexico. Other scientists, from astronomers to evolutionary biologists joined the fray (Raup 1986), and gradually, the `Alvarez' hypothesis was accepted by the best of them. The results of this development has been extraordinarily fruitful, "provoking new observations that no one had thought of making under old views" (Gould 1995, p.152). This led among other things, to a resolution, in evolutionary biology, of the ancient, but still acrimonious debate between *catastrophists* (Sedgwick, Cuvier, etc.) and *uniformitarianists* (Lyell, Darwin, etc.). Even popular culture was impacted (see the movies Meteor, Armageddon, and Deep Impact).

Indeed, the eventual creation of an international system for tracking potentially dangerous meteorites is not unlikely. Similarly, the evidence for a recent African origin of *Homo sapiens*, with subsequent dispersal to West Asia, Eurasia, Australia, the Americas, and finally Oceania is supported by archaeological and genetic evidence, with linguistics providing the clincher: an evolutionary tree that closely matches that generated by the physical disciplines (Cavalli-Sforza *et al.* 1988). The latter example shows that the hierarchy of sciences implied in consilience does not mean that the specific results of a more fundamental discipline are inherently more reliable that those of a more derived discipline. Rather, it only implies that the different sets of results must be mutually compatible. Thus, the linguistic evidence is, in this example, no less important than the evidence based on genetics. Similarly, when the physicist Lord Kelvin pronounced the Earth to be only a few thousands years old, based on the time required for a large sphere of burning coal to cool off, and evolution by slow natural selection thus impossible, it was he who was wrong, not Charles Darwin (Tort 1996).

The question now is whether we can make use of consilience in our work as oceanographers, marine biologists and fisheries scientists, i.e., in fields perhaps less glamorous than those in the above examples. Various concepts we may call consilient come to mind here, related to presentations at the workshop documented in this volume. The first of these is the mass-balance concept, i.e., the notion that, in a given system, mass must be conserved, irrespective of its movements and transformations. This principle is related to the First Law of Thermodynamics, which states that energy can be neither destroyed nor created (Gilmont 1959). For chemical reactions, this implies for example that "the sum of the masses of the reactants must equal the sum of the masses of the products" (Gilmont 1959, p. 146). Physical oceanographers also rely on mass-balance when calculating geostrophic flows from density fields (Sverdrup *et al.* 1942), or when calculating upwelling intensity from coastal wind stress, which implies water masses welling up to replace water blown off the coast (Bakun 1996).

On the other hand, one rarely hears biologists, or even ecosystem modellers explicitly invoke the principle of mass-balance, though it is also an absolute requirement for living things (Schrödinger 1992). An exception to this is the ecosystem modelling work of T. Laevastu and colleagues, in which mass-balance was used as a key structuring element for trophic interactions and migrations (Laevastu and Favorite 1977; Laevastu and Larkins

1981). J.J. Polovina emphasized this feature when he simplified Laevastu's model and formulated the Ecopath approach (Polovina 1984, 1985), thus giving it the feature which made it applicable to a wide range of system types (Christensen and Pauly 1993; Pauly, this volume). Ecopath uses the mass-balance approach to verify that the estimated production of the functional groups (exploited or not) of a given ecosystem matches the estimated consumption by their predators. Such verification is not ensured by the publication of individual estimates, however precise, even in the best journals catering to the different sub-disciplines of marine biology.

Rather, it is by incorporating such estimates into a mass balance ecosystem model that we render such estimates mutually compatible, and hence assure ourselves of their reliability and usefulness. This should have a beneficial impact on marine biology, whose work on different processes in an extremely wide variety of organisms is sometime perceived as lacking `relevance'. Moreover, ecosystem and mass-balance considerations should help renew fishery science as well, given that it has been too narrowly focused on the study of single species, and on industrial fisheries, usually overlooking by-catch discarding practices, non-commercial species, and other fisheries (artisanal, sport, etc.) as well.

The relation of these points to Wilsonian consilience should be obvious. Consilience also implies developing protocols for integrating the results of remote sensing studies (as illustrated here by Hardman-Mountford and McGlade or Roy *et al.*, this volume) into mass-balance trophic models of ecosystems. The key results relevant here are: (1) definition of geographic system boundaries, as in the case of the biochemical provinces of Longhurst (1998); and (2) synoptic estimates of primary production (Longhurst et al. 1995), i.e., of that which determines the boundaries of marine ecosystems in terms of the size of their biomass fluxes. Here, by constraining model size, remote sensing can link with ecosystem modelling, and thus work in consilient mode. Note also that both remote sensing and trophic modelling may be accused of being 'superficial': remote sensing because, quite literally, it cannot look deeper that a few decimetres into the sea, and trophic modelling because it does not consider interactions other than those generated by grazing and predation. Yet, when data from the two approaches are analyzed jointly, inferences can be drawn which go well beyond those based on more conventional approaches (see e.g., Pauly and Christensen 1995; Trites *et al.* 1997).

Perhaps we may infer from this issue of apparent superficiality that consilient work may suffer, at least for a while, from analogies with multidisciplinary work, wherein the methods of different disciplines are brought to bear on a given topic (e.g., as chapters in a book), without any of these methods being made to relate to each other. Such unconnected work is all too frequent, e.g., in that discipline called coastal area management, a theme to which we shall return. The concept of consilience, it seems to me, should also apply to the strengthening of inferences that results when the past is related to the present. This is what occurs when we use knowledge gathered by historians, or by scientists of past centuries, and often perceived as anecdotal, to establish stable baselines for biodiversity (Pauly 1995, 1996).

This is also what occurs when we draw inferences from time series, whose increasing length increases their contrast, and hence their usefulness for various analyses (Hilborn and Walters 1992). Thus, it is important that the physical and biological time series generated and/or used by the participants in this book be continued, and every effort should be made to ensure that this occurs. The last aspect of consilience to be covered here refers to its implications for the languages we use. Trivially, this means that we must speak the same vernacular language (English, as it mostly turns out), and only a bit less trivially, translate concepts in and out of our various discipline-specific jargons. Also, and this is where things start getting really complicated, we must identify concepts that cut across disciplines, and a multidisciplinary currency allowing for transactions between disciplines.

I shall illustrate this with reference to coastal area management (CAM), a discipline with many practitioners and applications, but whose defining tenet(s), topic(s) and technique(s) remain elusive. Indeed, some practitioners give the impression that anything that happens anywhere on or close to on any coastline is within the purview of CAM, and that any method ever used to investigate any of these things is appropriate for CAM. (The reader will understand that I could not provide references to back these claims without antagonizing people whose technical work I respect, in spite of the disorganisation of their discipline). Thus for CAM, as perhaps everywhere in science (Medawar 1967), the challenge is to identify tractable problems, related to its defining objects: coastlines.

Coastlines differ from other geographic features in that most of their different characteristics are arrayed in a single dimension, i.e., in the form of transects that are perpendicular to the coastline, and stretch from upland to the sea. On the other hand, fewer differences occur parallel to a coastline. Alexander von Humboldt, a founder of physical geography, may have been the first to use transects to document geographical variations along strong gradients (Gayet 1996). Conway (1985), on the other hand, introduced transects to agro-ecosystem analysis, as a tool to express in simplified manner the complex interactions within highly integrated farming systems. [Note, incidentally that such systems can also be described by trophic mass-balance models; Dalsgaard and Oficial 1997]. From this, it is straightforward to propose that multi-sectoral coastal transects should become a key concept in CAM, and that, suitably formalized, such transects could lead to the common currency required for comparison of coastal systems, and for comparative evaluation of various injuries to such systems (Pauly and Lightfoot 1992).
SimCoast™ (McGlade 1999; 2001) which was used in the EU Gulf of Guinea implements this coastal transect approach, and provides, via fuzzy logic, the currency that CAM had been lacking so far, enabling quantitative comparisons of impacts due to different, otherwise incommensurable agents (Nauen, this volume), ranging from upland erosion to fisheries policy. That this should lead to consilience among, and progress for, the different disciplines so far fruitlessly engaged in CAM needs little emphasis.

Another example of consilience requiring a common currency is FishBase, the electronic encyclopaedia of fishes (see www.fishbase.com), which works only because a standardised nomenclature (Eschmeyer 1998) is used to establish the links between the widely different data types included in FishBase (Froese and Pauly 1998). This is what enables FishBase to provide, among other things, a comprehensive coverage of the fishes of the Gulf of Guinea, and of their biology. As might have be noted, several of my examples (Ecopath, SimCoast, FishBase) are products of projects supported by the European Commission, particularly by its Directorate General concerned with development (i.e. VIII and XII INCO/DC), but not run by committees, or university consortia. Perhaps this indicates that such projects, providing support to participants only if they buy in the strong concept underlying such ventures, are more effective than the usual collaborative projects, where the partners agree only to share the available funds.

The ventures given as example here certainly have shown that participants from a variety of countries, both developed and developing, can contribute. And this is probably the neatest thing about consilience: it implies that we can all contribute, given some self-discipline.

Acknowledgements

I thank Dr. Cornelia Nauen, for useful discussions, and for all the support.

References

Alvarez, L. W. Alvarez, F. Asaro and H.V. Michel 1980 Extraterrestrial cause for the Cretaceaous-Tertiary extinction. Science 208: 1095-1106.

Bakun, A. 1996 Patterns in the Ocean: ocean processes and marine population dynamics. California Sea Grant, La Jolla, CA, and Centro de Investigaciones Biologicas del Noroeste, La Paz, BCS. 323p.

Cavalli-Sforza, L.L., A. Piazza, P. Menozzi, and J. Mountain 1988 Reconstruction of human evolution: bringing together genetic, archaeological and linguistic data. Proc. Acad. Sci. 85(16): 6002-6006.

Christensen, V. and D. Pauly (eds) 1993 Trophic models of aquatic ecosystems. ICLARM Conf. Proc. 26. 390p.

Conway, G.R. 1985 Agroecosystem analysis. Agric. Admin. 20:31-55.

Dalsgaard, J.P.T. and R.T. Oficial 1997 A quantitative approach for assessing the productive performance and ecological contribution of smallholder farms. Agricultural Systems 55(4): 503-533.

Eschmeyer, W.N. (ed.) 1998 *Catalog of fishes*. Special Publication, California Academy of Sciences, San Francisco, 3 volumes, 2905p.

Froese, R. and D. Pauly (eds) 1998 FishBase 98: concepts, design and data sources. ICLARM. 293 p. [distributed with 2 CD-ROMs]

Gayet, M. 1996 Humboldt, Friedrich Wilhelm Heinrich, Alexander von, 1769-1859. pp2284-2287. *In*: P. Tort (ed.) Dictionnaire du Darwinisme et de l'Evolution. Presse Universitaire de France. Paris.

Gilmont, R. 1959 *Thermodynamic principles for chemical engineers*. Prentice Hall, New York. 339p.

Gould, S.J. 1995 *Dinosaurs in a haystack: reflection in natural history*. Harmony Books, New York. 480p.

Hardman-Mountford, N.J. and J.M. McGlade (this volume, chapter 5) Variability of physical processes in the Gulf of Guinea and implications for fisheries recruitment. An investigation using remotely sensed sea surface temperature.

Hardman-Mountford, N. J. and J.M. McGlade (this volume, chapter 6) Defining ecosystem structure from natural resource variability: application of principal components analysis to remotely sensed sea surface temperatures.

Hilborn, R. and C.J. Walters 1992 *Quantitative fisheries stock assessment: choice, dynamics and uncertainty*. Chapman & Hall, New York. 570p.

Laevastu, T. and F. Favorite 1977 Preliminary report on a dynamic numerical marine ecosystem model (DYNUMES) for the Eastern Bering Sea. U.S. Natl. Mar. Fish./Northwest and Alaska Fisheries Center, Seattle. 81p.

Laevastu, T. and H.L. Larkins 1981 *Marine Fisheries Ecosystems: quantitative evaluation and management*. Fishing News Books, Farnham, Surrey. 162p.

Longhurst, A.R. 1998 *Ecological geography of the sea*. Academic Press, San Diego. 398p.

Longhurst, A.R.S. Sathyendranath, T. Platt, and C.M. Caverhill 1995 An estimate of global primary production in the ocean from satellite radiometer data. J. Plankton Research. 17: 1245-1271.

McGlade, J.M. 1999 Ecosystem analysis and the governance of natural resources. pp309-342. *In*: J.McGlade (ed.) *Advanced Ecological Theory. Principles and Applications.* Blackwell Science, Oxford.

McGlade, J.M. 2001 SimCoast™. A Fuzzy Logic Rule-Based Expert System. View the World, UK [distributed as CD-ROM]

Pauly, D. 1994 Resharpening Ockhams' Razor. Naga, the ICLARM Quarterly 17(2):7-8.

Pauly, D. 1995 Anecdotes and the shifting baseline syndrome of fisheries. Trends in Ecology and Evolution 10(10):430.

Pauly, D. 1996 Biodiversity and the retrospective analysis of demersal trawl surveys: a programmatic approach, pp1-6. *In*: D. Pauly and M. Martosuborto (eds). *Baseline studies in biodiversity: the fish resources of Western Indonesia.* ICLARM Studies and Reviews 23.

Pauly, D. (this volume, chapter 21) Spatial modelling of trophic interactions and fisheries impacts in coastal ecosystems: a case study of Sakumo Lagoon, Ghana.

Pauly, D. and V. Christensen 1995 Primary production required to sustain global fisheries. Nature 374:255-257.

Pauly, D. and C. Lightfoot 1992 A new approach for analyzing and comparing coastal resource systems. Naga, the ICLARM Quarterly 15(3):7-10.

Polovina, J.J. 1984 Model of a coral reef ecosystem I. The ECOPATH model and its application to French Frigate Shoals. Coral Reefs 3:1-11.

Polovina, J.J. 1985 An approach to estimating an ecosystem box model. U.S. Fish. Bull. 83(3): 457-460.

Raup, D.M. 1986 *The Nemesis affair: a story of the death of Dinosaurs and the Ways of Science.* W.W. Norton, New York. 219p.

Roy, C. P. Cury, P. Fréon and H. Demarq (this volume, chapter 10) Environmental and resource variability off northwest Africa.

Schrödinger, E. 1992 *What is Life?* Cambridge University Press, Cambridge. 184p.

Simonton, D.K. 1988 *Scientific genius: a psychology of science.* Cambridge University Press. Cambridge, 229p.

Sverdrup, H.U., M.W. Johnson and R.H. Fleming 1942 *The oceans: their physics, chemistry and general biology.* Prentice Hall, New York. 1087p.

Tort, P. 1996 Thompson, Sir William, Lord Kelvin, 1824-1907. pp4281-4283 *In*: P. Tort (ed.) Dictionnaire du Darwinisme et de l'Evolution. Presse Universitaire de France. Paris.

Trites, A., V. Christensen and D. Pauly 1997 Competition between fisheries and marine mammals for prey and primary production in the Pacific Ocean. J. Northw. Atl. Fish. Sci. 22: 173-187.

Wilson, E.O. 1998 *Consilience: the unity of knowledge.* Alfred A. Knopf. New York. 332p.

II

Environmental Forcing & Productivity

Environmental Forcing

The Gulf of Guinea Large Marine Ecosystem
J.M. McGlade, P. Cury, K.A. Koranteng and N.J. Hardman-Mountford (Editors)

5

Variability of Physical Environmental Processes in the Gulf of Guinea and Implications for Fisheries Recruitment. An Investigation Using Remotely Sensed SST

N.J. Hardman-Mountford and J.M. McGlade

Abstract

This study focuses on seasonal and interannual variability in SST (Sea Surface Temperature) for the Gulf of Guinea, with special attention given to specific features relating to the underlying dynamics of the system and relevant to fisheries recruitment. Remotely sensed AVHRR SST data were used for this investigation. Firstly, patterns of cloud contamination were assessed. Then, seasonal and interannual variability in SST patterns were investigated, spatially and temporally. Features identified were the Senegalese upwelling influence, coastal upwelling, tropical surface water, putative river run-off and fronts. These were all seen to vary seasonally. Of particular interest was the observed cooling along the coast of Liberia and Sierra Leone in February. Interannually, a quasi-cyclic pattern of warm and cool years was seen to exist, both generally and in a number of the features examined. The results are discussed in terms of existing theories and their relevance to fisheries recruitment.

Introduction

Physical environmental features of relevance to fisheries

Physical environmental variability is known to affect biological productivity (Mann and Lazier 1996, Bakun 1996). This is especially true for upwelling areas, where high levels of physical stress place significant biological constraints on marine life and ecosystem functioning. Pelagic fish species characteristic of these areas, such as the clupeiids, often have life history traits, such as serial spawning and early maturity; these are thought to be a mechanism for dealing with such environmental variability (Bakun, 1996; Cole and McGlade 1998a).

Bakun (1993, 1996) generalises three broad categories of oceanographic process considered to be important in influencing recruitment success; namely nutrient *enrichment* of the environment to support appropriate levels of primary production, *concentration* of food particles into denser aggregations to facilitate foraging and *retention/transport* of eggs and larvae within/to suitable nursery areas . Bakun refers to these three factors as a fundamental triad. Examples of enrichment processes are upwelling, river run-off and micro-scale turbulence. Concentration can occur in areas such as fronts, river plumes and the thermocline.

Features that form transport and retention mechanisms can include fronts, coastal boundaries, the thermocline, currents and local gyral circulation patterns.

Upwelling areas provide many of these factors. Cury and Roy's (1989) Optimal Environmental Window (OEW) theory states that rather than any particular environmental parameters being 'good' for recruitment *per se*, a balance between relevant parameters is required to provide optimal conditions for recruitment. For example, in a wind driven upwelling region, increase in wind stress will increase the offshore Ekman transport and hence the influx of nutrients to the euphotic layer to stimulate primary production. However, it will also lead to an increase in the level of turbulence, disrupting vertical stratification and the formation of chlorophyll maximum layers, and to offshore advection of eggs and larvae. Therefore, the optimal conditions for recruitment occur at intermediate levels of wind stress, where there is a balance between the upwelling of nutrient rich water and calm conditions required for successful first feeding and retention of eggs and larvae in the nursery area. This dome shaped relationship, predicted by OEW theory, has been shown to hold for all the major wind-driven, Ekman-type upwelling areas of the world (*ibid.*). For the Gulf of Guinea region, however, the lack of correlation between wind strength and upwelling strength means that the risk of high levels of wind stress disrupting concentration and retention mechanisms is minimised. Therefore, the wind limitation does not hold and optimal conditions for recruitment success occur when enrichment, through upwelling, is at its strongest. This fits with the prediction of OEW theory of a linear correlation between upwelling intensity and recruitment in non Ekman-type upwelling situations where wind remains moderate (*ibid.*).

A number of physical oceanographic features that have relevance to fisheries dynamics have been identified in the Gulf of Guinea LME. One of these is the seasonal upwelling along the coast of Ghana and Côte d'Ivoire, previously described (Demarcq and Aman this volume). Other features include a large, seasonal, southerly extension of cold water from the Senegalese upwelling region, the presence of warm, low salinity Guinean waters outside the upwelling area (Binet and Marchal 1993) and a seasonal cooling in the late boreal summer off Sierra Leone (Bakun 1996).

Remote sensing of the West African region

Remote sensing has been used extensively for monitoring the global oceans and a number of studies have used satellite derived SST (Sea Surface Temperature) data to study coastal regions along the coast of West Africa. These include studies on the north-west African region (Hernández-Guerra and Nykjaer 1997), the coastal area of Mauritania (Maus 1997), the Senegalese upwelling (Demarcq and Citeau 1995), the Northern Benguela off Namibia (Cole and McGlade 1998b) and the Gulf of Guinea (Hardman and McGlade this volume). CZCS data have been used to study the Benguela off Namibia and South Africa (Weeks and Shillington 1994). Additionally, compound remote sensing studies, combining SeaWiFS data with *in situ* measurements, have taken place on recent cruises along the north-west and south-west African coastlines as part of the Atlantic Meridional Transect (AMT) program (Robins and Aiken 1996, Aiken *et al.* 2000).

In the Gulf of Guinea, Aman and Fofana (1995,1998) used Meteosat SST data to identify the location of upwelling and study seasonal variability along the coast of Côte d'Ivoire. They identified the major upwelling season and two periods of minor upwelling. The first of these minor events occurred in January during the *harmattan* period and the second event took place in March. The harmattan can cause deviations of the ITCZ, which may lead to upwelling (Allersma and Tilmans 1993). Demarcq and Aman (this volume) have attempted to supplement Meteosat data with two *in situ* data sets. They report considerably improved spatial and seasonal precision with the blended product.

This paper focuses on seasonal and interannual variability in SST and the specific features related to the underlying physical dynamics of relevance to fisheries recruitment.

Data acquisition and pre-processing

The satellite data used are from the Cloud and Ocean Remote Sensing around Africa (CORSA) AVHRR data set produced by the Space Applications Institute, Joint Research Centre (JRC) of the European Commission. The data are level-2 and level-3 pre-processed by the Marine Environment Unit of the Space Applications Institute from level 1 GAC AVHRR data obtained under a scientific collaboration agreement with NASA. This processing includes masking out land and cloud pixels, validating the images against SST data from the Combined Ocean Atmosphere Data Set (COADS), NASA Multi-Channel SST (MCSST) and the Global Ocean Surface Temperature Atlas (GOSTA) and processing the daily images into weekly composites. Four weekly composites are produced for each month of the time series; the first three composites being seven day averages and the final composite being an average of all the remaining days in the month (Cole 1997). The time series used extend from week 2 of July 1981 to week 4 of December 1991 with two missing weekly composites (week 1 of November 1989 and week 1 of January 1991), giving a total of 501 weekly composite images. The CORSA-AVHRR images have a spatial resolution of 4km at the equator (GAC), which is sufficient for the investigation of shelf processes. Three sub-regions were used: *SGLP* Sierra Leone-Guinea Plateau; *CWAU* Central West African Upwelling; and *EGOG* Eastern Gulf of Guinea (Hardman and McGlade this volume).

The AVHRR sensor can measure SST to an accuracy of 0.1K and a comparative analysis between the CORSA-AVHRR product and NASA 'Pathfinder' 9km resolution AVHRR data showed the two data sets to agree within 1°C in 96% of cases (C. Villacastin, JRC, Ispra, Italy, pers. comm.).

Cloud contamination

The tropical location of the Gulf of Guinea means that the region has particularly high cloud cover and high levels of atmospheric water vapour content, which can influence the accuracy of satellite-borne radiometers. Therefore, to assess the usability of the data set for this region, the degree of cloud contamination was derived. The term 'cloud contamination' is used to describe non-land pixels which had previously been recoded during the cloud-removal process of the JRC to have a value of zero. Firstly, a reclassification routine was applied to the images, which converts the original weekly composites into binary images. Non-cloud pixels were given a value of one and cloud pixels remained at zero. The percentage of cloud pixels for each weekly composite layer was then calculated by counting the proportion of pixels in each layer with a value of zero, excluding land areas. This calculation was performed for the whole Gulf of Guinea region and its three sub-systems (see Figure 5-1a, colour plate). The spatial distribution of cloud contamination was generated by summing the values for each pixel site from the Boolean layers. This procedure was carried out for the whole time series to show the overall spatial distribution of cloud (Figure 5-1a, colour plate) and on a monthly basis to show the seasonal cycle of cloud contamination (Figure 5-1b, colour plate). The spatial distribution of cloud contamination throughout the entire data series clearly shows the area of highest cloud contamination lying in a central band around 3°N. This area of heavy cloud is associated with the mean position of the Inter-Tropical Convergence Zone (ITCZ).The monthly spatial distribution of cloud contamination shows the seasonal meridional migration of the ITCZ. The frequency distribution of the degree of cloud contamination in 5% bands for the weekly composite layers is shown in Figure 5-2a for the whole Gulf of Guinea region. Almost 80% of layers had less than 50% cloud contamination. Figure 5-2b shows the proportion of layers in each 25% band of cloud contamination for each of the subsystem areas. For the Sierra Leone-Guinea Plateau (SLGP) and Central West African Upwelling (CWAU) subsystems, between 75% and 80% of the layers have less than 50% cloud contamination. This figure is lower for the Eastern Gulf of Guinea (EGOG) subsystem (60.3%).

On average, for the whole Gulf of Guinea, cloud cover is highest from May to September, however, each subsystem shows a different seasonal cycle. The EGOG has the highest cloud cover in most months and peaks in August, as does the SLGP. Cloud cover in the CWAU peaks in June and then drops in July before increasing again slightly until September. Seasonal cycles of cloud cover are shown in Figure 5-3.

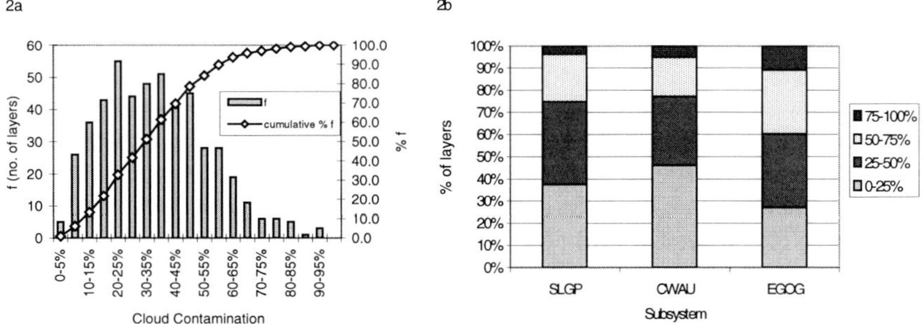

Figure 5-2a Frequency of input images (layers) with different levels of cloud contamination (5% bands); **b** Proportion of input images (layers) showing different levels of cloud contamination (25% bands) in each subsystem.

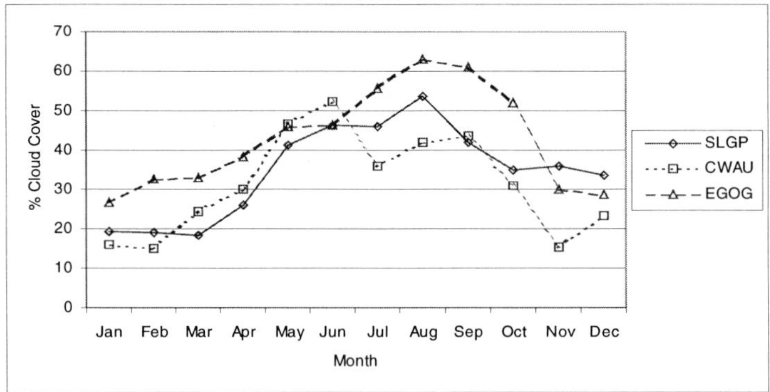

Figure 5-3 Seasonal cycle of cloud contamination in the Gulf of Guinea by subsystem.

Data processing and analysis methods

Temporal statistics (spatial variability)

Descriptive statistical images (mean SST, standard deviation) were calculated across time from the weekly SST composite images to show the mean spatial pattern for the whole time series and areas of greatest variance, as well as the seasonal and interannual spatial variability. These images were examined to identify important oceanographic features, as well as large scale patterns of seasonal and interannual variability.

Spatial statistics (temporal variability)

The weekly composite images were averaged spatially for each subsystem to show temporal trends in SST. Time series charts were constructed to show temporal variability in mean SST, both of a seasonal and interannual nature. The resulting mean SST time series was subjected to a 49 point moving average smoothing routine (modified from Chatfield 1989) to emphasise interannual variability.

Standardised SST anomalies were calculated to show unsmoothed seasonal and interannual variability on a weekly time scale. Interannual variability was shown by standardising data with respect to the week of the year (1-48), thereby eliminating seasonal variability. Seasonal variability was shown by standardising with respect to the year (1981-1991), thereby eliminating interannual variability.

Spectral Analysis

Power spectra of the time series, filtered with a Hanning window (see Chatfield 1989) with five degrees of freedom, were generated, using a Fast Fourier Transform (FFT) routine, to show the period of any cycles present in the data. SST signals relating to specific spectral peaks were extracted using a rectangular band pass filter (Vandenhouten *et al.* 1996).

Indices of Interannual Variability in SST Features

- Senegalese Upwelling Influence

Interannual variability in the dominance of the Senegalese upwelling influence (henceforth known as the SUI) over the SLGP subsystem was measured as the anomaly from the latitude of the most southerly position reached by the feature. This occurred during either March or April for each year throughout the time series. The southern boundary is seen in the monthly composite images as a region of rapid temperature change (temperature gradient $>1^{\circ}$C per 100 km).

- Coastal Upwelling

Interannual variability in upwelling along the coast of Ghana and Côte d'Ivoire was assessed quantitatively by calculating measures of both upwelling intensity and spatial extent for both the major (Jul-Sep) and minor (Jan-Mar) upwelling seasons. These measures were calculated for a strip of 1°x1° boxes stretching along the coast from Cape Palmas to Cotonou (Figure 5-4, colour plate). Upwelling intensity (I) was calculated for each box by subtracting the mean SST value for each box ($\overline{\text{SST}}$) from a SST of 25°C, i.e. SST below 25°C have positive values.

$$I = 25.0 - \overline{SST} \qquad\qquad \textbf{Eq.5-1}$$

The value of 25°C was chosen as this was the contour used to define the upwelling by Bakun (1978). The overall value for the season was then taken as the box with the maximum upwelling intensity value.

Spatial Extent (E) was calculated as the proportion of the total area (all boxes) having SST values less than a threshold level (T).

$$E = \frac{no.\ of\ pixels < T}{total\ no.of\ pixels} \qquad\qquad \textbf{Eq.5-2}$$

For the major season the threshold level was SST below 23°C and for the minor season it was SST less than 24.5°C. These threshold limits were determined from examination of the SST contours associated with each of the upwelling seasons in the weekly SST composite images.

The two indices were multiplied together to give a combined upwelling index.

Results

Overall

The overall mean SST image (Figure 5-4a, colour plate), averaged through time for the whole series, shows the southerly intrusion of water from the Senegalese upwelling region and the major upwelling areas along the coast of Ghana and Côte d'Ivoire to be the dominant cold water features of the system, superimposed on the warm waters of the tropical Atlantic. This image also shows a cold water influence, probably of equatorial upwelling origin, around the southern limit of the image. Additionally there is a small, cool strip of water along the coast of Sierra Leone. Warm water, probably originating from river run-off, is visible around the mouths of some of the major rivers of the Niger Delta and Cameroon coasts. A strong SST gradient, suggestive of a front, exists at the interface between the SUI and warmer, coastal waters around the Bijagos Islands.

The similarity in patterns between the mean percentage cloud contamination of weekly SST composites for the region (Figure 5-1a, colour plate) and the mean SST image (Figure 5-4a, colour plate) reflects the high degree of coupling between atmospheric and oceanographic conditions in the region. The overall standard deviation image (Figure 5-4b, colour plate) shows the greatest variance to be associated with the SUI. The coastal and equatorial upwelling areas also show a reasonably high degree of variability. This is expected given the seasonal nature of the upwellings.

Seasonal Variability

Seasonal SST variability in the Gulf of Guinea is shown in Figure 5-5 (colour plate).
* Senegalese Upwelling Influence
The SUI is composed of cold water ($<19^{\circ}$C at its core) originating from the Senegalese and Mauritanian coastal upwellings. It usually becomes visible in the Gulf of Guinea region during December and continues to move south until March or April. It then begins to recede.

During this period, the coldest water extends in a narrow finger close to the coast of Senegal and Gambia, but then moves away from the coast of Guinea-Bissau, appearing to be trapped along the shelf break rather than the coast. The result is a cold tongue of SUI water encircling the Bijagos Islands. During May, the tongue of cold water is still present around the Bijagos Islands, but by June only a narrow cold water front remains. By July, the SUI is absent from the whole coast as far north as Cap Vert, Senegal and by August, the SUI in also absent from offshore waters.

- Tropical Surface Water

In the SLGP subsystem, warm Tropical Surface Waters (TSW) appear to be furthest south in March/April and then move north until November, when they begin to recede towards the south again. This coincides with the commencement of the southerly movement of the SUI. The movement of these warm tropical waters appears to be associated with the seasonal migration of the ITCZ. The area of warm tropical waters appears to increase in size and temperature from January to May. Although they then appear to keep pushing north and growing in area, their temperature drops markedly from June to September. They have their greatest spatial extent, and are at their warmest (26°-27°C), during October and November.

In the CWAU subsystem, warming of the offshore area appears to start in February but is most noticeable from March to May. The warmer water tends to move in a north-easterly direction, associated with the seasonal movement of the ITCZ. The warm signal is damped from June to October by the summer upwelling along the Ghana-Côte d'Ivoire coastline to the north and by the equatorial upwelling to the south. This is especially visible during July to September when the warm signal is almost totally overridden. In November and December warmer waters appear over the whole area with the warmest waters in the north-easterly corner of the subsystem, towards the EGOG subsystem. These are related to the southerly movement of the ITCZ but are also contributed to by river run-off from the EGOG subsystem.

In the EGOG subsystem, during February to March, warm water pushes in from the south. During April, the major warm water areas appear to be coming more from the east and north-east of the subsystem. In May, warm water areas are also present on the north coast. In June, the warm water extent diminishes and cold water appears in the south. The extent of warm water areas is minimal from July to September. Small warm water patches begin to occur along the Cameroon coast, in the north-east of the subsystem, in October, probably associated to some degree with river run-off. In November and December these areas spread and are also present along the northern coast. This southward expansion of warm water appears, again, to be associated with the seasonal movement of the ITCZ.

- Minor Coastal Upwelling

The minor upwelling season along the Ghanaian and Ivorian coastlines appears to start in January, when upwelling appears over the continental shelf from Cape Palmas to the Volta delta, with the largest area over the wide shelf to the east of Cape Three Points. In February, upwelling appears to be absent from the area to the east of Cape Three Points, although is present around this headland. The spatial extent of upwelling along the concave coast to the east of Cape Palmas is approximately the same as in January, however, SSTs are slightly cooler. A cool water strip over the narrow continental shelf from Cape Palmas to Sierra Leone has also developed over this month. This appears coolest around the shelf break (50-200m). The cool strip turns offshore at Sherbro Island, following the continental shelf edge, and forms a 'pool' of cool water. It then appears to connect with the most southerly extent of the SUI. In March, it appears that cold waters are present along the entire coast from Togo to

Sierra Leone. Along the Ghanaian coast, SSTs are warmer than in January and the spatial upwelling extent is reduced. Along the Ivorian coast, upwelling is reduced compared to February, with the main upwelling centre to the east of Cape Palmas. The cool strip is still present along the Liberia and Sierra Leone coasts, following the shelf edge at Sherbro Island to move offshore and meet up with the SUI. A cold water plume off Sherbro Island appears intense. In April, some upwelling is still observed off Ghana, along the coast to the east of Cape Three Points. A cool strip is also present along the coast west of Cape Three Points up to Sierra Leone.

- Major Coastal Upwelling

The major upwelling season appears to start in June. Upwelling in this month is greatest to the east of Cape Three Points. In July, there is a large drop in SST and strong upwelling occurs along the entire coast from Cotonou, Benin to Cape Palmas. The upwelling is greatest to the east of Cape Three Points, where it has a considerable offshore extent, and there is strong putative frontal development between the upwelled water and offshore water. The most intense upwelling appears to occur in August along the entire coast from Cape Palmas to Cotonou. The upwelled waters extend well beyond the continental shelf edge. SST are reduced from the coast to the equator (equatorial upwelling) throughout the whole of the CWAU subsystem. As such, frontal development between upwelled waters and offshore waters appears weakened. In September, upwelling is weaker than in August but still strong, extending to or beyond the shelf edge. In October, upwelling is greatly reduced in both spatial extent and intensity. The main upwelling area is seen on the Ivorian coast to the east of Cape Palmas, although some cold water is still visible around Cape Three Points. In November, warm water covers the whole coastal area. Cool waters are still observed in the equatorial area, but these are detached from coastal SST.

- Fronts

As the SUI annually migrates south, it is seen to encircle the Bijagos Islands of Guinea-Bissau for a few months every year. In the average monthly SST images, this first becomes apparent in December. By January, the cold waters of the SUI have encircled the islands. During February, warmer waters begin to appear around the islands. As the extent of warm water around the islands increases throughout March, a strong temperature gradient, suggestive of frontal development, is apparent between the two water masses. In April, the SUI appears to encircle the islands even further. During May, the SUI begins to retreat but is still present in June as a thin plume around the islands, although the strength of the front is much reduced.

The near connection of SUI waters with the Sherbro Island upwelling in February forms a strip of cool water that separates the warm shelf waters from the warm offshore oceanic waters. This front is also apparent in March and April, however, the extent of warm shelf waters is much reduced in these months.

The strongest frontal development associated with the major upwelling appears to be in July. Strong temperature gradients are also well developed in June when the upwelling is not as intense and the upwelled water not as cool, but the putative front is closer to the shore. Additionally, strong temperature gradients exist in August but the putative fronts between water masses appear to break down due to cooling offshore. During the minor upwelling, the coolest temperatures and strongest putative frontal development appear to occur in January.

Interannual Variability
- General

Figure 5-6 (colour plate) shows the mean SST images for each year. In general 1982-1986 was a cool period and 1987-1991 was warm. Also, 1984 appears to have been a warm year. The monthly composites of mean SST for each year show some change in the influence of the Senegalese upwelling throughout the time series, however, the warming visible in the central Atlantic and eastern Gulf of Guinea appears to dominate the interannual variability. The time series of spatial means for each subsystem shows a trend of alternating warm and cool years from 1981 to 1986 then an extended warm period from 1987 to 1991 (Figure 5-7). This pattern is also clearly shown for each subsystem in the charts of interannual variability in global SST anomaly values for each week in the time series (Figure 5-8, colour plate). Additionally, unusually warm conditions are visible in 1984, as is an anomalously cool period from mid-1982 to early 1983.   Towards the end of 1991, cool temperatures, comparable with those generally seen in the first half of the time series, become visible again. In the SLGP subsystem, the first half of 1985 also appears unusually cool.

Power spectra generated for the spatial mean SST time series for each subsystem (Figure 5-9) show three peaks: two explaining the seasonal cycle (1 cycle per year and 2 cycles per year) and one explaining interannual variability (less than 1 cycle per year). The SST signal represented by the interannual peak shows the same trend as the smoothed time series for each subsystem, but more clearly (Figure 5-10). The oscillations have a period of approximately 3 years. The amplitude of the oscillations varies between the subsystems with the CWAU and EGOG showing the greatest similarity. The SLGP shows the cooler trend from 1981 to 1987 most clearly, however, all three subsystems show a large peak around mid 1987 to 1988. In 1991, the SLGP appears to be re-entering a warm phase, whereas the CWAU and EGOG appear to be cooling.

- Coastal Upwelling

In the annual mean SST images, the spatial extent of coastal upwelling appears increased in 1983 and reduced in 1984 and 1987-1990. Table 5-1 gives a qualitative measure of interannual variability in the strength of the upwelling along the coasts of Ghana and Côte d'Ivoire using an index that takes both intensity and spatial extent into account. This index shows the same pattern of variability as the SST images. The combined upwelling index for the major season shows strong upwelling in 1982 to 1983, weak upwelling in 1984 and an extended period of weak upwelling from 1987 to 1990 (Figure 5-11). The minor upwelling shows a similar pattern from 1984 to 1989 but then diverges with a continued weakening of the upwelling from 1990 to 1991. It is mainly the spatial component of the upwelling index that follows this pattern. The upwelling intensity shows strong interannual variability with peaks in 1983 and 1986 and troughs in 1984 and 1987, with this interannual scale variability also dominating from 1987 to 1991 rather than the longer-term pattern of an extended warm period seen in other indices.

- SUI

In the mean annual SST images, the SUI appears to be reduced in 1983 and 1984 and from 1987 to 1991 and to increase slightly in 1985. Table 5-2 shows interannual variability in the most southerly extent (latitude) of the SUI from the monthly mean SST images for each year. The SUI was always seen to be furthest south during March or April. The pattern of reduced and strong years for the SUI observed qualitatively in the SST images is also seen using this index, plotted in Figure 5-11 as the interannual anomaly from the mean most southerly position, standardised with respect to the standard deviation.

Year	Spatial extent		Intensity		Combined	
	Minor	*major*	*minor*	*Major*	*Minor*	*major*
1981	No data	0.30	No data	53	No data	16.09
1982	0.06	0.82	20	52	1.11	42.88
1983	*	0.69	*	70	*	48.32
1984	0.40	0.13	25	41	9.98	5.51
1985	0.61	0.48	37	46	22.61	22.16
1986	0.51	0.61	37	52	19.00	31.65
1987	0.10	0.08	15	30	1.51	2.33
1988	0.11	0.11	36	39	3.92	4.24
1989	0.18	0.10	33	44	5.90	4.23
1990	0.14	0.28	19	39	2.69	10.79
1991	0.01	0.46	10	53	0.12	24.44

Table 5-1 Upwelling Index combining spatial extent and intensity. * represents seasons severely contaminated by atmospheric water vapour.

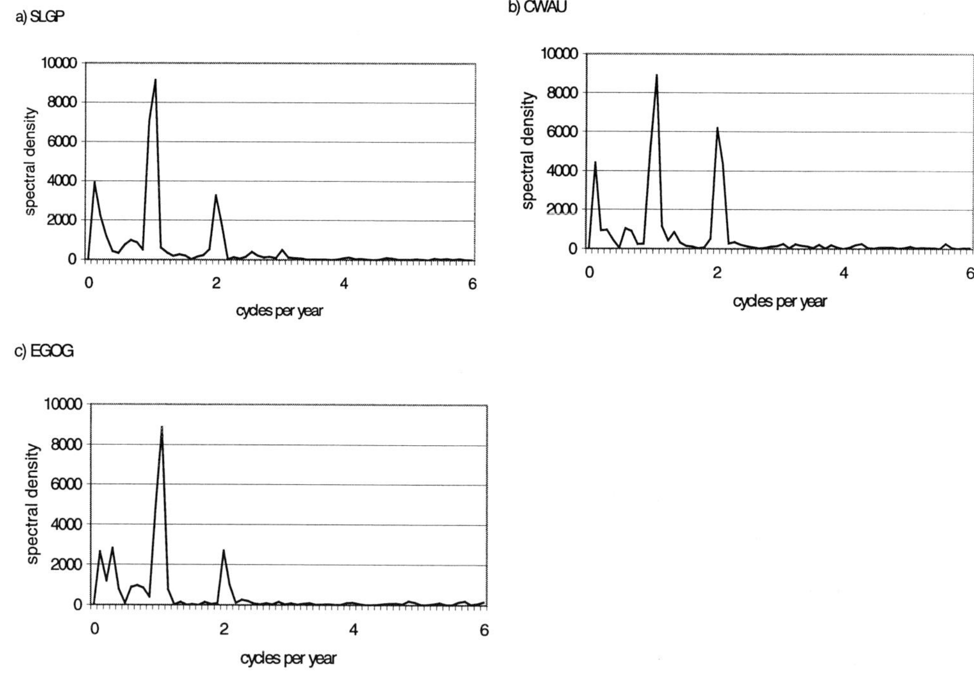

Figure 5-9 Power spectra of the time series of spatial mean SST for each subsystem: **a** SLGP, **b** CWAU and **c** EGOG. The spectra are filtered with a Hanning window, with five degrees of freedom.

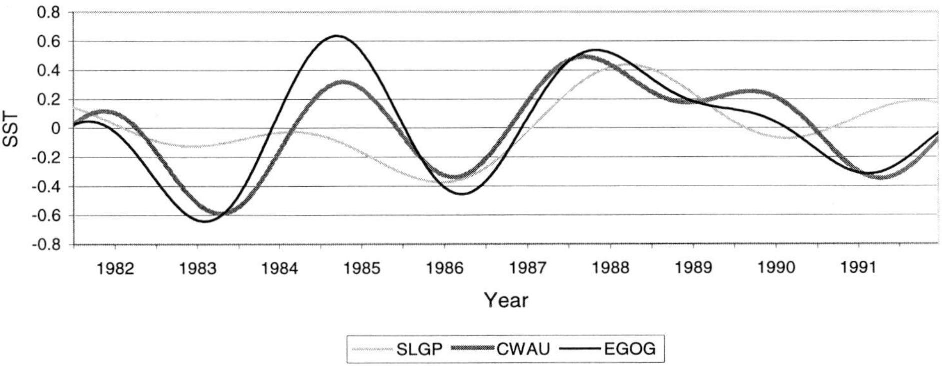

Figure 5-10 SST signal extracted from the time series of weekly spatial means using a rectangular bandpass filter, for frequencies corresponding to the interannual spectral peaks for each subsystem.

Year	Lat.(N)	Lat. Anomaly	Month
1982	5.719	-0.91	March
1983	7.071	0.37	April
1984	7.483	0.75	April
1985	4.425	-2.13	March
1986	6.189	-0.47	April
1987	7.894	1.14	April
1988	6.777	0.09	March
1989	7.777	1.03	April
1990	6.307	-0.35	April
1991	7.189	0.48	April

Table 5-2 Most southerly latitude of SUI, standardised anomaly from the mean most southerly position and month of reaching this limit.

Discussion

Seasonal Pattern

The temporal mean SST images clearly show the SUI and the upwelling seasons, as does the seasonal SST anomaly. These dominant features of the subsystem are also the most variable, with most of the variance being explained by the seasonal cycle. The cold SUI is present in the Gulf of Guinea from December to May. The SUI almost encircles the Bijagos Islands

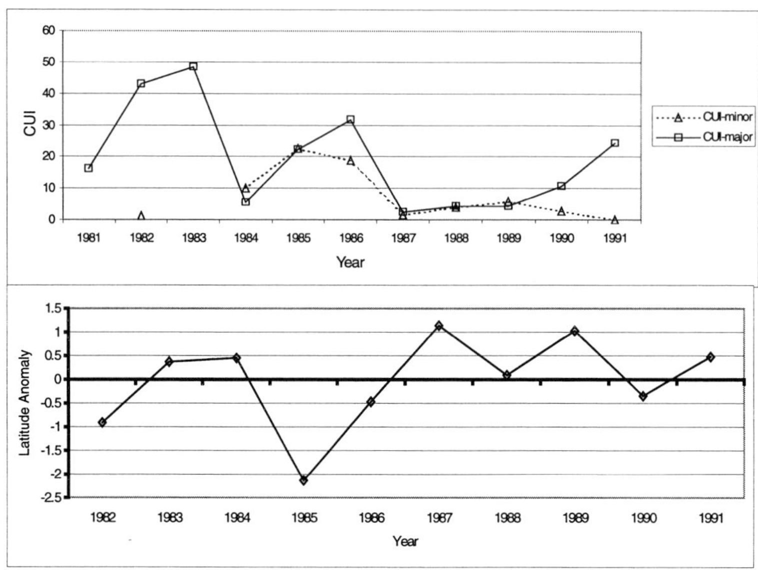

Figure 5-11 Combined upwelling index (CUI) values for the major (CUI-major) and minor (CUI-minor) upwelling seasons. Standardised anomaly from the mean most southerly position of the Senegalese upwelling influence (SUI) for each year.

from February to May with a strong temperature gradient, suggesting strong frontal development, between the warm and cold water masses. Demarcq and Citeau (1995) also showed this pattern. The SUI then recedes to the north and is absent from the Gulf of Guinea from June to November.

The major upwelling season occurs during June to October but is dominant from July to August; the minor season appears strongest in January but is also visible during February and March and cool temperatures have been observed in some areas both in December and as late as May. Both upwelling seasons appear to commence to the east of Cape Three Points. This also appears to be the area of greatest upwelling strength. The upwelling appears then to shift westwards towards Cape Palmas. The position of the major upwelling cells to the east of the capes, downstream in the Guinea current, provides support for the cape effect theory of Marchal and Picaut (1977). This effect is also seen to a lesser degree around the mouth of the Volta Delta. The observed westward shift in dominance from Cape Three Points to Cape Palmas is consistent with the results of Picaut (1983) who observed a westward propagation of the upwelling, counter to the dominant surface flow. Such observations support the theory of the upwelling propagating via a coastally trapped wave (Moore *et al.* 1978; Clarke 1979).

The timing of these seasons seen in the CORSA-AVHRR data is in agreement with other observations (Aman and Fofana 1995, 1998, Roy 1995), as is the position of the major upwelling. However, the position of the minor upwelling contradicts the results seen in the Côte d'Ivoire and Ghana coastal station SST time series, which recorded its presence only along the Ivorian coast (Arfi *et al.* 1991). The appearance of cold SST along the coast of Liberia and Sierra Leone in early February suggests that the minor upwelling may extend up the coast from Cape Palmas during February to April. This cooling also turns offshore at Sherbro Island, following the continental shelf edge and connecting with the SUI during

February and March.   Whether the cooling along this coast is strengthened by this connection, perhaps with cold SUI water pushing south along the coast, is not clear. The rapid appearance of this coastal cooling is consistent with the speed of a coastally trapped Kelvin wave travelling at 0.7 ms^{-1} (Picaut 1983) to 1.7 ms^{-1} (Houghton 1983) along a coast of approximately 650 km in length (Cape Palmas to Sherbro Island) over a period of one to two weeks. The continental shelf is very narrow in this area and the shelf break is quite shallow. Interaction between a trapped wave in the shoaling thermocline and the shelf break may also contribute to the observed upwelling in this area. Bakun (1996) also describes a seasonal cooling observed off the coast of Sierra Leone in the 1970s, which he attributed to a coastally trapped wave, but this feature was present in August and September, when no cool feature is observable in the seasonal SST images. Bakun's observation could be related to, or masked by, a general cooling in the central SLGP subsystem during July to September, however, this observed cooling coincides with the maximal level of cloud cover over this area, associated with the overhead presence of the ITCZ. Thus, the cooling may be caused by contamination of the images by atmospheric water vapour or high cloud albedo leading to reduced insolation of surface waters. Whether the observed large scale cooling is a real oceanographic phenomenon or not, no upwelling signal is observed to propagate along the coast of Liberia and Sierra Leone at this time of year.

Both the major and minor upwelling seasons appear to commence with rapid drops in SST in January and July. These events, combined with the rapid propagation of the upwelling signal along the coast suggest that the onset of upwelling is triggered by coastally trapped, upwelling Kelvin waves causing shoaling of the thermocline. However, this does not explain why coastal cooling is also seen more weakly in June. This change could be explained by changes in the local and cross-equatorial winds and intensification of the Guinea current, as demonstrated in the models of Philander (1979) and Philander and Pacanowski (1981). Additionally, Ingham (1970) suggested that upwelling along the coasts of Ghana and Côte d'Ivoire occurred all year round but was not strong enough to penetrate the strong thermocline in the area. Instead, minor cooling took place by mixing in the Guinea current.

Fronts associated with the seasonal upwellings are visible in the SST record. Along the coasts of Ghana and Côte d'Ivoire during the major upwelling, the strongest frontal development tends to occur in July and occurs beyond the shelf edge. The offshore position of this front explains why no such structure was observed in the coastal studies of Houghton (1976). During the minor upwelling, the strongest frontal development is associated with the areas of lowest SST and appears to occur in the vicinity of the shelf edge along the coast of Ghana and Côte d'Ivoire in January and Côte d'Ivoire in February. The connection of the cool waters along the coasts of Liberia and Sierra Leone with the SUI in February and March forms a cold water front along the shelf edge between the warm shelf waters of the wide Guinea continental shelf and the warm, offshore oceanic water.

Interannual Pattern

The annual mean SST images show a cool period from 1982 to 1986 and a warm period from 1987 to 1990. During the initial cool period, 1982 to 1983 was cooler than average and 1984 was exceptionally warm. This pattern is also observed in the spatial mean SST time series, the spatial mean SST anomalies, and the SST signal represented by the interannual peak of the power spectra. The initial cycles have a period of around 3 years, approximately the same period as El Niño events.   Indeed, the unusually warm conditions in 1984 have been attributed to an Atlantic Niño event (Hisard *et al.* 1986; Hisard 1988; Philander 1986).   This suggests that the cool periods in the years either side of this warm event could be the Atlantic equivalent of Pacific la Niñas or cold events.   The elongated warm period, from 1987 to 1991, could be due to a shift in the system or the presence of a longer, perhaps decadal, scale cycle.   Changes in coastal sea temperatures and salinity, consistent with this prolonged warm period, were also noted by Koranteng and McGlade (this volume).

The combined upwelling index shows the same pattern, with strong upwelling in 1982 to 1983, weak upwelling in 1984 and an extended period of weak upwelling from 1987 to 1990.   This is more obvious for the major upwelling season.   The interannual behaviour of the seasons appears to diverge from 1990 to 1991 with the major upwelling beginning to strengthen but the minor upwelling continuing to weaken.   It is mainly the spatial component of the upwelling index that follows this pattern.   The upwelling intensity shows strong interannual variability with a period of approximately 3 years throughout the time series rather than giving way to the longer-term pattern of an extended warm period observed in other indices.   This suggests that an interannual mode of variability, perhaps related to an Atlantic Niño cycle, may influence the intensity of upwelling.   Underlying both components of the upwelling index is a long-term linear trend of weakening in the upwelling.

Binet and Marchal (1993) reported considerable changes in the spatial distribution of the major upwelling season seen in the coastal station SST record (Arfi *et al.* 1991).   From 1969 to 1980 the observed spatial pattern agreed with the concept of two upwelling cores, seen on the eastern side of Cape Palmas and Cape Three Points.   Temperature dropped abruptly on the downstream side of each cape, then increased progressively eastward as the waters were carried by the Guinea current (Morlière and Rébert 1972).   However, in 1982 the observed decrease in temperature was almost the same along the entire coast.   Furthermore, from 1983 to 1986 the pattern was reversed with the coldest waters found to the western side of Cape Three Points.   This study finds no evidence for such a shift, rather the upwelling is seen most strongly to the east of Cape Palmas and Cape Three Points in all years.   However, the observation by Herbland and Marchal (1991) of very weak upwelling at all stations in 1987 is supported by the SST images used in this study.

Another trend has also been observed in the strengths of the coastal upwellings. Pézennec and Bard (1992) noted an intensification of the minor upwelling season and reduction in strength of the major upwelling season over the period 1970 to 1990.   This trend is not seen in the combined upwelling index values from this study, although the strength of the minor upwelling is seen to weaken at a lesser rate than that of the major upwelling.   This pattern is mainly contributed to by the intensity component of the index.   Allowing for the methods used by Pézennec and Bard (*ibid.*), which removed any long term trend from seasonal components of the signal, these two studies are in agreement, i.e. the intensity of the minor season has increased relative to that of the major season).

The between-year index of the most southerly position of the SUI also shows a high degree of interannual variability, however, the pattern of this is different to that observed in SST and upwelling indices, described above.   Rather, peaks occur in 1984 and 1987 (further north) and there is a trough in 1985 (further south).   On the other hand, there is an extended period of high index values from 1987 to 1991, resembling the pattern in SST and upwelling index values.

Relevance of Features Observed to Fisheries Recruitment

Oceanographic features of relevance to fisheries recruitment, according to the triad hypothesis (Bakun 1993, 1996), must affect enrichment, concentration and retention/transport mechanisms.   In the Gulf of Guinea, several features have been identified from the SST images that may influence these parameters.   The Ghana and Côte d'Ivoire coastal upwelling appears to be the largest putative enrichment mechanism in the images.   Both within and outside of the upwelling areas, river run-off may also provide nutrient enrichment.   Both river plumes, observed in the eastern Gulf of Guinea, and fronts, such as are seen at the edge of the upwelling area, may act by concentrating the nutrients supplied by such enrichment processes.

Binet and Marchal (1993) described the relationship between enrichment from upwellings and river flows and plankton biomass in the coastal upwelling area.   Primary production throughout the boreal summer is first stimulated by nutrient input from the run-off of the first rains over the land in June, then by the major upwelling season, and finally by the flood of the larger rivers in September and October.   The number of phytoplankton cells is, therefore, at a peak from June to September.   This seasonal peak in primary production has also been noted by John *et al.* (this volume).   Zooplankton (copepod) biomass follows the same seasonal pattern but with a two week lag time.   The taxonomic composition of phytoplankton and zooplankton is also highly related to the different hydrological seasons (Binet and Marchal 1993).

Fronts can also act as retention mechanisms.   The strong frontal development between the cold SUI and the warm Guinea-Bissau shelf waters around the Bijagos Islands may be a powerful example of this.   The front observed at the edges of the coastal upwelling area is also a potential retention mechanism.

Conclusions

The results of this study show the usefulness of remote sensing for investigating seasonal and interannual variability in SST in the Gulf of Guinea.   Additionally, many of the features observed are interpretable in terms of physical processes known to exist in the region and analysis of these data can aid understanding of the underlying physical dynamics.   Of particular interest are observations associated with the coastal upwellings.   Some of these corroborate previous observations, such as the westward movement of the upwelling (Picaut 1983), the "cape effect" (Marchal and Picaut 1977) and the relative intensification of the minor season (Pézennec and Bard 1992).   Additionally, the observed rapid onset of upwelling lends some support to the remote forcing theory of Moore *et al.* (1978) and the coastally trapped wave theory of Clarke (1979).   Other observations contradict earlier studies, such as the observed presence of an offshore front during the major upwelling season (Houghton 1976) and the absence of a shift in position of the main upwelling cores (Binet and Marchal 1993).   Furthermore, the previously unrecorded observation of coastal cooling in Liberia and Sierra Leone, coinciding with the minor upwelling season in Ghana and Côte d'Ivoire, raises more questions regarding the different mechanisms forcing the two upwelling seasons. Quasi-cyclic behaviour of SST on interannual scales has also been observed.   The possible presence of an El Niño scale (3-5 year) cycle and the presence of warmer than average temperatures throughout the region in 1984, coinciding with an Atlantic Niño event, suggest that global scale climate interactions may be forcing SST in the Gulf of Guinea to some extent. However, the length of the time series used limits any further speculation. Finally, this study has shown how satellite derived SST time series can be used to resolve oceanographic features of putative relevance to fisheries recruitment, and their seasonal and interannual variability. Further details surrounding this research are given in Hardman-Mountford (2000).

Acknowledgements

The authors wish to express their gratitude to the European Commission for their support of this project. J.M.McGlade would also like to acknowledge the support provided by NERC (GT5/00/MS/1).

References

Aiken J, N. Rees, S. Hooker, P. Holligan, A. Bale, D. Robins, G. Moore, R. Harris and D. Pilgrim 2000 The Atlantic Meridional Transect: overview and synthesis of data. Progress in Oceanography 45(3-4): 257-312.

Allersma, E. and W.M.K. Tilmans 1993 Coastal conditions in West Africa - a review. Ocean and Coastal Management 19: 199-240.

Aman, A. and S. Fofana 1995 Coastal sea surface temperature as detected by Meteosat satellite and received at the University of Abidjan. Pp52-59. *In* F.X. Bard and K.A. Koranteng (eds.). *Dynamics and Use of Sardinella Resources from Upwelling off Ghana and Ivory Coast.* ORSTOM Editions, Paris.

Aman, A. and S. Fofana 1998 Spatial dynamics of the upwelling off Côte d'Ivoire. pp 139-147. *In* M.H. Durand, P. Cury, R. Mendelssohn, C. Roy, A. Bakun and D. Pauly (eds). *Global versus local changes in upwelling systems.* Editions ORSTOM, Paris.

Arfi, R., O. Pézennec, S. Cissoko, and M. Mensah 1991 Variations spatiales et temporelles de la résurgence ivoiro-ghanéenne. pp162-172. *In*: P. Cury and C. Roy (eds.). *Variabilité, instabilité et changement dans les pêcheries ouest africaines.* ORSTOM Editions. Paris.

Bakun, A. 1978 Guinea Current upwelling. Nature 271: 147-150.

Bakun, A. 1993 The California Current, Benguela Current and Southwestern Atlantic Shelf Ecosystems: A Comparative Approach to Identifying Factors Regulating Biomass Yields. pp 199-221. *In* K. Sherman, L.M. Alexander, and B. Gold (eds.). *Large Marine Ecosystems – Stress, Mitigation and Sustainability.* American Association for the Advancement of Science, Washington, DC.

Bakun, A. 1996 *Patterns in the Ocean: Ocean Processes and Marine Population Dynamics.* California Sea Grant College System, NOAA, La Jolla, California.

Binet, D. and E. Marchal 1993 The Large Marine Ecosystem of the Gulf of Guinea: long-term variability induced by climatic changes. pp 104-118. *In*: K. Sherman, L.M. Alexander, and B. Gold (eds.). *Large Marine Ecosystems – Stress, Mitigation and Sustainability* . American Association for the Advancement of Science, Washington DC.

Chatfield, C. 1989 *The Analysis of Time Series: An Introduction.* 4th Edition. Chapman and Hall, London. 241 p.

Clarke, A.J. 1979 On the generation of the seasonal coastal upwelling in the Gulf of Guinea. Journal of Geophysical Research 84 (C7): 3743-3751.

Cole, J.F.T. 1997 *The surface dynamics of the Northern Benguela upwelling system and its relationship to patterns of clupeoid production.* PhD Thesis, University of Warwick, UK.208p.

Cole, J. and J.McGlade 1998a Clupeoid population variability, the environment and satellite imagery in coastal upwelling systems. Reviews in Fish Biology and Fisheries 8: 445-471.

Cole, J.F.T. and J.McGlade 1998b Temporal and spatial patterning of sea surface temperature in the northern Benguela upwelling system: possible environmental indicators of clupeoid production. South African Journal of Marine Science 19: 143-157

Cury, P. and C. Roy 1989 Optimal environmental window and pelagic fish recruitment success in upwelling areas. Canadian Journal of Fisheries and Aquatic Sciences 46 (4): 670-680.

Demarcq, H. and A. Aman (this volume, chapter 7) A multi-data approach for assessing the spatio-temporal variability of the Ivorian-Ghanaian coastal upwelling: interest for understanding pelagic fish stock dynamics.

Demarcq, H. and J. Citeau 1995 Sea surface temperature retrieval in tropical area with METEOSAT: the case of the Senegalese coastal upwelling. International Journal of Remote Sensing 16 (8): 1371-1395.

Hardman-Mountford, N. J. 2000 *Environmental variability in the Gulf of Guinea Large Marine Ecosystem. Physical features, forcing and fisheries.* PhD Thesis, University of Warwick, UK. 251p.

Hardman-Mountford, N.J. and J.M. McGlade (this volume, chapter 6) Defining ecosystem structure from natural resource variability: application of principal components analysis to remotely sensed sea surface temperature.

Herbland, A. and E. Marchal 1991 Variations locales de l'upwelling, répartition et abondance des sadinelles en Côte d'Ivoire. *In* : P. Cury et C. Roy (eds.) *Variabilité, instabilité et changement dans les pêcheries ouest africaines*, ORSTOM, Paris.

Hernández-Guerra, L. and L. Nykjaer 1997 Sea surface temperature variability off northwest Africa: 1981-1989. International Journal of Remote Sensing 18: 2539-2558.

Hisard, P. 1988 El Niño response of the tropical Atlantic Ocean during the 1984 year. pp273-290. *In*: T.Wyatt and M.G. Larrañeta (eds). *International Symposium on Long Term Changes in Marine Fish Populations.* Vigo, Spain. Consejo Superior de Investigaciones Cientificas.

Hisard, P., C. Hénin, R. Houghton, B. Piton, and P. Rual 1986 Oceanic conditions in the tropical Atlantic during 1983 and 1984. Nature 322: 243-245.

Houghton, R.W. 1976 Circulation and hydrographic structure over the Ghana continental shelf during the 1974 upwelling. Journal of Physical Oceanography 6: 909-924.

Houghton, R.W. 1983 Seasonal variations of the subsurface thermal structure in the Gulf of Guinea. J. of Physical Oceanography 13: 2070 - 2081

Ingham, M.C. 1970 Coastal upwelling in the Northwestern Gulf of Guinea. Bulletin of Marine Science 20: 1-34.

John, A.W.G., P.C. Reid, S.D. Batten and E.R. Anang (this volume, chapter 11) Monitoring levels of 'phytoplankton colour' in the Gulf of Guinea using ships of opportunity.

Koranteng, K.A. and J. M. McGlade (this volume, chapter 8) Physico-chemical changes in continental shelf waters of the Gulf of Guinea and possible impact on resource variability.

Mann, K.H. and J.R.N. Lazier 1996 *Dynamics of Marine Ecosystems: Biological-Physical Interactions in the Oceans.* 2nd Edition. Blackwell Science Inc., Boston. 394p.

Marchal, E. and J. Picaut 1977 Répartition et abondonce évaluées par échointégration des poissons du plateau continental ivoiro-ghanéen en relation avec les upwellings locaux. J. Res. Océanogr. 2: 39-57.

Maus, J.W.V. 1997 *Sustainable fisheries information management in Mauretania: Implications of institutional linkages and the use of remote sensing for improving the quality and interpretation of fisheries and biophysical data.* PhD Thesis, University of Warwick. 269p.

Moore, D.W., J.P. McCreary, P. Hisard, J. Merle, J.J. O' Brien, J. Picaut, J.-M. Verstraete and C. Wunsch 1978 Equatorial adjustment in the eastern Atlantic Ocean. Geophysical Research Letters 5: 637-640.

Morlière, A. and J.P. Rébert 1972 Etude hydrologique du plateau continental Ivoirien. Documents Scientific de Centre de Recherche Oceanographique Abidjan 3 (2): 1-30.

Pézennec, O. and F.X. Bard 1992 Importance écologique de la petite saison d'upwelling ivoiro-ghanéenne et changements dans la pêcherie de *Sardinella aurita.* Aquatic Living Resources 5: 249-259.

Philander, S.G.H. 1979 Upwelling in the Gulf of Guinea. Journal of Marine Research 37 (1): 23-33.

Philander, S.G.H. 1986 Unusual conditions in the tropical Atlantic Ocean in 1984. Nature 322: 236-238.

Philander, S.G.H. and R.C. Pacanowski 1981 The oceanic response to cross-equatorial winds (with application to coastal upwelling in low latitudes). Tellus 33: 201-210.

Picaut, J. 1983 Propagation of the seasonal upwelling in the eastern tropical Atlantic. Journal of Physical Oceanography 13(1): 18-37.

Robins, D.B. and J. Aiken 1996 The Atlantic Meridional Transect: an oceanographic research programme to investigate physical, chemical, biological & optical variables of the Atlantic Ocean. Underwater Technology 21: 8-14.

Roy, C. 1995 The Cote d'Ivoire and Ghana Coastal Upwellings: Dynamics and Changes. p346-361. *In*: F.X. Bard and K.A. Koranteng (eds). *Dynamics and Use of Sardinella Resources from Upwelling off Ghana and Ivory Coast.* ORSTOM Editions, Paris.

Vandenhouten, R., G.Goebbels, M. Rasche and H. Tegtmeier 1996 SANTIS – a tool for Signal ANalysis and TIme Series processing Version 1.1: User Manual.

Weeks, S.J. and F.A. Shillington 1994 Interannual scales of variation of pigment concentrations from coastal zone color scanner data in the Benguela Upwelling system and the Subtropical Convergence zone south of Africa. Journal of Geophysical Research 99 (C4): 7385-7399.

The Gulf of Guinea Large Marine Ecosystem
J.M. McGlade, P. Cury, K.A. Koranteng and N.J. Hardman-Mountford (Editors)

6

Defining Ecosystem Structure from Natural Variability: Application of Principal Components Analysis to Remotely Sensed SST

N. J. Hardman-Mountford and J.M. McGlade

Abstract

It has been proposed that the Gulf of Guinea Large Marine Ecosystem can be divided into three subsystems, however, the limits of this system and its constituent subsystems have only been defined arbitrarily. This study uses principal components analysis to investigate the variance structure of remotely sensed SST data for the Gulf of Guinea in an attempt to define important ecosystem structures and boundaries using the natural variability of the system. The results are interpreted in terms of underlying physical dynamics and boundaries of the large marine ecosystem and its subsystems are determined. Inter and intra-comparison of SST in these new subsystems is then used to successfully validate the new subsystems.

Introduction

Ecosystem definition

The definition of meaningful boundaries for marine ecosystems continues to be an area of active research, and heavily debated in both scientific and political arenas (McGlade 1999). For whilst in terrestrial ecology, different biomes have been defined using relatively simple factors such as latitude, altitude, rainfall, slope, underlying soil or rock type and vegetation, in marine science below the scale of regions, i.e. polar, temperate and tropical regions, there no clear consensus has emerged. Longhurst (1998) and Sherman and co-workers (AAAS 1986; Alexander 1986; Sherman 1986) have elucidated two systems of classification, one based on biogeochemical provinces and the other on the productivity and associated socio-economic activities of Large Marine Ecosystems: in the Gulf of Guinea these two systems have identical boundaries (Pauly *et al.* 2000). This study looks at whether or not these boundaries are constant and tests for the presence of any persistent structures within them, using observations of the natural variability in remotely sensed sea surface temperature data.

Systems and Subsystems of the West African Coast

The coastal marine environment of West Africa has been classified into three Large Marine Ecosystems (LMEs): the Canary Current LME (NW Africa), the Gulf of Guinea LME (Central West Africa) and the Benguela Current LME (SW Africa) (Binet and Marchal 1993).

The Gulf of Guinea LME, as defined by Binet and Marchal (*ibid.*) lies between the

67

Bijagos Islands (Guinea-Bissau, ~ 11° N) and Cape Lopez (Gabon, ~ 1° S). This area includes the maritime waters of 12 coastal states (Guinea-Bissau, Guinea, Sierra-Leone, Liberia, Côte d'Ivoire, Ghana, Togo, Benin, Nigeria, Cameroon, Equatorial Guinea and Gabon) as well as the island states of Equatorial Guinea (Bioko, formerly Fernando Po) and Sao Tome and Principe. The system is generally defined by the flow of the Guinea Current so is sometimes referred to as the Guinea Current LME. It is bounded to the north by the Canary Current and to the south by the South Equatorial Current (*ibid.*).

Because of its heterogeneous nature, Tilot and King (1993) divided the Gulf of Guinea LME into three arbitrary subsystems. Each is defined by particular characteristics, which nevertheless contribute to the functioning of the ecosystem as a whole and interact with each other. These subsystems are:

a) *Sierra Leone and Guinea Plateau (SLGP)*: from the Bijagos Islands (Guinea-Bissau) to Cape Palmas (Liberia/Côte d'Ivoire). This area is characterised by the largest continental shelf in West Africa and has large riverine inputs, giving thermal stability. It is also in the seasonal passage of the Inter-Tropical Convergence Zone (ITCZ).

b) *Central West African Upwelling (CWAU)*: from Cape Palmas to Cotonou (Benin). This thermally unstable subsystem is characterised by seasonal upwelling of cold, nutrient rich, subthermocline water, which dominates its annual cycle and drives the biology of the subsystem. Variability in upwelling strength leads to variability in productivity.

c) *Eastern Gulf of Guinea (EGOG)*: from Cotonou to Cape Lopez (Gabon), including the offshore islands of Bioko and Sao Tome and Principe. This area is characterised by thermal stability and a strong pycnocline. Its productivity depends on nutrient input from land drainage, river flood and turbulent diffusion through a stable pycnocline. Variability of river input depends on climatic fluctuations of the monsoon.

Use of PCA in remote sensing studies

PCA has become a commonly used technique for remote sensing image analysis. It is generally used as a technique for data compression between highly correlated spectral bands (Lillesand and Kiefer 1987). However, it has also been used for change detection studies. Fung and LeDrew (1987) used PCA for detecting land-cover change in Canada from Landsat MSS data and Eastman (1992) assessed the usefulness of the technique for studying changes in vegetation cover over Africa using Normalised Difference Vegetation Index (NDVI) images. Eastman (*ibid.*) concluded that the technique has excellent capabilities in the detection of change, regardless of whether it is cyclic, aperiodic repeating or isolated. In addition, it appears to be equally sensitive to changes that are additive or multiplicative in nature. Gallaudet and Simpson (1994) used the technique to investigate oceanographic processes off Baja California using SST from AVHRR data. More recently, both Cole and McGlade (1998) and Maus (1997) used PCA to show spatial structure and temporal variability in upwelling along the West African coast from the CORSA-AVHRR SST product. Cole and McGlade (1998) investigated the Benguela region off Namibia while Maus (1997) studied the Canary Current upwelling off Mauritania. This study attempts to apply this technique to the Gulf of Guinea region.

Rationale for use of PCA in this study

Analysis of the current data set, described in Hardman-Mountford and McGlade (this volume), showed a high degree of variability through time with a strong seasonal cycle and a large degree of interannual variability. Additionally, the images were seen to have a spatially heterogenous structure, with three main areas or subsystems. Furthermore, within the images, a number of oceanographic features were identified. Thus, the SST time series has a high degree of both geometric (space and time) and geophysical (underlying physical dynamics) dimensionality. Therefore, because PCA works on the variance structure of the data, it seemed an appropriate technique to investigate further the different modes of temporal and spatial variability in this highly variable data set.

Methods

Data scales and areas

Composite SST images from the CORSA-AVHRR time series, described in Hardman-Mountford and McGlade (this volume), were used for this analysis. To ensure that the analysis covered the appropriate scales for relevant oceanographic features and natural variability, two different spatial scales were used. The temporal scale used was that of monthly composites as this allowed for the temporal sequence of the data to be best maintained when eliminating time steps with excessive cloud cover.

Firstly, CORSA-AVHRR data for the Gulf of Guinea, was combined with data for the Canary current region, as used by Maus (1997), the Benguela region, as used by Cole and McGlade (1998) and previously unused data for the Angolan offshore area. Monthly composites for each area were joined together using the mosaic routine of the ERDAS *Imagine*® software (ERDAS 1997). Data were available for all areas over the 9.5 year time series from July 1981 to December 1990. This combination of areas gave spatial coverage of tropical and subtropical West Africa, thus allowing significant features at the basin scale to be examined. In the second analysis, the 10.5 year time series, from July 1981 to December 1991, of CORSA-AVHRR data for the Gulf of Guinea region only was used.

Selection of layers and interpolation

For PCA to be performed, the input layers must have any pixels classified as cloud (DN=0) removed. Input layers for the analyses are the monthly SST composite images, in temporal sequence. To achieve this removal of cloud contamination, the whole Gulf of Guinea area was first divided into its three subsystems (Figure 6-1). Each of these subsystems was then subdivided into smaller blocks and the percent cloud cover of each layer was calculated for each block. Any layers with greater than 50% cloud cover in a block or with excessively large areas of cloud were removed from the data set. A total of 10 layers were removed for the Gulf of Guinea analysis. Remaining patchy cloud cover was then eliminated by interpolating SST values from surrounding pixels using the Focal Analysis routine in the software. A square 7x7-pixel kernel was used with all pixels weighted equally. This allowed the interpolation procedure to retain the structure of mesoscale features without being overly dependent on single neighbouring pixel values. The interpolation procedure was repeated until all cloud cover was removed from the data set. No layers were removed for the West

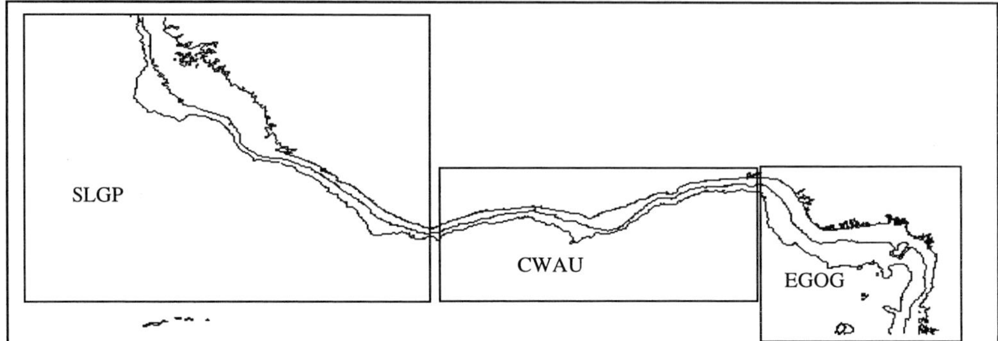

Figure 6-1 Map showing potential subsystem boundaries.

African analysis. Instead, focal analysis was used to interpolate all missing pixels. This was acceptable because only coarse scale features are of relevance to this analysis.

Spatial standardisation

The next stage of data preparation involved spatially standardising the data set so that each layer had zero mean and unit standard deviation. Although a correlation matrix was used for the PCA, spatial standardisation was still performed to facilitate interpretation of the output principal component images by presenting them as positive and negative deviations from the mean.

Principal Components Analysis

Standardised PCA was performed on the prepared data using a modification of the PCA routine in the ERDAS *Imagine*® software package. Modifications included replacing the calculation of the variance-covariance matrix with calculation of the correlation matrix. The mathematical basis of the analysis was taken from Cole (1997) and Cole and McGlade (1998), and modified using Gallaudet and Simpson (1994), Fung and LeDrew (1987) and Kendall and Stuart (1966).

The output from this analysis consisted of principal component (PC) images, eigenvalues and eigenvectors for each image. The PC image gives the spatial output from the analysis and the eigenvalues are a measure of the percent variance explained by each PC image. The eigenvector gives the loadings through time for each PC image, i.e. the temporal output of the analysis.

Interpretation of Principal Components

To be able to interpret the patterns observed in component images and loadings in terms of features of the system, it is necessary to understand how the data are presented. The patterns in the component images are represented in standardised units. For convenience, positive values and negative values are labelled by colour or code; however, the sign of these units is arbitrary and only important for the relative relationship of pixels in the image. Therefore, in some images, positive pixels represent relatively cool areas and negative relatively warm areas, whereas in others the opposite pattern applies. The relative intensity of the colour or grey-scale indicates the strength to which a certain pattern applies in certain areas. Hence, darker tomes show the pattern is strong in a particular area and white represents areas where the pattern is weakly represented. These are often, but not always, close to boundaries between opposite patterns. Black pixels indicate the boundaries between opposite patterns. The relative importance of a particular pattern to each individual time step is given by the loadings. As with the component images, the sign of the loadings is arbitrary. Therefore, large peaks or troughs represent time steps when a pattern in the corresponding component image is either strongly represented or its opposite pattern is strongly represented. Loading values around zero represent time steps when the patterns seen in the corresponding component image are not relevant.

Spectral Analysis

Power spectra were calculated for the loadings to show the temporal modes associated with each of the principal components.

Results and Discussion

Interpretation of PCA for West Africa

Inspection of the eigenvalues for the West African PCA shows only the first two to have easily interpretable patterns, explaining 79.8% and 13.6% of the total variance, respectively (Figure 6-2). These were retained for interpretation. The other components were discarded on the basis that individually they explained very little of the remaining variance. PC I (Figure 6-3a) shows SST values in the tropics to be warm relative to the subtropics. PC I loadings (shown in Figure 6-4) are always positive indicating that this dominant temperature structure is remarkably constant. Thus, PC I appears to relate to the global SST structure based on the fact that the area of greatest solar irradiance occurs in the tropics.

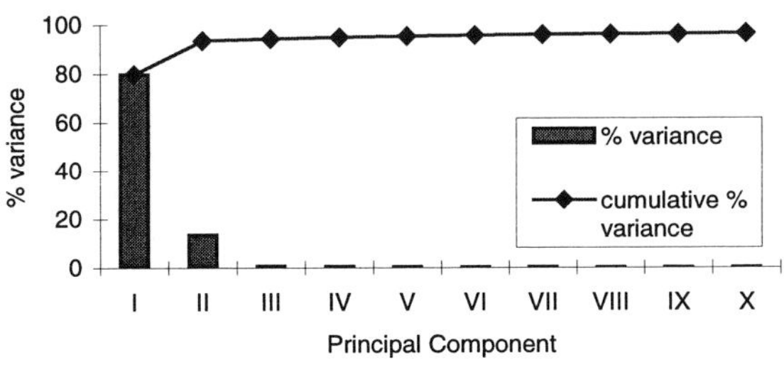

Figure 6-2 Percent variance (eigenvalues) explained by each of the first 10 principal components of the West African PCA.

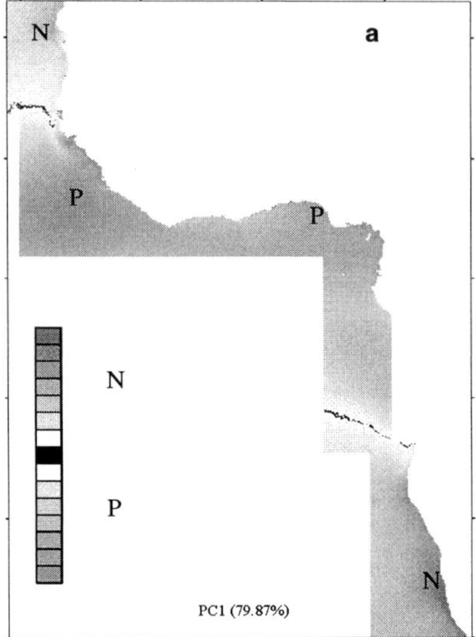

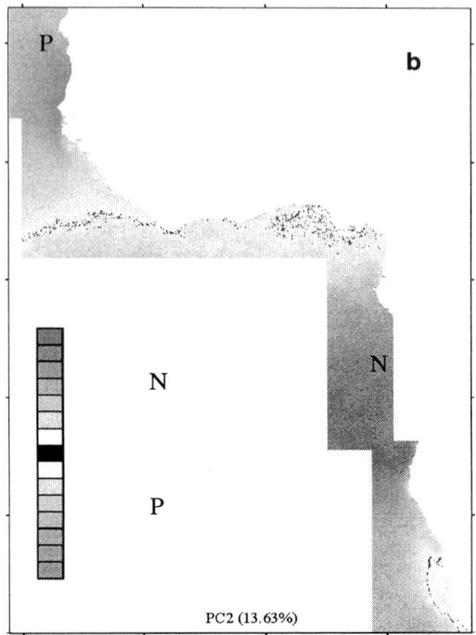

Figure 6-3a PCI and **b** PCII for SST data for West Africa. N represents strong negative correlation with the loadings, P represents strong positive correlation, white represents weak correlation and black represents the boundary between areas of positive and negative correlation (zero correlation).

PC II (Figure 6-3b) indicates that the North Atlantic is warm with respect to the South Atlantic, but with the boundary between the two regions extending westward, perpendicular to the Liberian coast. The loadings for PC II (Figure 6-4) show a strong annual cycle. Comparison of the loadings with the monthly SST composites shows that the peaks occur during September or October of each year. Similarly, troughs, representing the inverse of the pattern seen in the component image, generally occur during February or March, although have been seen in January.

Because the loadings have both positive and negative values, the relative SST pattern can be interpreted both ways. When loadings are positive SST values are warmer north of the boundary than south of it. In the same way, when the loadings are negative, the SST pattern has the opposite structure.

The north-south dipole in SST structure and strong seasonal cycle suggest that the observed pattern represents the reversed seasonal polarity of the different hemispheres. The position of the boundary, however, is situated around 5°N rather than at the equator. This implies that the South rather than North Atlantic dominates the area from the equator to around 5°N and, therefore, the observed pattern is related to oceanic circulation rather than solar irradiance. Additionally, the position of the boundary between the areas appears to correspond to the mean position of the ITCZ and migrates accordingly. This has two consequences: firstly, the boundary may change position seasonally like the ITCZ and secondly, the oceanic dominance of an area may be linked to climate circulation patterns.

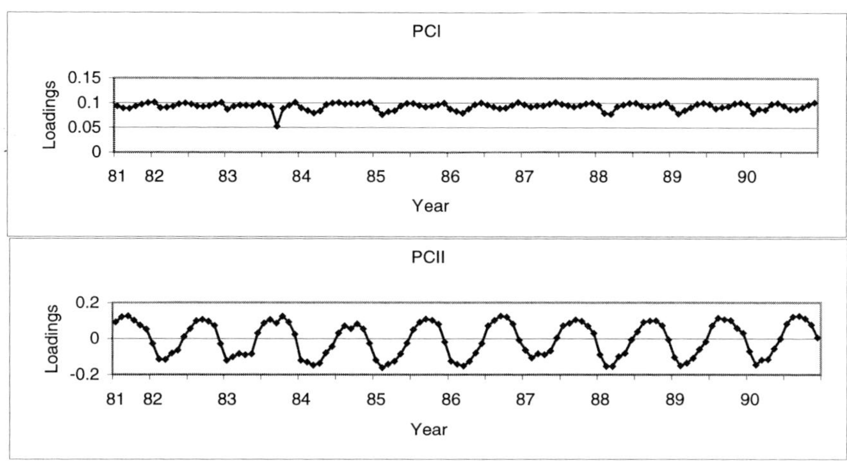

Figure 6-4 Loadings for PCI and PCII for West Africa PCA.

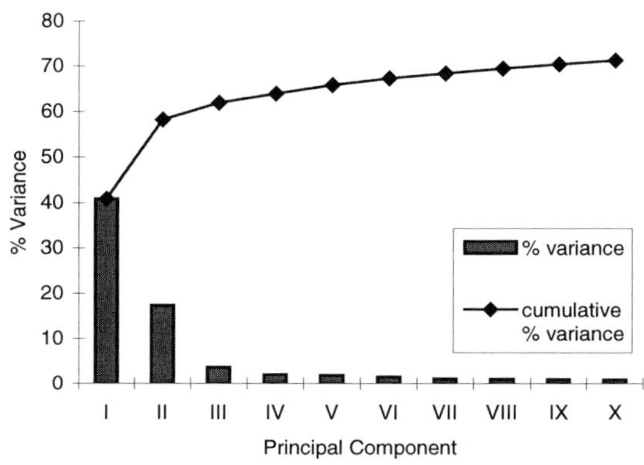

Figure 6-5 Percent variance (eigenvalues) explained by each of the first 10 principal components of the Gulf of Guinea PCA.

Interpretation of PCA for the Gulf of Guinea

The first two Principal Components of the Gulf of Guinea analysis appeared to show meaningful patterns and were retained for interpretation. Between them, they account for approximately 58% of the total variance in the standardised input data set (Figure 6-5). As with the previous analysis, the other components were discarded on the basis that individually they account for very little of the remaining variance. The first two components are discussed individually below.

- Gulf of Guinea PC I

PC I (Figure 6-6) accounts for 40.9% of the total variance. It represents the dominant pattern of variability in the dataset and shows a meridional temperature gradient with the break between relative warm and cold temperatures extending westward from the Liberian coast. PC I loadings show a strong seasonal cycle (Figure 6-7). Power spectra of the PC I loadings (Figure 6-8) clearly show the annual peak and additional peaks with frequencies of 2 cycles per year and 4 cycles per year. The peak at 4 cycles per year and part of the peak at 2 cycles per year are probably harmonics of the annual peak, however, the biannual (2 cycles per year) peak is larger than the annual peak. This suggests that it is more than just a harmonic of the annual cycle and that if the harmonic effects were removed a real cycle of this frequency would remain with reduced amplitude.

This component appears to have the same spatial and temporal structure as PC II for the West African PCA, however, a slight difference is seen in the position of the boundary between the relatively warm and cold areas. This could be an artefact of processing, based on the relative areas of the inputs for the two analyses. Comparison of the loadings with the monthly SST composites showed that the peaks, representing the pattern seen in the component image, occur each October or November. In these months, the central area of the SLGP subsystem is warm whereas the area to the south, is relatively cool. The SUI is not present in the Gulf of Guinea at this time. Troughs in the SST loadings generally occur during March to April, although have also been seen in January and February. They occur when the SUI is extended far to the south, making the northern area much cooler than the south. The explanation for this principal component is the same as for PC II of the West African PCA. The movement of the SUI and tropical surface water in the SLGP subsystem is consistent with the seasonal migration of the ITCZ

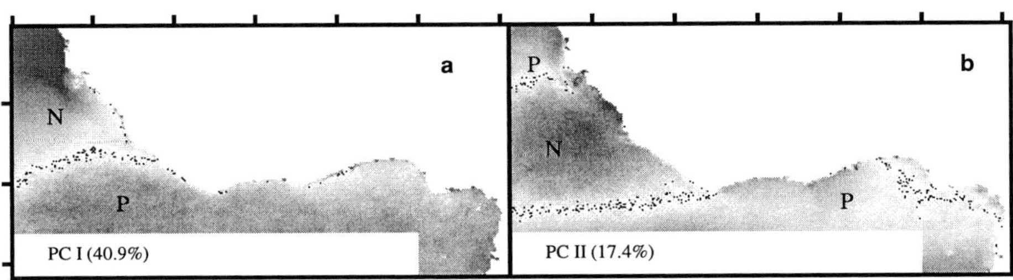

Figure 6-6 a PCI and **b** PCII for the Gulf of Guinea. N represents strong negative correlation with the loadings, P represents strong positive correlation, white represents weak correlation and black represents the boundary between areas of positive and negative correlation (zero correlation).

- Gulf of Guinea PC II

PC II (Figure 6-6b) accounts for 17.4% of the total variance. It represents the main residual pattern of variability when the variability explained by PC I is removed. The pattern shows four areas of contrasting relative temperatures. Area 1 corresponds to the mean position of the Senegese upwelling influence in the Gulf of Guinea. Area 2 extends from the boundary with the SUI in the north to Cape Palmas in the south. Area 3 covers the equatorial Atlantic and Ghana/Côte d'Ivoire coastal upwelling area. Area 4 lies close to the coasts of Nigeria and Cameroon. The SST structure for Areas 1 and 3 varies in phase, as does the SST structure for Areas 2 and 4. Areas 1 and 3 have the opposite structure and phase to Areas 2 and 4.

The loadings for PCII (Figure 6-7) appear to show two annual cycles as well as an interannual trend. This pattern is also seen in the power spectra of the PCII loadings (Figure 6-8) which show peaks at <1 (interannual), 1 (annual) and 2 (biannual) cycles per year. Despite their strong seasonality, the loadings are nearly always positive, suggesting this structure is present, either strongly or weakly, most of the time.

The characteristic shape and the position of *Area 1* clearly identify it as the cold SUI, as already stated. As with PC I, the troughs in the loadings coincide with this feature being near the limit of its southerly extension, during March to April. When the loadings peak, the SUI is either in the far north or absent from the Gulf of Guinea. *Area 2* corresponds to an area of ocean space not identified with any one particular process, but characterised by generally warm temperatures for most of the year. However, it can also be quite a variable area, under the seasonal influence of both the migration of cold Canary Current waters from the north and the influx of warm water from the tropical central Atlantic to the west. It is typically warm when the loadings are at a peak, between October and December, and cold when the loadings are at a trough and the SUI is extended. *Area 3* represents the coastal upwelling areas along the coast, but offshore there is no boundary, instead the area extends both east and west and to the southern limit of the scene. This pattern is strongest in the coastal and equatorial upwelling areas, with a weaker band between them. Thus, Area 3 appears to represent equatorial dynamics, characterised by coastal and equatorial upwelling of cold water. The monthly SST composites show this pattern to be strongest after the end of the major upwelling, when the loadings are at a peak. This is because during this period Area 3 is relatively cool compared to Area 2.

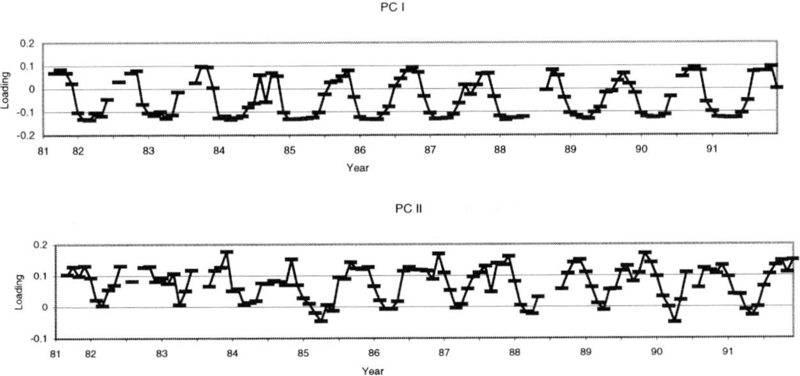

Figure 6-7 Loadings for PC I and PC II of the Gulf of Guinea PCA.

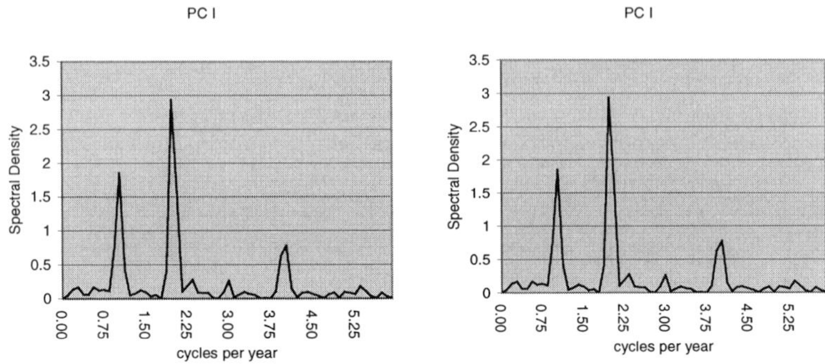

Figure 6-8 Power spectra for PC I and PC II of the Gulf of Guinea PCA.

The position of *Area 4*, close to the coast of Nigeria and Cameroon, appears to identify it with the large areas of warm, low salinity water that exist in this location, originating from river discharge. Again, this pattern is observed when the loadings are at a peak.

The characteristics of these areas appear to be similar to those of Tilot and King's (1993) subsystems, with the exception of Area 1, which represents a boundary feature for the Gulf of Guinea LME. Area 2 coincides with the major part of the SLGP subsystem and Area 4 appears to be dominated by river run-off, a characteristic of the EGOG subsystem. The coastal section of Area 3 corresponds to the CWAU subsystem, However, PCA identifies this zone with equatorial dynamics offshore. This is consistent with the remote forcing theory of coastal upwelling (Moore *et al.* 1978). It appears, therefore, that the areas identified by this principal component can be used to redefine the subsystems of the Gulf of Guinea based on physical dynamics of the system rather than arbitrary boundaries.

Redefining and validating new subsystems

The PC II image for the whole Gulf of Guinea PCA appears to have structured the LME into its constituent subsystems according to patterns of SST variability. The location in coastal areas of the boundaries between these subsystems is remarkably close to those described by Tilot and King (1993), although the offshore extensions of these boundaries vary a great deal. This is not surprising as the offshore limits were defined arbitrarily for the purposes of this study. The loadings for this component show that although the pattern varies seasonally, it is nearly always present to some degree. The northern boundary of the LME is also delineated, in accordance with previous observations.

Additionally, PC I of the West African PCA shows a constant pattern of warm tropical waters compared with cooler subtropical waters. The northern boundary of the warm tropical waters seen in this component corresponds closely to the northern limit of the Gulf of Guinea LME, as defined in this study and that of Binet and Marchal (1993). The southern boundary, however, is situated along the Angolan coast, far south of the predefined limit of the Gulf of Guinea LME at the equatorial upwelling. This raises important questions as to whether the Gulf of Guinea LME extends south of the equatorial upwelling, perhaps including the area around the Angola basin as another subsystem, or whether the tropical system to the south of the equatorial upwelling is a separate LME, perhaps in some ways mirroring the Gulf of Guinea to the north. Furthermore, the structure described in PC I of the West African PCA describes the Gulf of Guinea LME in its tropical context. Thus, these studies appear able to describe the physical structure of the Gulf of Guinea LME in nested spatial scales, from the global context, to the level of the LME and even into its constituent subsystems.

The conclusion that the structures identified in this study correspond to different subsystems of the Gulf of Guinea LME is based on their interpretation in terms of physical features known to exist in the overall system. If these conclusions are correct, then the new subsystems should also appear both similar in nature to each other and different from surrounding systems. To investigate these relationships, time series of SST were calculated for the new areas by taking their spatial means for each week of the original dataset. Area 1, because not part of the Gulf of Guinea LME under this description, was included as a control area. These time series were then used to explore the properties of the new subsystems.

Comparison of Areas

- SST Time Series

Visual comparison and linear regression between the SST time series (Figure 6-9) for each of the new areas showed Areas 2 to 4 to be closely related to each other but not to Area 1. Areas 3 and 4 were the most closely related (r^2=0.66). Area 2 was almost equally related to Areas 3 and 4 (r^2=0.42 and r^2=0.40, respectively). No evidence of a relationship was found between Area 1 and Areas 2, 3 and 4 (r^2=0.01, r^2=0.09, r^2=0.001, respectively). All results are highly significant (F-test, P<0.01), except between Area 1 and Area 4, which is still significant (F-test, P<0.05).

Visual comparison and linear regression of two SST series from within each Area showed a stronger relationship than that between subsystems, except for Area 4, which still showed a strong relationship. This relationship was strongest for Area 1 (r^2=0.89). Area 2 and Area 3 showed approximately the same level of coherence (r^2=0.69 and r^2=0.68, respectively). Area 4 was also significantly coherent (r^2=0.52). All results are highly significant (F-test, P<0.01).

- Seasonal Cycles

Charts of the seasonal cycles in each area show three distinct seasons for Areas 2, 3 and 4, but only two seasons for Area 1 (Figure 6-10). Power spectra for each of the areas show

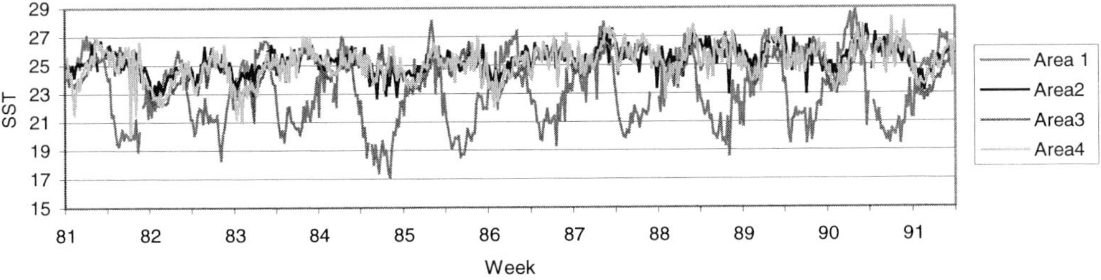

Figure 6-9 SST time series for each of the Areas identified in PC II of the Gulf of Guinea PCA.

peaks with frequencies at both 1 cycle per year and 2 cycle per year for Areas 2, 3 and 4, however, only an annual peak (1 cycle per year) is seen for Area 1 (Figure 6-11). Charts of the seasonal cycles for the two SST series from within each Area show the seasonal cycle of each Area to be consistent throughout that Area.

• Interannual Variability

To investigate interannual variability between the areas, the interannual peaks from the power spectra were used to extract the interannual SST pattern from the SST data (Figure 6-12). SST monthly anomalies were also calculated for the time series of monthly SST data and these were used for linear regression between the areas.

The interannual pattern from power spectra for Areas 2, 3 and 4 showed an alternating cycle of warm and cool SSTs, with a period of about three years, with a superimposed prolonged warm period from 1987 to 1990. The prolonged warm period was seen most strongly for Area 2 and was least evident in Area 4. This pattern is very similar to that described in chapter 4. The interannual pattern from power spectra for Area 1 was very different. SSTs appeared to be declining for the first half of the time series and increasing for the second half, with an additional increase in 1988.

Linear regression of the SST monthly anomalies showed stronger relationships than for the SST weekly values, however, the relationships were essentially the same in nature. Area 3 and Area 4 were most strongly correlated (r^2=0.75), Area 2 was more strongly correlated with Areas 3 and 4 (r^2=0.65, r^2=0.53, respectively) than Area 1 (r^2=0.22). Area 1 was least strongly correlated with Areas 3 and 4 (r^2=0.11, r^2=0.14, respectively). All results are highly significant (F-test, P<0.01).

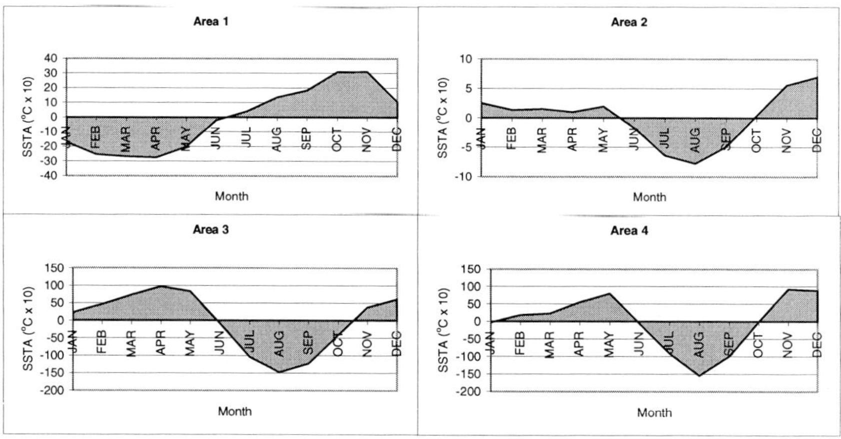

Figure 6-10 Seasonal mean SST anomaly for each of the Areas identified in PC II of the Gulf of Guinea PCA.

Conclusion on new subsystems

The SST signals for Areas 2 to 4 show that they belong to the same system as each other and a different system from Area 1. This is apparent in the similarities between the signals on a seasonal and interannual basis. Areas 3 and 4 appear to be the most closely related. The fixed boundaries used for the subsystems in these validation studies mean that the seasonal contraction and expansion of Areas 1 and 2 has not been taken into account. This means that Area 2 actually has a mixed nature, with cold SUI waters penetrating into the area on a seasonal basis. Nonetheless, the seasonal signal of Area 2 is still very distinct from Area 1. Perhaps, however, on an interannual basis, the closer relationship of Area 1 with Area 2 than with other Areas may be due to variability between years in the most southerly extent of the SUI.

The strong correlation seen for SST within each Area, compared with those between Areas, shows each Area to be a coherent subsystem. The weak relationship within Area 4 compared to its relationship with Area 3 for weekly SST data may be due to high frequency, small scale variability in this Area caused by its dependence on meteorological conditions (rainfall and river run-off). The much stronger relationship seen in the monthly anomalies shows the Area is coherent on an interannual basis, although this is still weaker than the relationship between Area 3 and 4. Nonetheless, the fact that this Area can be interpreted in terms of physical processes known to occur justifies it being defined as a subsystem of the Gulf of Guinea.

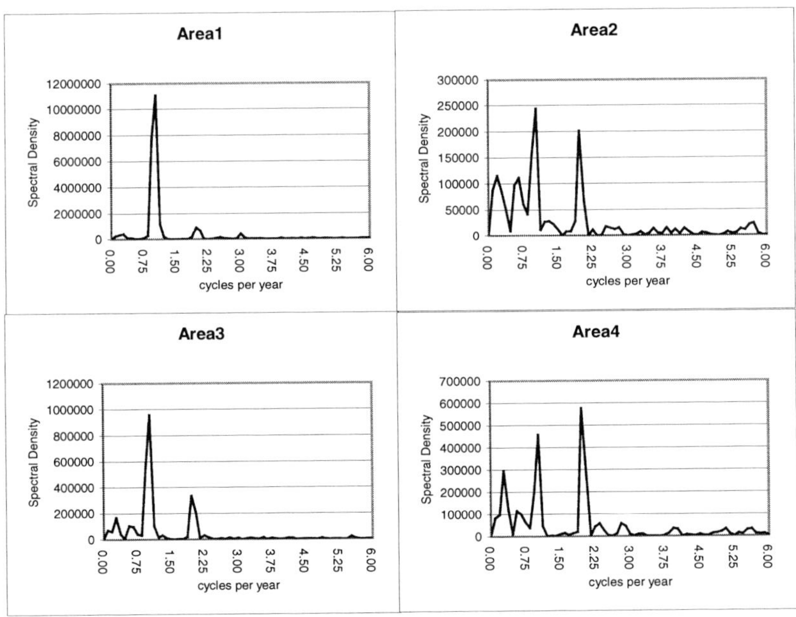

Figure 6-11 Power spectra of SST for each of the Areas identified in PC II of the Gulf of Guinea PCA.

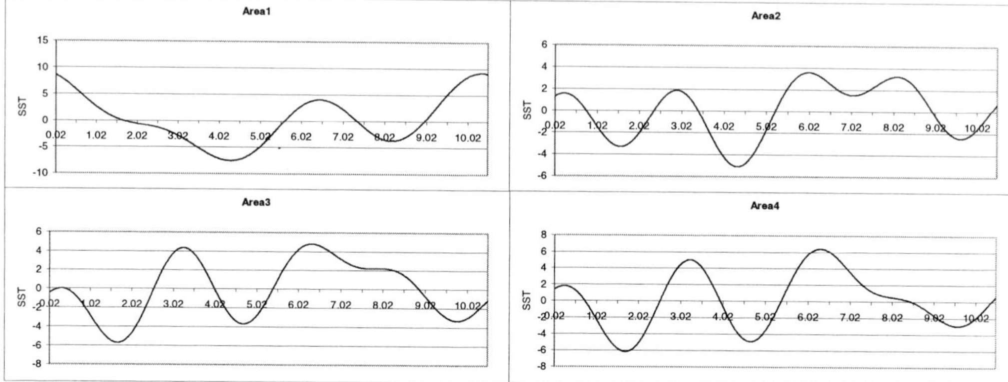

Figure 6-12 Interannual SST cycles from power spectra for each of the Areas identified in PC II of the Gulf of Guinea PCA.

Overall Conclusions

From the interpretations and discussion in the previous sections, the following conclusions can be drawn. Firstly, the Gulf of Guinea LME is situated within the tropical equatorial Atlantic. It's northern boundary is defined by the interface between the SUI and warmer tropical Atlantic waters to the south, with the SUI lying outside the Gulf of Guinea. The southern boundary of the LME cannot be determined from this study. Secondly, the Gulf of Guinea LME can be divided into three subsystems, identified by Areas 2 to 4 of PC II for the whole Gulf of Guinea PCA. These areas correspond closely to the coastal subsystems defined by Tilot and King (1993) and are validated by investigation of their SST signals, both seasonally and interannually. The importance of these boundaries is that they are delineated by the variance structure of physical data, rather than being arbitrarily determined as has previously been the case.

Finally, the seasonal SST dipole picked out in both analyses shows the boundary between areas of North and South Atlantic influence to be situated around 5°N and correspond to the mean position of the ITCZ. This implies that ocean-atmosphere interactions exist in determining seasonal circulation patterns and that depending on the position of the ITCZ, areas of the Gulf of Guinea can be under the influence of either the North or South Atlantic at different times of the year.

Acknowledgements

The authors wish to express their gratitude to the European Commission for their support of this project. J.M.McGlade would also like to acknowledge the support provided by NERC (GT5/00/MS/1).

References

AAAS (American Association for the Advancement of Science) 1986 *Variability and management of large marine ecosystems.* AAAS Selected Symposium 99. Westview Press, Inc., Boulder, CO.

Alexander, L.M. 1986 Large marine ecosystems as regional phenomena. pp239-240. *In*: K. Sherman and L.M. Alexander (eds). *Variability and management of large marine ecosystems. AAAS Selected Symposium* 99. Westview Press, Inc., Boulder, CO.

Binet, D. and E. Marchal 1993 The Large Marine Ecosystem of the Gulf of Guinea: Long-Term Variability Induced by Climatic Changes. pp104-118. *In*: K. Sherman, L.M. Alexander, and B. Gold (eds.). *Large Marine Ecosystems – Stress, Mitigation and Sustainability.* American Association for the Advancement of Science, Washington DC.

Cole, J.F.T. 1997 *The Surface Dynamics of the Northern Benguela Upwelling System and its Relationship to Patterns of Clupeoid Production.* PhD Thesis, University of Warwick. 208p.

Cole, J. and J. McGlade 1998 Temporal and spatial patterning of sea surface temperature in the northern Benguela upwelling system: possible environmental indicators of clupeoid production. South African Journal of Marine Science 19: 143-157

Eastman, J. 1992 Time series map analysis using standardized principal components. ASPRS ACSM RT technical Papers 1: 195-204

ERDAS 1997 Erdas Imagine® version 8.3.

Fung, T. and E. LeDrew 1987 Application of principal components analysis to change detection. Photogrammetric Engineering and Remote Sensing 53 (12): 1649-1658.

Gallaudet, T.C. and J.J. Simpson 1994 An empirical orthogonal function analysis of remotely sensed sea-surface temperature variability and its relation to interior oceanic processes off Baja-California. Remote Sensing of the Environment 47 (3): 375-389.

Hardman-Mountford, N.J. and J. McGlade (this volume, chapter 5) Variability of Physical Environmental Processes in the Gulf of Guinea and Implications for Fisheries Recruitment. An Investigation Using Remotely Sensed Sea Surface Temperature.

Kendall, M.G. and Stuart, A. 1966 *The advanced theory of statistics* (vol. 3). Charles Griffin and Company Ltd, London. 552 p.

Lillesand, T.M. and R.W. Kiefer 1987 *Remote Sensing and Image Interpretation.* (2nd Edition). John Wiley and Sons, New York. 721 p.

Longhurst, A. 1998 Ecological Geography of the Sea. Academic Press, San Diego. 397 p.

Maus, J.W.V. 1997 *Sustainable fisheries information management in Mauretania: Implications of institutional linkages and the use of remote sensing for improving the quality and interpretation of fisheries and biophysical data.* PhD Thesis, University of Warwick. 269p.

McGlade, J. 1999 Ecosystem analysis and the governance of natural resources. pp309-341. *In*: J.M.McGlade (ed.) *Advanced Ecological Theory: Principles and Applications.* Blackwell Science, Oxford.

Moore, D.W., J.P. McCreary, P. Hisard, J. Merle, J.J. O' Brien, J. Picaut, J.-M. Verstraete and C. Wunsch 1978 Equatorial adjustment in the eastern Atlantic Ocean. Geophysical Research Letters 5: 637-640.

Pauly, D., V. Christensen, R. Froese, A. Longhurst, T.Platt, S. Sathyendranath, K. Sherman and R. Watson 2000 Mapping fisheries onto marine ecosystems: a proposal for a consensus approach for regional, oceanic and global integrations. International Council for the Exploration of the Seas Annual Science Conference C.M.2000/T:14

Sherman, K. 1986 Introduction to Parts One and Two: large marine ecosystems as tractable entities for measurement and management. pp3-7. *In*: K. Sherman and L.M. Alexander (eds). *Variability and management of large marine ecosystems.* AAAS Selected Symposium 99. Westview Press Inc., Boulder, CO.

Tilot, V. and A. King 1993 *A Review of the Subsystems of the Canary Current and Gulf of Guinea Large Marine Ecosystems.* IUCN Marine Programme.

The Gulf of Guinea Large Marine Ecosystem
J.M. McGlade, P. Cury, K.A. Koranteng and N.J. Hardman-Mountford (Editors)

7

A Multi-Data Approach for Assessing the Spatio-Temporal Variability of the Ivorian-Ghanaian Coastal Upwelling: Understanding Pelagic Fish Stock Dynamics

H. Demarcq and A. Aman

Abstract

The majority of coastal pelagic fish species are known to be affected by strong interannual environmental variations. This variability is recognised as being largely due to the impact of the environment on recruitment processes, however, a complex balance exists between physical and biological processes which does not allow a comprehensive understanding of the observed stock dynamics. Habitats along the Côte d'Ivoire and Ghana coast are mainly characterised by the presence of seasonal or permanent coastal upwellings. Sea Surface Temperature (SST) monitoring is the most suitable parameter to quantify these upwellings. A specific approach has been developed here using remote sensing observations of the METEOSAT satellite series. Two data sets from *in situ* sources are combined with daily satellite observations in order to quantify the seasonal intensity and spatial extent of coastal upwelling off Côte d'Ivoire and Ghana.

A monthly climatology was produced for the region, with a considerably more accurate spatial and seasonal precision than shown in currently available data. This has been made possible to estimate more reliably the SST anomalies found along the coast. This is particularly useful in this region where studies have shown the importance of subtle environmental variations on fish stocks such as changes in upwelling seasonality.

Introduction

To manage fisheries effectively a broad knowledge of the mechanisms which control stock dynamics is needed. In coastal upwelling areas and especially for short lived species, such as small pelagic fishes, these dynamics are linked to climate forcing via recruitment success (Sissenwine 1984; Cury and Roy 1989).

In the Gulf of Guinea region, ocean dynamics are mainly linked to the seasonal presence of the Guinea current, to equatorial upwelling and to the upwellings along the coasts of Côte d'Ivoire and Ghana (Ingham 1970; Morlière 1970; Morlière and Rebert 1970; Colin *et al* 1993). Enrichment caused by the upwellings is strong enough to support regional stocks of several economically important pelagic fish species (Herbland 1983).

The coastal upwelling in the Gulf of Guinea region is characterised by coastal anomalies in the SST signal (Colin *et al* 1993). This signal can be picked up via the existing

83

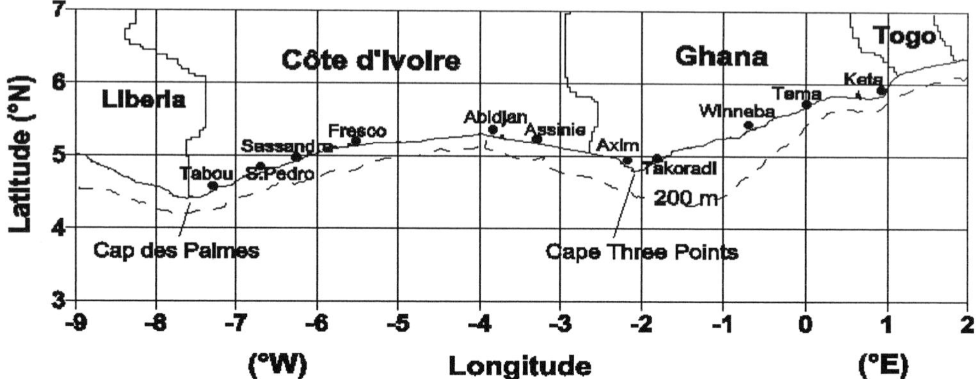

Figure 7-1 Working area with the location of the oceanographic coastal stations.

network of oceanographic coastal stations that cover Côte d'Ivoire and Ghana and supply high quality coastal SST samples (Figure 7-1). Comparatively lower quality coastal SST samples may be obtained from the 'ship of opportunity' meteorological data set (called 'SHIP' data), on a coarser space-time scale. Finally, remote sensing data provide excellent, complementary information on the spatial structure of the upwelling, however, the spatio-temporal sampling rate is low and irregular due to the very high cloud cover of the region.

Remote Sensing

SST reconstruction over the Gulf of Guinea through passive remote sensing techniques is severely limited by the quasi-permanent cloud cover associated with the high water vapour content of the atmosphere. The satellites in the METEOSAT series have the advantage of a very high observational repeat rate of 30 minutes, giving a spatial resolution of 5 km. This level of specificity is used to process daily synthesis images of radiative temperature by minimising the atmospheric absorption (Demarcq and Citeau 1995). Daily images were processed from July 1989 to December 1997 but few of them allow the reconstruction of sea surface thermal fields. Previous studies have shown the capabilities of METEOSAT for quantifying the upwelling off Mauritania and Senegal (Demarcq and Citeau 1995) and to localise the surface cooling off Côte d'Ivoire and Ghana (Aman and Fofana 1994).

Figure 7-2 displays the mean number of monthly observations with partially clear atmospheric conditions, making it possible to observe an upwelling related SST gradient for the 1989-1997 period. January to February and July to August are the best observational periods, as they are associated with the minor and major upwelling seasons. During these periods, the relative cooling of the sea surface has a "back-drying" effect on the atmosphere by reducing evaporation from the sea.

Typical coastal upwelling situations can be described off Cote d'Ivoire and Ghana (Figure 7-3) using radiative temperature fields analysis (uncorrected temperature). These fields show the main spatial differences between the minor upwelling structure - restricted

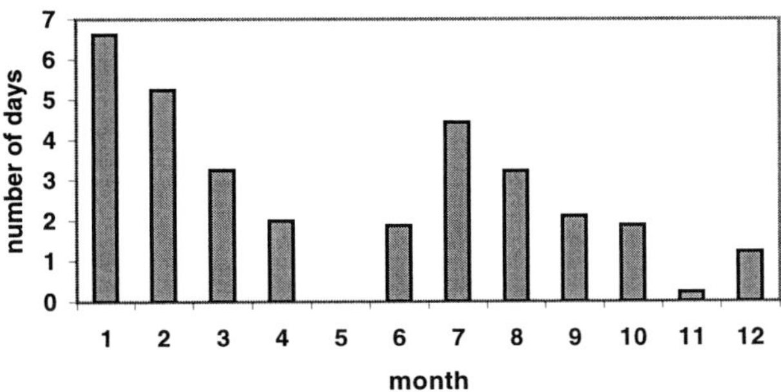

Figure 7-2 Number of days with cloud-free upwelling structures observed with METEOSAT.

more to coastal areas (Figure 7-3a to 3c), and the major upwelling structure (Figure 7-3d). The extent of upwelling offshore is generally limited to less than 50 km during the minor upwelling season on both the Côte d'Ivoire and Ghana coasts.   During the major upwelling season the extent is similar to that for the minor season in the west part of Côte d'Ivoire but regularly increases eastwards with an area greater than 100 km off Ghana.

The simultaneous appearance of regional upwelling during the minor upwelling season is worth noting (Figure 7-4, colour plate). Four independent coastal upwelling points can be observed. Three of these are to the east of the three capes: Cape Palmas, Cape Three Points and Keta. More surprisingly is the presence of an upwelling source point to the west of Abidjan. This may be generated by the submarine canyon close to Abidjan. This pattern can also be observed in the oceanographic data collated by Colin *et al.* (1993) for the major upwelling season.

External data: SHIP records and coastal oceanographic measurements

The spatial coverage of the SHIP data is inadequate to supply precise spatial information on upwelling. Nevertheless, averaged spatial information on the extent of the upwelling can be obtained. Approximately 25 000 records were compiled for the 1984-1997 period covering each upwelling season (Figure 7-5, colour plate).

The thermal signature of the major upwelling (mainly during July and August) is highly pronounced (Figure 7-5a, colour plate), with a clear continuous minimum SST along the coast from 7°W (Tabou) to 1°E (Winneba). This observation is consistent with the main structures observed via the satellite data (Figure 7-3d, colour plate). For the minor upwelling season (January to February), the resulting signature of the upwelling is very weak. Only the coldest areas can be observed, mainly to the east of Cape Palmas and to the west of the Cape Three Points. These features are not consistent with the satellite data and are most likely due to the high degree of noise in the SHIP data and the relatively poor sampling of merchant ships near the coast. As seen in the satellite features, the offshore extent of the upwelling during this season is weak, and the SHIP data significantly underestimates the upwelling intensity. The mean coldest value on the continental shelf is 24.5°C, but this temperature is more than 4°C above the coastal temperature usually recorded.

Coastal oceanographic measurements of nearshore SST are continuously provided by six coastal stations located along the open Ivorian Gulf from Tabou to Assinie and six other stations in Ghana from Axim to Keta (Figure 7-1). These stations are sampled daily by ORSTOM and the Centre de Recherches Océanologiques d'Abidjan (CROA) for Côte d'Ivoire

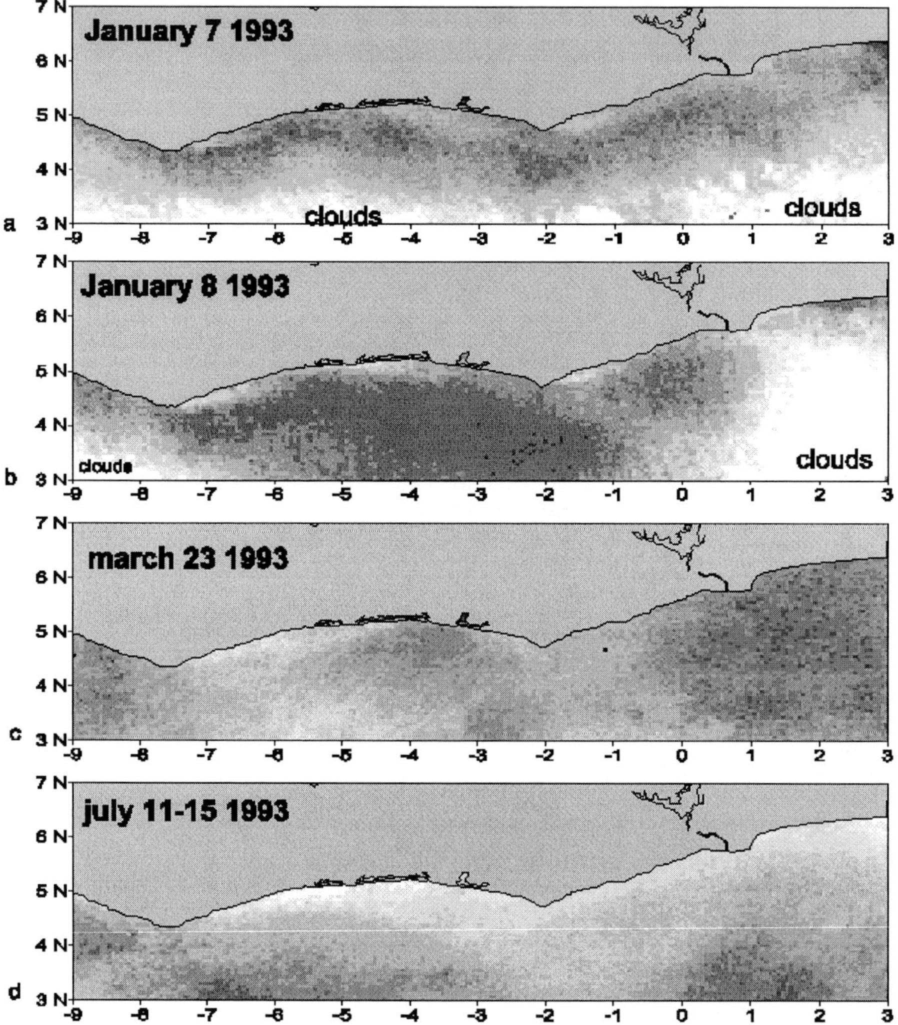

Figure 7-3 Typical SST fields as depicted by METEOSAT during the minor upwelling season **a** to **c** and during the major upwelling season **d**.

(at about 09.00) and daily by the Fisheries Research & Utilisation Branch (FRUB) for Ghana. These measurements have been taken since 1977 for Côte d'Ivoire and since 1985 for Ghana. Compared to the SHIP data, the regular sampling of these stations gives a precise absolute value of the zonal SST gradient. It is generally possible to detect the gradients on the satellite images but impossible to obtain with the SHIP data.

Figure 7-6 (colour plate) shows an example of the mean SST sampling provided during one week for these two *in situ* data sets. The measurement precision is about 0.8 °C for the SHIP data (Gohin 1987)and about 0.1°C for the coastal measurements. For reasons of sampling and precision, the coastal data are the only convenient data to provide a precise and regular local description of the SST dynamics.

Integration of Satellite/SHIP/coastal data

The amount of cloud-free satellite data is insufficient to provide an inter-annual description of the upwellings dynamics, however, remote sensing can supply adequate data for an objective spatial approach. A specific processing method has been developed for this highly cloudy area to restore accurate SST fields using a combination of satellite data, coastal SST measurements and SHIP data. Raw infrared imagery of the geostationary European satellites of the METEOSAT series is available from 1988 from the receiving station of the Centre de Recherches Océanographiques de Dakar-Thiaroye (Dakar, Sénégal) and from 1996 from the physical department of the University of Cocody at Abidjan, via CROA. The processing methods are based on those previously develop by Demarcq and Citeau (1995) for the coastal upwelling areas and adapted to the regional specificity of Côte d'Ivoire and Ghana, in particular for the integration of the coastal measurements.

After calibration and geometric correction, daily composite images of "brightness temperature" (including the atmospheric alteration of the signal) are processed from the 48 half-hourly images of the earth disk. This pre-processing is performed to reduce the cloud cover by minimising the atmospheric disturbance (see Demarcq and Citeau 1995 for details). A total of 270 daily images containing a part of a valid SST gradient were processed for the period July 1989 to December 1997. The intimate mixing of oceanic and atmospheric features on the images enables the user to select interactively usable images using oceanographic criteria. The main criteria are the identification of upwelling related structures and a preliminary oceanographic knowledge of the area.

The challenge is then to perform adequate atmospheric correction to obtain validated SST fields. It is widely acknowledged that even the most powerful methods, using split-window algorithms on NOAA/AVHRR imagery, are not satisfactory for equatorial regions because of the very high water vapour contents of the atmosphere (Sobrino *et al.* 1993, Mathur and Agarwal 1992, Steyn Ross *et al.* 1993). In this situation, the residual SST departure due to uncorrected atmospheric effects (despite the use of two infrared channels) can potentially reach 2°C.

The METEOSAT infrared imagery is characterised by the presence of a unique infrared channel so the use of a direct calculation of atmospheric absorption is not possible, unlike with the two infrared channels of AVHRR. The particularly high atmospheric absorption due to water vapour in this region means that processing of the atmospheric attenuation is of crucial importance. Therefore, the only adequate way of performing an accurate atmospheric correction, in order to restore SST fields using METEOSAT, is to use external *in situ* SST data sets. In the present case, two kinds of data are available for this purpose: the SHIP data and the daily oceanographic coastal measurements. These data are used in combination with the satellite radiative temperature fields to obtain corrected SSTs. Two areas are considered for the calculation of the atmospheric absorption (Figure 7-6a, colour plate).

The coastal coverage of the SHIP data is insufficient to recover SST gradients on a daily or weekly scale (Figure 7-5, colour plate). The typical SHIP data coverage (Figure 7-6b, colour plate) allows processing of the mean offshore atmospheric absorption but does not enhance the SST gradient between the coastal and offshore area. The coastal SST is then severely over-estimated (Figure 7-7b, colour plate). On the other hand, the coastal measurements represent a very good sampling of the coastal area (Figure 7-6c, colour plate) and the SST processed with only this data set (Figure 7-7c, colour plate) gives a satisfying SST reconstruction of the coastal area, though it is not able to account for the offshore atmospheric absorption. The offshore SST is consequently under-estimated. The coastal data are used preferentially in the coastal area, when available, because of their greater precision. This is achieved by assigning them a "protection perimeter" to avoid mixing with the noisy SHIP data. This perimeter is logically assimilated as the mean extension of the upwelling (Figure 7-6d, colour plate).

The processing of the initial radiative temperature field (Figure 7-6a, colour plate) with this specific *in situ* data mixing (Figure 7-6d, colour plate) shows a better result in terms of coastal/offshore SST gradient (Figure 7-7d, colour plate) than those obtained with only one data set (Figures 7-6b,c, colour plate). The specific interest of each *in situ* data set is taken into account particularly well. The calculated absorption field, principally due in these latitudes to the integrated water vapour content of the atmosphere, shows a strong offshore gradient range of between 3°C for the coastal area and 6° C for the offshore area, alongside the clouds (Figure 7-7d, colour plate). As generally observed in other upwelling areas, the upwelling's drying effect on the atmosphere (by minimising the water evaporation in the coastal area) is clearly visible. This drying effect is generally visible in the radiative temperature daily synthesis as a 10-15 km band of clear sky above the continent in front of the upwelling areas (Figure 7-8, colour plate). This effect is also an indicator of the upwelling presence and consequently does not exist, on the particular day shown, to the east of Cape Palmas.

Monthly climatologies

Monthly SST climatologies were calculated from the mean of the 270 clearest METEOSAT daily synthesis images processed using to the above method. Figure 7-9 (colour plate) displays the mean SST fields obtained on a monthly scale, with contours superimposed. The number of daily images included in the processing varies according to the mean seasonality of cloud cover in the region. This cloud cover is not constant throughout the year (Figure 7-2) and is linked partly to the presence of the coastal upwellings themselves and partly to the latitudinal migration of the Inter-Tropical Convergence Zone. No images have been processed for May as during this month the upwelling intensity is minimum (from *in situ* data sets), so cloud cover reaches a maximum. Therefore, SST has been interpolated from the months of April and June accounting for the coastal measurements recorded during this month.

Both the major and minor upwelling seasons are clearly visible. The major season occurs from July to November with a very clear maximum in August. The upwelling spatial pattern shows a local maximum of the upwelling intensity off the west part of Côte d'Ivoire, particularly in August when the upwelling reaches its maximum intensity. During this maximum of intensity, the mean SST is around 22.5°C in Côte d'Ivoire and 23.5°C in Ghana.

In contrast to the main upwelling, the minor upwelling is very weak in Ghana and has to be considered as an indirect extension of the Côte d'Ivoire upwelling as its spatial signature does not effect the eastern part of Ghana. The duration of the minor upwelling is approximately four months, from January to April. Because upwelling is generally considered to be weaker in April than in February and March, special attention is given to results from this month. Twelve satellite SST situations during four different years were processed for the month of April; the results show low SSTs at the coast, as in the coastal measurements. Therefore, the relative importance of this period, especially when compared to the slightly higher SSTs obtained for the months of February and March, has to be carefully examined.

A comparison with the Reynolds SST climatology (Reynolds and Smith 1995) for the months of January, August and October (Figure 7-10, colour plate) shows a considerable improvement of the SST reconstruction for the coastal areas. The minor upwelling is almost absent on the Reynolds climatology (Figure 7-10a, colour plate) and the August SST minimum is seen situated off Ghana in the Reynolds climatology when it is clearly situated in the western part of Côte d'Ivoire in our climatology. The SST departure from the Reynolds climatology in both upwelling seasons is greater than 2°C - a significant value in any satellite

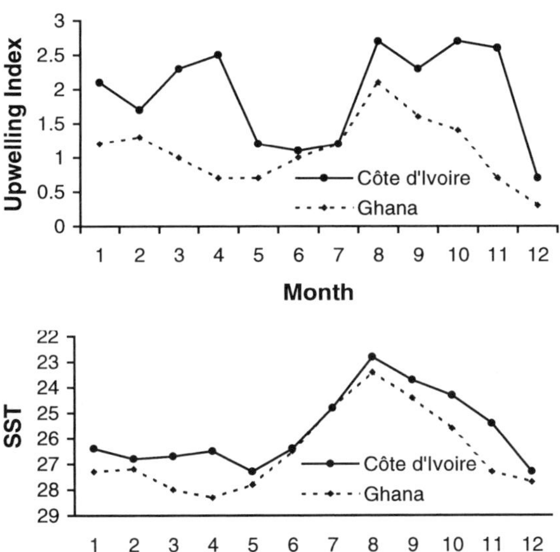

Figure 7-11 Mean coastal SST (top) and Upwelling Index (offshore SST - coastal SST) calculated from the climatology off Côte d'Ivoire and Ghana.

Mean coastal SST values were extracted from this climatology for the two main upwelling areas, from Cape Palmas to Cape Three Points and to the east of Cape Three Points (Figure 7-11, top). The amplitude of the seasonal SST variation reached 5°C (from 28°C to 23°C) and the SST was always slightly lower (approximately 1°C) for the western region (Côte d'Ivoire and western Ghana) from October to April.

A satellite upwelling index was calculated from the SST deficit between the offshore area (taken as the maximum of SST from 0° to 4° of latitude) and the coastal area for the same regions (Figure 7-11). The upwelling index shows considerably higher values than expected during the minor upwelling season, particularly off Côte d'Ivoire. The upwelling index shows a stable minimum from May to July, whereas the mean duration of the minor and major upwelling season seems to be approximately the same (4 months) with some reservations concerning the high values obtained in March and April. Compared to previous upwelling indices calculated only with coastal measurements, this result shows the potentially great importance of the minor upwelling season that appears to have been underestimated in previous studies. However, the relative importance of the minor upwelling season and the temporal changes depicted in its seasonality (Pézennec and Bard 1992) reinforces the hypothesis of the potentially high impact of this season in local biological processes concerning the pelagic species dynamic.

Conclusions and perspectives

The infrared imagery supplied by the European METEOSAT satellites is an essential data set which supplies sparse but useful information on the mean seasonal cycle of the upwellings and their associated spatial structure. The integration of complementary *in situ* data in the processing of infrared satcllitc images over a very cloudy equatorial area shows the ability to compute high value data in the form of a spatial SST climatology for the Ivorian-Ghanaian upwellings, despite *a priori* severe atmospheric limitations.

The processing of this preliminary climatology was based on the clearest satellite images of the 1989-1997 time series in order to take account of the spatial structure of the upwelling from daily data where satellite/SHIP/coastal data were available. This processing could be refined to obtain a better adjustment for the seasonal upwelling intensity and the main SST gradient. The inconvenience of this method is the irregular amount of data on a monthly basis because of satellite restrictions. Future refinements of this climatology will require systematic use of the whole time series from the coastal stations in order to avoid possible local and seasonal artefacts due to a seasonal irregularity in the amount of processed data. On the other hand, the data from the coastal stations of Ghana were not available for this study after 1991 and this may have led to an overestimate of the Ghana SST values in any subsequent period.

Concerning the temporal approach, the remote sensing data from meteorological satellites does not supply information on a regular temporal basis on the Gulf of Guinea, despite the high initial observation frequency of the METEOSAT satellites, because of the high degree of cloud cover. A direct observation of the dynamics of the upwelling structure is thereby made impossible from satellite data alone. Moreover, the time lag between two successive exploitable satellite data images is sometimes too long, even if compared to the mean duration of the upwelling seasons.

The hypothesis of an intensification of the minor upwelling noticed by some authors (Pézennec and Bard 1992) needs to be taken into consideration in the face of the relatively high upwelling indices obtained for the minor upwelling season from the remotely sensed data.

References

Aman, A. and S. Fofana 1994 Spatial dynamics of the upwelling off Côte d'Ivoire. Colloques Et Séminaires, ORSTOM Editions: 139-147.

Colin, C., Y. Gallardo, R. Chuchla and S. Cissoko 1993 Environnements climatique et océanographiques sur le plateau continental de Côte d'Ivoire. pp75-110. *In*: P. Le Lœuff, E. Marchal and J. B. Amon Kothias (eds). *Environnement et ressources aquatiques de Côte-d'Ivoire*. Tome I - Le milieu marin. ORSTOM éditions ed. ORSTOM, Paris.

Cury, P. and C. Roy 1989 Optimal environmental window and pelagic fish recruitment success in upwelling areas. Canadian Journal of Fisheries and Aquatic Sciences, 46(4): 670-680.

Demarcq, H. and J. Citeau 1995 Sea surface temperature retrieval in tropical area with METEOSAT: the case of the Senegalese coastal upwelling. International Journal of Remote Sensing. 16(8):1371-1395.

Gohin, F. 1987 *Analyse géostatistique des champs thermiques de surface de la mer*. Ecole nationale supérieure des Mines de Paris. 103p.

Herbland, A. 1983 Le maximum de chlorophylle dans l'Atlantique tropical oriental: description, écologie, interprétation. Océanographie Tropicale 18(2):295-318.

Ingham, M. C. 1970 Coastal upwelling in the northwestern Gulf of Guinea. Bulletin of Marine Science: 20:2-34.

Mathur, A. K. and V. K. Agarwal 1992 A quantitative study on the effect of water vapor on estimation of sea surface temperature using satellite IR observations. pp673-680. *In: Int. Symp. on the Oceanography of the Indian Ocean*. 14-16 Jan 1991, Natl. Inst. Oceanogr., Goa.

Morlière, A. 1970 Les saisons marines devant Abidjan. Doc. Scient. Centre Rech. Océanogr. Abidjan.; 1(2):1-15.

Morlière, A. and J.P. Rebert 1970 *Etude hydrologique du plateau continental ivoirien*. Doc. Scient. Centre Rech. Océanogr. Abidjan; 3(2):1-30.

Pézennec, O. and F. X. Bard 1992 Ecological importance of the Ivorian and Ghanaian minor upwelling season and changes in the *Sardinella aurita* fishery. Aquatic Living Resources 5(4):249-259.

Reynolds, R. W. and T. M. Smith 1995 A high-resolution global sea surface temperature climatology. Journal of Climatology 8(6):1571-1583.

Sissenwine, M. P. 1984 Why do fish populations vary? pp59-94. *In* R.May (ed.) *Exploitation of marine communities*. Dahlem Konferezen, Springer-Verlag, Berlin.

Sobrino, J. A., Z.L. Li and M. P. Stoll 1993 Impact of the atmospheric transmittance and total water vapor content in the algorithms for estimating satellite Sea Surface Temperatures. IEEE Transactions of Geoscience-Remote Sensing 31(5):946-952.

Steyn Ross, M. L., D.A. Steyn Ross, P.J. Smith, J.D. Shepherd, J. Reid and P. Tildesley 1993 Water vapour correction method for advanced very high resolution radiometer data. Journal of Geophysical Research 98 (C12): 22817-22826.

The Gulf of Guinea Large Marine Ecosystem
J.M. McGlade, P. Cury, K.A. Koranteng and N.J. Hardman-Mountford (Editors)
© 2002 Elsevier Science B.V. All rights reserved.

8

Physico-Chemical Changes in Continental Shelf Waters of the Gulf of Guinea and Possible Impacts on Resource Variability

K.A. Koranteng and J.M. McGlade

Abstract

Changes in the continental shelf ecosystem of Ghana were examined, based on sea surface and bottom temperature (SST, SBT), salinity and dissolved oxygen. The de-composed trend of temperatures exhibited a phase of cooling from the beginning of the series until 1976/77 and warming thereafter. The trend of salinity and dissolved oxygen showed different tendencies but in terms of timing, they were consistent with the pattern of change of temperature. These changes resulted in the division of the observational period into three blocks, namely the period before 1972, from 1972 to 1982 and the period after 1982. In the first block, sea surface temperature (both coastal and offshore) and bottom temperature declined, coastal salinity was low. The second block was colder, with less than average SST and SBT. The mixed layer was narrow with the thermocline remaining shallower than its long-term average position. Coastal and bottom salinity (measured at 100 m deep) were relatively high but the seasonal variation was minimal. This was a period of significant change in the physical components of the ecosystem of the Ghanaian shelf waters which to date has not been so clearly documented in the literature. In the final phase, temperatures were high, and salinity was low and erratic.

Introduction

The waters of the Gulf of Guinea, situated in the Eastern Central Atlantic, are defined by the Guinea Current Large Marine Ecosystem (LME) (Sherman, 1993). Situated between the Canary Current LME to the north and the Benguela Current LME to the south, the Guinea Current LME extends from the Bijagos Islands (approximately latitude 11° N, longitude 16° W) to Cape Lopez (latitude 0° 41′S, longitude 8° 45′E) (Binet and Marchal 1993) (Map 2, colour plate). It includes the maritime waters of all countries between Guinea Bissau and Gabon in West Africa.

Off Ghana and Côte d'Ivoire situated in the central part of the LME, two upwelling seasons, major and minor, occur annually with differing duration and intensities. Small pelagic fisheries in the area are sustained by these seasonal upwellings (FRU/ORSTOM 1976; Pézennec and Koranteng 1998). In the last decade, there were remarkable changes in the distribution, abundance and reproductive strategy of the round sardinella (*Sardinella aurita* Clupeidae) in this ecosystem. These have been attributed to the intensification of the minor upwelling (Pézennec and Bard 1992; Pézennec 1994).

In the same period, there were significant increases and decreases of some demersal fish abundances; most significant among these was the complete disappearance of the triggerfish (*Balistes capriscus* Balistidae) from the sub-region where it had once accounted for over 60 % of total fish biomass assessed in bottom trawl surveys (Koranteng 1984) and about 83 % of total pelagic biomass assessed in acoustic surveys (Koranteng 1998). The motivation behind this paper is to seek an explanation for the variation in demersal fish stocks including possible response to the observed environmental changes.

Data and Analysis

Several data types were used in this investigation. These included daily sea surface temperature (SST) recorded at seven locations along the Ghanaian coastline between 1962 and 1992 and offshore SSTs from the Comprehensive Ocean Atmosphere Data Sets (COADS) (Woodruff *et al.* 1987). The COADS database was created from marine surface observations made on various platforms including ships of opportunity and moored buoys. Using the CODE program (Mendelsshon and Roy 1996), sub-sets of the COADS data were extracted for four areas off Côte d'Ivoire - Ghana (i.e. between latitudes 1 and 6° N and from longitude 7°W to 1°E and for the period 1946 to1992.

At the coastal stations, water samples were also collected daily and sent to the MFRD laboratories in Tema for the determination of salinity using an inductive salinometer. Collection of salinity records from stations other than Tema was discontinued in 1982, hence only salinity measurements from Tema were used in this work. Also used are subsurface temperature, salinity, and dissolved oxygen values measured at 30 m, 50 m, and 100 m on a hydrographic transect off Tema between 1968 and1992.

Each time series of environmental data was decomposed using a model of the form:

$$Y_i = T_i + S_i + R_I \qquad \textbf{Eq. 8-1}$$

where Y_i represent the observed data (e.g. SST), T_i is the trend, S_i is the seasonal variation and R_i represents the remainder random elements of the series after the trend and seasonal variation have been accounted for. The analysis was carried out using the STL routine incorporated in the S-PLUS computer software (MathSoft 1995). "STL is a **S**easonal-**T**rend decomposition procedure consisting of a sequence of smoothing operations that employ locally weighted regression or **L**oess"; a detailed mathematical treatise of the method may be found in Cleveland *et al.* (1990).

Following Laevastu (1993) and Becker and Pauly (1996), standardised departures of the monthly averages of each environmental parameter (e.g. SST) were calculated from:

$$X_j = \frac{r_j - \bar{r}}{s} \qquad \textbf{Eq. 8-2}$$

where r_j is the monthly mean value of the parameter, and $\bar{r}$ and s are the mean and standard deviation respectively of the series. Such standardised departures (or anomalies) are distributed with a zero mean and a variance of one.

Where there was more than one series of the same parameter (e.g. the coastal SSTs), standardised departures were calculated for each series and a combined anomaly was calculated from:

$$X_j = \frac{1}{N_j} \sum_{i=1}^{N_j} \frac{r_{ij} - \overline{r_i}}{s_i}$$

Eq. 8-3

where r_{ij} is the mean of the parameter for year j at station i, $\overline{r_i}$ and s_i are the mean and standard deviation respectively of the ith station's series, and N_j is the number of stations with complete records in year j. Thus, information from all the stations was combined in this index. A similar index was used to monitor rainfall in the Sahel region of West Africa (Kraus 1977; Katz 1978; Lamb 1982). Statistical properties of such an index are examined by Kraus (1977). A 13-point moving average was applied as a low-pass filter to remove the seasonal variation in the time series of anomalies (Chatfield 1996).

Temperature values recorded weekly at one of the stations (depth of 100 m) on the Tema hydrographic transect between 1968 and 1992 were examined. The monthly mean depth of the thermocline, represented by the depth of the 21°C isotherm (Merle 1978; Koranteng and Pézennec 1998) was calculated and the depth variation examined.

Preliminary results showed that between 1963 and 1992, there were distinct time blocks within each of which environmental conditions were different. To clearly identify the environmental time blocks, new temperature and salinity anomalies were calculated using the monthly values of the decomposed trend of each series as the data values.

Results

Figure 8-1a,b shows the extracted trend and seasonal variation of the coastal SST series recorded at Tema in Ghana. The decomposed trend for the other series follow a similar pattern. From the plot of the seasonal component of each series it is possible to follow the intra-annual (seasonal) variation in sea surface cooling. Within each year, the lowest point on the graph corresponds to the peak of cooling (major upwelling) and the highest point corresponds to the peak of warming. Between these are points corresponding to the first cold period in the year (January/February) and the warm period towards the end of the year (October/November).

A linear regression of the derived trend values (for the Tema series) versus time (years) shows a highly significant ($p < 0.001$) decline of SSTs (signifying cooling) between 1963 and 1975 and a rise (warming) between 1975 and 1992. The trend of the COADS SST series (Figure 8-2) also exhibits periods of cooling and warming.

Monthly mean temperature, salinity and dissolved oxygen measured at 100m off Tema are presented in Table 8-1. The seasonal component of the temperature series shows the following:

a) consistently increasing temperatures during the second warm season (i.e. October/November) between 1975 and 1992;
b) general reduction of temperatures during the minor upwelling period signifying an intensification of this upwelling; and
c) slight increases in temperatures during the main warm season signifying a relative weakening of the major upwelling.

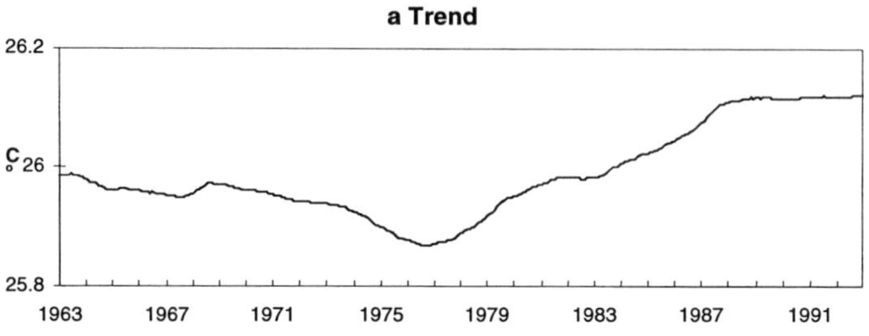

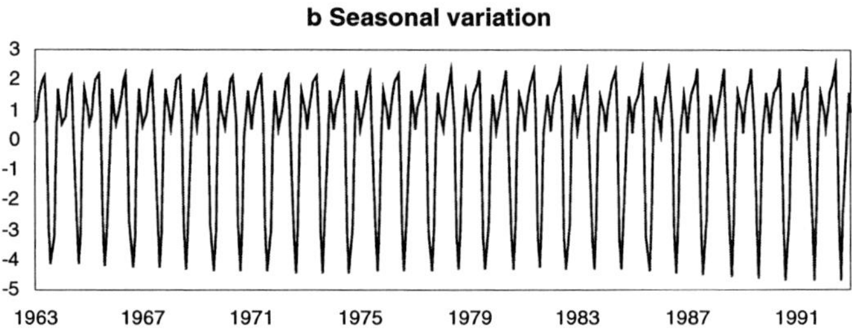

Figure 8-1 a Trend and **b** seasonal variation of a series of sea surface temperature measured off Tema, Ghana.

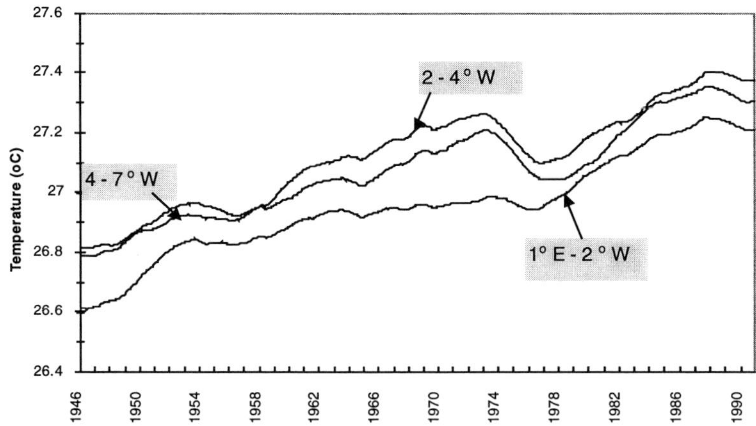

Figure 8-2 Trend component of COADS SST data for three areas off Ghana and Côte d'Ivoire. Areas are between shoreline and latitude 4°S and within the longitudes indicated.

The trend in the Tema coastal sea surface salinity (Figure 8-3a) shows a period of increasing or high salinity (up to about 35.45 ‰) between 1970 and 1978, a period of rapid reduction of salinity between 1978 and 1982, and a period of relatively stable salinity of around 35.2 ‰ between 1982 and 1992. The striking feature of the salinity series is the decline (in the first period) and rise (in the second period) in amplitude of the seasonal component (Figure 8-3b). Mean values of dissolved oxygen (DO) ranged between 2.8 and 2.9 ml l^{-1} in 1968 - 1978, increased to over 2.9 ml l^{-1} between 1980 and 1985 and declined between 1985 and 1990 (Table 8-1). During the period of investigation, the 21°C isotherm (used here to represent the depth of the thermocline) was encountered at depths from just below the surface to over 100 m deep. The annual mean thermocline depth generally reduced between 1968 and 1975 where the smallest annual mean depth (29.4m) was calculated. The largest annual mean depth (51.0m) was in 1978 and the long-term annual mean depth was 41 m. The large standard deviations associated with the means signify large variation in the depth of the isotherm. Using the trend values, the resultant anomalies of the bottom and surface temperatures and salinity are shown in Figure 8-4a,b).

Year	Surface Temperature	Bottom Temperature	Coastal Salinity	Bottom Salinity	Bottom Dissolved Oxygen
1963	26.6				
1964	25.3				
1965	26.2				
1966	26.0				
1967	25.0				
1968	26.7	16.6		35.60	2.99
1969	26.0	16.8		35.26	3.02
1970	26.2	16.7	34.83	35.57	2.74
1971	25.8	16.8	35.22	35.74	2.73
1972	25.9	16.3	35.36	35.74	2.85
1973	26.4	16.7	35.33	35.56	2.88
1974	26.0	16.2	35.25	35.70	2.62
1975	25.7	16.0	35.22	35.66	2.76
1976	25.1	15.9	35.51	35.73	2.50
1977	25.3	16.2	35.44	35.75	2.90
1978	25.7	16.3	35.33	35.83	3.28
1979	26.5	17.2	35.28	35.87	3.08
1980	26.0	16.5	35.25	35.78	2.82
1981	26.5	16.3	34.8	35.81	2.95
1982	25.8	16.9	35.25	35.86	3.06
1983	25.4	16.8	35.09	35.64	3.17
1984	26.4	17.0	34.98	35.68	3.06
1985	25.8	16.7	35.21	35.66	2.93
1986	25.4	16.6	35.16	35.36	2.74
1987	27.0	16.8	35.17	35.80	2.59
1988	26.5	17.2	35.22	35.52	2.78
1989	26.3	17.1	34.87	35.43	2.65
1990	25.9	17.0	35.46	35.79	2.59
1991	26.1	17.3	35.17	35.65	2.71
1992	26.1	17.2	35.18	35.63	2.94

Table 8-1 Mean values of environmental parameters measured off Tema, Ghana. Bottom parameters were measured at 100 m deep.

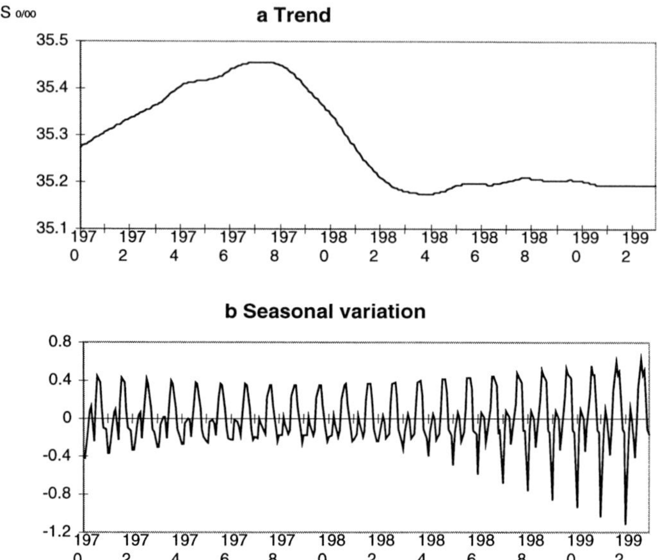

Figure 8-3a Trend and **b** seasonal variation of coastal salinity measured off Tema, Ghana.

Discussion

The results of the analysis of sea surface temperatures clearly show the seasonal and inter-annual patterns of cooling in the coastal and shelf waters off Ghana as described by Longhurst (1998) for the entire Gulf of Guinea. The two cold seasons (January-February and July-September) as well as the two warm seasons (April-June and October-December) are clearly identifiable in all the series of both sea surface and bottom temperatures.

The period of decline of the coastal SSTs, from the 1960s to the mid-1970s, and a subsequent rise, are in phase with the behavior of the COADS SST series (Figures 8-1 and 8-2). The latter series shows a general underlying increase since 1946 with some decline in 1954-55, 1964-65, 1975-76 and 1987-88, thus appearing to exhibit decadal variability. The decline in 1975-76 was the most pronounced within the 47-year period, and this was picked up in the coastal and bottom temperature series recorded in Ghanaian waters.

Also evident from the plot of the seasonal component of the SST series (Figure 8-1) is the intensification of the minor upwelling and increased secondary warming from the mid-1980s. These events have been picked up more clearly in the time series decomposition analysis than from the use of an upwelling index (e.g. Arfi *et al.* 1991; Pézennec 1994; Koranteng and Pézennec 1998). The change in seasonal variation is not very evident from the COADS data, implying less significant change in the intensity of the two upwelling events in offshore areas. This is understandable as the upwelling is coastal.

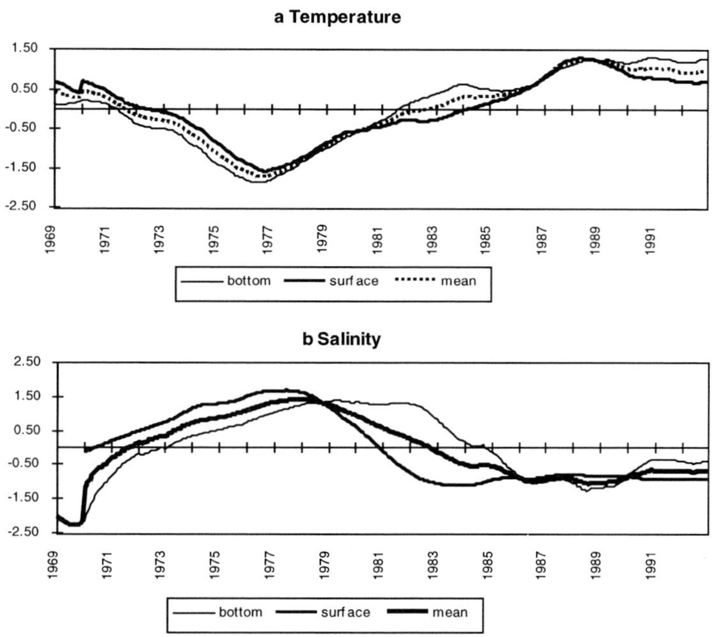

Figure 8-4 Plots showing time blocks of **a** temperature and **b** salinity.

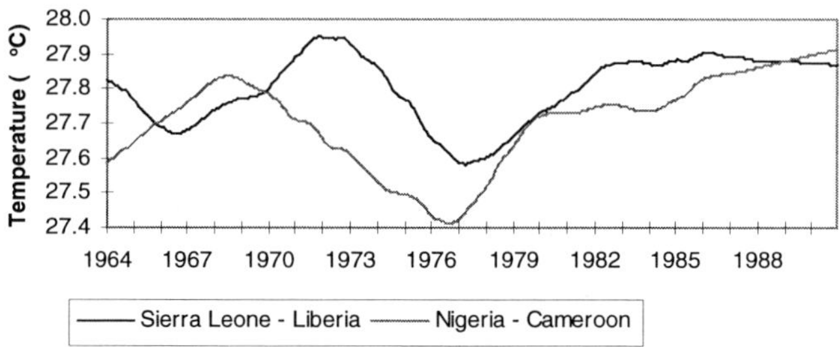

Figure 8-5 SST trends in areas to the north-west and south-east of Ghana.

The re-calculated trend anomalies for temperature and salinity (Figure 8-4a,b) clearly divide the period under consideration into three regimes, namely the period until 1972, 1972-1982 and 1982-1992. These are referred to as Environmental Time Blocks (ETBs).

Figure 8-5 gives trends of SSTs obtained from COADS for the areas off Sierra Leone to Liberia and Nigeria to Cameroon, to the north and south of Ghana respectively. The figure also shows periods of decline of temperatures (between 1970 and 1982) and rising temperature from 1982. The behaviour of these SST trends are in aggreement with the situation off Ghana.

Koranteng (1998) showed a relationship between standardized anomalies of SST measured at Tema, Ghana (05° N, 00° 04′ E) and at Cananeia (south of Sao Paulo) in Brazil (25° S, 47° 55′ W). It was shown that similar specific cold and warm events existed in the two series with the Tema series usually lagging behind the Cananeia series. For example, the years of sustained cold SSTs that occurred off Tema between 1975 and 1978 is also seen in the Cananeia series but for 1973-76. Similarly, the warm events of 1987-89 off Tema occurred during 1985-87 off Cananeia. These observations suggest that the specific environmental changes observed off Ghana and Côte d'Ivoire as described in this work could be generalised for the whole Gulf of Guinea and possibly for the whole southern Atlantic basin. Interconnections between the western and eastern parts of the Atlantic Ocean have been noted in the past. For example, Hisard *et al.* (1986) noted eastward movement of positive SST anomalies from the Brazilian coast towards the Angolan and Namibian coasts during the 1984 warm events in the Atlantic.

Koranteng (1998) also noted that the patterns of change of rainfall and river discharge in the nearshore area off Ghana were similar to those describe for physical parameters of the shelf waters. Rainfall and river discharge generally declined between 1970 and 1983 with the exception of a particularly wet year of 1975.

The literature is quite clear on the warm climatic events that occurred in the Atlantic over the last three decades and attempts to draw a parallel with ENSO (El Niño Southern Oscillation) events in the Pacific (e.g. Shannon, 1986; Bakun, 1996). On the contrary, the cold period of 1975-79, as identified in this work, has not been seriously discussed before in the literature.

Infrequent perturbations in the aquatic environment could affect the internal dynamics of a system (Margalef 1986; Baird *et al.* 1991). These events could have significant impacts on the behaviour of some species of fish, especially on their growth, distribution and abundance as have been noted by Pézennec and Bard (1992), Pézennec (1994), Pézennec and Koranteng (1998) for *Sardinella aurita* in the western Gulf of Guinea and by Cury and Roy for fishery resources off Côte d'Ivoire and Ghana (this volume). For example, there is a remarkable synchrony between the temperature and salinity results and the events that have occurred in the sardinella and triggerfish fisheries in this ecosystem. In the case of the sardinella, the environmental regimes closely match the relatively 'healthy' phase of the stock before the high landings of 1972, the 'collapsed' phase between 1973 and 1982 and the recovered or 'prosperous' phase between 1983 and 1992 as described by Pézennec (1994). The environmental regimes also match the years of appearance and disappearance of *Balistes capriscus* in this ecosystem. Proliferation of the species started around 1972-1973, and flourished in the mid-to-late 1970s until the late 1980s. Its decline was observed between 1987 and 1988 (Pézennec 1994; Koranteng 1998).

Conclusion

The period between 1963 and 1992 had within it time blocks with distinct environmental characteristics in the Gulf of Guinea. In the first time block (i.e. before 1972) sea temperatures (surface and bottom) were relatively high, salinity was low and the thermocline was below its long-term average depth.

The second time block was characterised by low temperatures and high but stable salinity. The peak of the changing events was between 1975 and 1979. The period was characterised by sustained cold temperatures in the coastal area and offshore. The mixed layer was shallow, due to the persistence of the thermocline at depths shallower than the long-term mean. Coastal (surface) and deep-water (sub-surface) salinities were relatively high, but quite stable in terms of seasonal variation.

In the third time block, a rising trend in sea surface and bottom temperature was observed. Salinity decreased but was erratic. These changes appear to have influenced the dynamics of fishery resources in this ecosystem with significant events coinciding with the environmental time blocks described here.

Acknowledgements

The authors wish to express their gratitude to the European Commission for their support of this project. J.M.McGlade would also like to acknowledge the support provided by NERC (GT5/00/MS/1).

References

Arfi, R., O. Pézennec, S. Cissoko and M.A. Mensah 1991 Variations spatiale et temporelle de la résurgence ivoiro-ghanéenne. pp162-172. *In*: P. Cury and C. Roy (eds). *Variabilité, instabilité et changement dans les pêcheries ouest africaines.* ORSTOM Editions, Paris.

Baird, D., J.M. McGlade and R.E. Ulanowicz 1991 The comparative ecology of six marine ecosystems. Philosophical Transactions of the Royal Society, London B 333: 15 - 29.

Bakun, A. 1996 Patterns in the ocean: ocean processes and marine population dynamics. University of California Sea Grant, San Diego, California, USA, in co-operation with Centro de Investigaciones Biológicas de Noroeste, La Paz, Baja California Sur, Mexico. 323p.

Becker, G.A. and M. Pauly 1996 Sea surface temperature changes in the North sea and their causes. ICES Journal of Marine Science 53: 887 - 898.

Binet, D. and E. Marchal 1993 The Large Marine Ecosystem of the shelf areas in the Gulf of Guinea: Long-term variability induced by climatic changes. pp104-118. *In*: K. Sherman, L.M. Alexander and B. Gold (eds). *Large Marine Ecosystems - Stress, Mitigation, and Sustainability.* American Association for the Advancement of Science, Washington DC.

Chatfield, C. 1996 *The analysis of time series, an introduction.* Chapman and Hall. 283 p.

Cleveland, R.B, W.S. Cleveland, J.E. McRae and I. Terpenning 1990 STL: A seasonal-trend decomposition procedure based on Loess. Journal of Official Statistics 6 (1): 3 - 73.

FRU/ORSTOM 1976 Rapport du Groupe de Travail sur la sardinelle (*Sardinella aurita*) des cotes ivoiro-ghaneenes, 28 juin - 3 juillet, 1976, Abidjan, Cote d'Ivoire. ORSTOM Editions, Paris. 62 p.

Hisard, P., C. Hénin, R. Houghton, B. Piton, and P. Rual 1986 Oceanic conditions in the tropical Atlantic during 1983 and 1984. Nature 322: 243-245.

Katz, R.W. 1978 Persistence of subtropical African droughts. Monthly Weather Review 106: 1017 - 1021.

Koranteng, K.A. 1984 A trawling Survey off Ghana, CECAF/TECH/84/63. Dakar, Senegal: CECAF Project (FAO). 72p.

Koranteng, K.A. 1998 *The impacts of environmental forcing on the dynamics of demersal fishery resources of Ghana*. PhD Thesis, University of Warwick. 377p.

Koranteng, K. A. and O. Pézennec 1998 Variability and Trends in Environmental Time Series along the Ivorian and Ghanaian Coasts. pp167-177. *In*: M.-H. Durand, P. Cury, R. Mendelssohn, C. Roy, A. Bakun and D. Pauly (eds). *Global versus Local Changes in Upwelling Systems*. ORSTOM Editions, Paris.

Kraus, E.B. 1977 Subtropical droughts and cross-equatorial energy transports. *Monthly Weather Review* 105: 1009 - 1018.

Laevastu, T. 1993 *Marine Climate and Fisheries. The effects of weather and climatic changes on fisheries and ocean resources*. West Byfleet: Fishing News Books. 204 p.

Lamb, P.J. 1982 Persistence on Subsaharan drought. Nature 299: 46 - 48.

Longhurst, A.1998 *Ecological Geography of the Sea*. Academic Press, San Diego. 397p.

Margalef, R. 1986 Reset successions and suspected chaos in models of marine populations. pp321–343. *In*: T. Wyatt and M.G. Larrañeta (eds). *Proceedings of International Symposium on Long Term Changes* in *Marine Fish Populations*. Vigo, Spain. Consejo Superior de Investigaciones Cientificas.

MathSoft Inc. 1995 S-PLUS for Windows. Version 3.3 Release 1.

Mendelssohn, R. and C. Roy 1986 Environmental influences of the French, Ivory-coast, Senegalese and Moroccan tuna catches in the gulf of Guinea. pp170-88. *In*: *Proceedings of the ICCAT Conference on the International Skipjack Program. Comptes-rendus de la Conference ICCAT sur le programme de l'année internationale du listao. Actes de la Conferencia ICCAT sobre el programa de Ano International del listao*. ICCAT, Madrid.

Merle, J. 1978 Atlas hydrologique saisonnier de l'océan atlantique intertropical. Travaux et documents de l'ORSTOM: 82.

Pézennec, O. 1994 *Instabilité et changements de l'écosystème pélagique côtier ivoiro-ghanéen. Variabilité de la ressource en sardinelles : faits, hypothèses et théories*. Thèse de Doctorat, Université de Bretagne Occidentale, Brest.

Pézennec, O. and F.-X. Bard 1992 Importance écologique de la petite saison d'upwelling ivoiro-ghanéenne et changements dans la pêcherie de *Sardinella aurita*. Aquatic Living Resources 5: 249 - 259.

Pézennec, O. and K. A. Koranteng 1998 Changes in the dynamics and biology of small pelagic fisheries off Cote d'Ivoire and Ghana - the ecological puzzle. pp329-343. *In*: M.-H. Durand, P. Cury, R. Mendelssohn, C. Roy, A. Bakun and D. Pauly (eds). *Global versus Local Changes in Upwelling Systems*. ORSTOM Editions, Paris

Shannon, L.V., A.J. Boyd, G.B. Brundrit and J. Taunton-Clark 1986 On the existence of an El Niño phenomenon in the Benguela system. Journal of Marine Research 44: 495 - 520.

Sherman, K. 1993 Large Marine Ecosystems as global units for marine resources management - an ecological perspective. pp3-14. *In*: K.A. Sherman, L.M Alexander and B.D. Gold (eds). *Large Marine Ecosystems-Stress, Mitigation, and Sustainability*. AAAS Selected Symposium, Westview Press Inc., Boulder, CO.

Woodruff, S.D., R.J. Slutz, R.L. Jenne and P.M. Steurer 1987 A comprehensive ocean-atmosphere data set. Bulletin of the American Meteorological Society 68 (10): 1239 - 1250.

The Gulf of Guinea Large Marine Ecosystem
J.M. McGlade, P. Cury, K.A. Koranteng and N.J. Hardman-Mountford (Editors)

9

Environmental Variability at a Coastal Station near Abidjan: Oceanic and Continental Influences

R. Arfi, M. Bouvy and F Ménard

Abstract

Hydrobiological conditions (light penetration, temperature, salinity, chlorophyll biomass at six depths and bacterioplankton abundance at the surface) were measured weekly for six years (1992-1997) at a station located in a coastal area of the Gulf of Guinea (Abidjan, Côte d'Ivoire). Nutrient concentrations completed the data set from December 1994. This coastal area is strongly influenced by a major upwelling (July-September) and by a minor upwelling (short cold events during January-February). Continental inputs induced by local rainfalls (May-June and October) and river floods (September-November) have also pronounced hydrological effects. SST varied from 20.6°C (August) to 30.7°C (May), while surface salinity showed an obvious annual cycle with a minimum of 30.12 psu in June and a maximum of 35.87 psu in September. Bacterial abundance and phytoplankton biomass show seasonal cycles, with simultaneous peaks noted during the main upwelling (maximum: 1.9 10^6 cell ml^{-1} and 5.6 μg l^{-1} respectively). Interannual fluctuations of upwelling intensity and of continental inputs explain the hydrological variability. Freshwater inputs are associated with oligotrophy, while upwellings contribute to the enrichment of the euphotic layer. As a consequence of the drought in the Sahel and of the decreasing rainfall on the coastal area, freshwater inputs are now considerably reduced, and the related impoverishment is less pronounced. At the opposite, the increasing duration of upwellings (and the importance of the short cold events) allows a higher primary productivity (and therefore a more active bacterial compartment). Combined, these two factors would explain the marked outburst of small pelagic fishes in this part of the Gulf of Guinea.

Introduction

Neritic ecosystems show high variability because they are influenced by major hydrological and biological fluctuations occurring on various time scales. Major upwellings or short cold events (SCE) induced by transient inputs of deep water within the euphotic layer increase seasonally the productivity of a coastal zone for months or days. Inputs of continental origin (large river floods or local rain) modify the coastal environment through salinity variation, organic enrichment and light attenuation. Such conditions are observed along the Gulf of Guinea, which is alternately under the influence of upwellings and continental inputs (Roy 1995). The forcing functions of these coastal upwellings (i.e. equatorial Kelvin waves, with local current and wind enforcement) are still under discussion. During the FOCAL/SEQUAL experiment, Colin and Garzoli (1988) reported high frequency variability for temperature and currents in the eastern equatorial Atlantic. From data recorded between 1964 and 1990, Binet and Servain (1993) have shown the interannual variation of SST and the progressive warming of the surface layer near Abidjan (0.07°C y^{-1}). They also reported a change in the apparent westerly wind stress in the Northern Gulf of Guinea, with increasing speeds and slight change in direction. However, Roy (1995) showed that the intensification of the wind is weaker than previously thought and is observed during the minor upwelling. Continental influence is linked to the freshwater inputs from numerous small rivers in the forest zone and large rivers in the Sahel zone. In the last decade, continental flows have sharply decreased, and even the humid forest area along the Gulf of Guinea shoreline can, in some years, be regarded as semi-arid (Mahé 1991; Servat *et al* 1996).

Oceanic water circulation in the northern Gulf of Guinea is dominated by the Guinea Current (GC), an eastward surface flow with high speeds usually observed from April to July (Colin 1988). This current overlays the Guinea Under Current (GUC) flowing westward. When the GC speed is low, the GUC becomes predominant and reversals of surface circulation are common. These currents show large fluctuations in speed and direction, both on a temporal and a spatial scale, but little data on a long term-scale are available. Binet and Servain (1993) explained that the change in the current system observed since 1980 is compatible with other major events, such as the Atlantic Niño (Philander 1986) or the circulation anomaly noted along the Namibian coast (Shannon *et al* 1986).

Along the Côte d'Ivoire continental shelf, environmental patterns were investigated using data collected from 1966 to 1984 (Morlière and Rebert 1972; Hisard 1973; Colin 1988). Characteristics of coastal upwelling and their interannual variability are well documented (Morlière 1970; Voituriez 1981; Colin 1988; Arfi *et al* 1993; Colin *et al* 1993; Pézennec 1994). Along the Côte d'Ivoire shoreline, this seasonal enrichment supports pelagic and demersal fisheries, both very sensitive to environmental change (Binet *et al* 1991; Pézennec and Bard 1992; Binet 1993). Continental influence is linked to four major rivers. Cavally, Sassandra and Bandama Rivers flow directly into the Gulf of Guinea, while the Comoé River flows seaward through the Ebrié Lagoon and the Vridi Canal. These large river inputs are high during the flood season, from October to December. Rainfalls in the coastal forest area induce local river floods during the rainy seasons, from April to June and from October to November (Binet 1983; Mensah 1991).

Several studies have been conducted on relating the effects of upwellings on the structure of the marine ecosystem, particularly on bacterioplankton and phytoplankton (Hanson *et al.* 1986; Fiala and Delille 1992; Painting *et al.* 1993; Wiebinga *et al.* 1997). In the coastal area of the Gulf of Guinea, no recent information is available on plankton productivity. Along the Côte d'Ivoire, coastline, phytoplankton (Dandonneau 1973; Binet 1993; Sevrin-Reyssac 1993) and zooplankton (Leborgne and Binet 1979) data were collected during the 1965-1979 period. Therefore, very little is known on the recent planktonic productivity and on the relationships existing between planktonic communities, and especially, how bacteria biomass responds seasonally to the changes in phytoplankton abundance.

A sea sampling programme was initiated in the mid-sixties in the Abidjan coastal zone. But the station location has changed several times, although globally, the same area was studied. Since 1982, a weekly hydrological sampling (temperature, salinity and Secchi disk measurements) has been maintained at the same site (Bakayoko 1990; Cissoko *et al.* 1995, 1996).

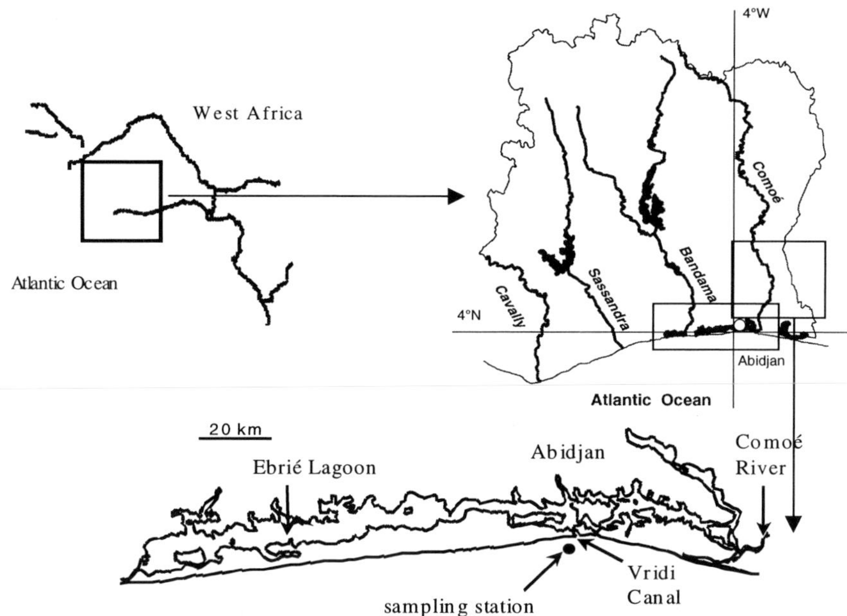

Figure 9-1 Location of the sampling station in the Gulf of Guinea.

This station is located four miles southwest of the Canal de Vridi (5°11N 4°04W, 80m depth, Figure 9-1). From May 1992, bacterial and phytoplanktonic biomasses were added to the parameters studied, while nutrients (PO_4-P, NH_4-N, NO_2-N and NO_3-N) completed the data set from December 1994. Local rainfall and monthly discharges of the main rivers were also collected, when available.

The aims of the study were to describe the seasonal and interannual fluctuations of physical parameters in relation to major continental (rain, river floods) and oceanographic events (upwellings) in the northern Gulf of Guinea during the 1992-1997 period and to compare these data to older information; and to assess the respective importance of these hydrological factors on the pelagic system (bacteria and phytoplankton) in that coastal station.

Materials and Methods

Rainfall was recorded daily at a meteorological station located 17 km west of Abidjan. River discharges (monthly average flow at the last gauge station before the estuaries) were obtained from the Côte d'Ivoire Water Authority.

Due to the many processes contributing to the phenomenon, upwelling indices could not be derived from wind and Ekman transport relation (Picaut 1983). Because upwelling results in a drop of SST, a fortnightly index based on the daily deviation from 26°C of surface shore temperatures was calculated (Arfi et al 1991, 1993; Pézennec 1994).

Seawater was collected using Niskin bottles at different depths (-1, -10, -20, -30, -50 and -75 m). Temperature was recorded using reversing thermometers and salinity was determined with a salinometer. Water transparency was measured with a 30-cm diameter Secchi disk. From repeated measurements using a LiCor underwater quantum sensor, a relation between Secchi disk values (Z_s in m) and light attenuation coefficient (k in m^{-1}) was calculated:

$$\ln(k) = -0.64*\ln(Z_s) - 0.27 \ (r^2 = 0.77, n = 41) \qquad \textbf{Eq. 9-1}$$

Thickness of the euphotic layer (Z_{eu} in m) was estimated from the 1% light level (Z_{eu} = 4.605/k). Nutrient concentrations of GF/F filtered water were analyzed according to the methods proposed by Aminot and Chaussepied (1983): ammonia was measured with a spectrophotometer, while soluble reactive phosphate (SRP), nitrite and nitrate concentrations were measured using an AutoAnalyzer. Bacterial abundance at -1m was estimated from direct counts by the epifluorescence method after staining cells by DAPI (Porter and Feig 1980). Chlorophyll a, considered as an index of phytoplankton biomass, was analyzed fluorometrically (Yentsch and Menzel 1963). Each profile of nutrients and chlorophyll biomass was integrated over the euphotic layer and expressed in mg m^{-2}.

It was not possible to retrieve the original oceanographic data collected in the sixties and the seventies. Therefore, figures illustrating temperature and salinity at -10 m and Secchi disk values recorded at the coastal station (fortnight average for the 1966-1971 period) were digitized from Morlière and Rebert (1972). The same process was used for the chlorophyll fortnight average values corresponding to the 1966-1969 period (Sevrin-Reyssac 1993).

Results and Discussion

Freshwater Inputs

Rainfall shows an obvious annual cycle (Figure 9-2a), with high values recorded during the main (May and June, 300 to 500 mm) and the secondary (October and November, around 200 mm) rainy seasons. Low values (less than 50 mm) are usually observed in August and in January-February. In this coastal area, rainfall shows a decreasing trend, embedded within large interannual variations. Drought characterises the end of the 1955-1997 period (Figure 9-3), even though high precipitation rates were still recorded in the later years (1987, 1993 and 1996). Annual values calculated for the 1988-1997 period (average 1551 mm) are lower than those calculated for the 1955-1965 (2278 mm), 1966-1976 (1844 mm) and the 1977-1987 (1633 mm) periods. When the two extreme periods are compared, the main rainy season of the recent sequence shows a marked rainfall decrease, while an increase is observed

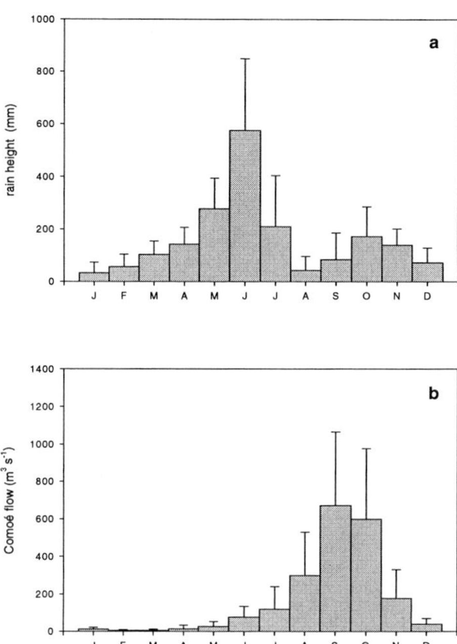

Figure 9-2 a Monthly rainfalls averaged over the 1948-1997 period and **b** Comoé flow averaged over the 1955-1995 period.

during the secondary rainy season. On several years during the 1990s, the main and the secondary rainy seasons showed comparable rainfalls.

The Comoé River, close to the station, exhibits a period of low water from January to May and a flood period in September-October (Figure 9-2b). These waters arrive in the ocean after a gap close to one month corresponding to the transit in the Ebrié lagoon (Durand and Guiral 1994). The same seasonal alternation of low flow and flood periods characterizes other large rivers along the Gulf of Guinea (see Mensah, 1991 for the Ghana shoreline).

River discharges show marked interannual fluctuations. The recent drought in the Sahel, and dam construction in the 1970s for the Sassandra and Bandama Rivers explain the sharp decline of freshwater inputs reaching the Côte d'Ivoire coastal area during the last two decades (Table 9-1). For the Comoé River (not dammed), the annual flow (Figure 9-3) is much lower during the 1977-1987 and the 1988-1995 periods (respective averages: 120 and 128 $m^3 s^{-1}$) than during the 1955-1965 period (224 $m^3 s^{-1}$) or the 1966-1976 period (197 $m^3 s^{-1}$). The main difference between these sequences is due to the flood flow decrease, but low waters recorded recently are nearly half those recorded 25 years ago. The annual average

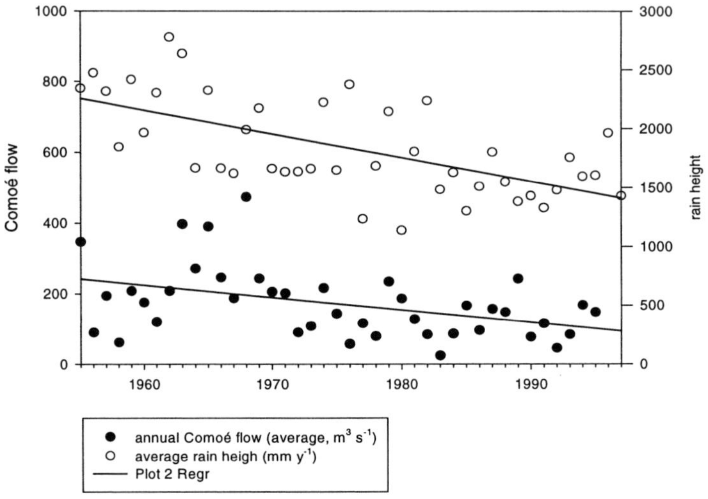

Figure 9-3 Annual rainfalls near Abidjan and Comoé flow for the 1955-1997 period.

Period	Cavally	Sassandra	Bandama	Comoé
1958-1969	Nd	Nd	Nd	248
1966-1969	689	663	372	287
1970-1981	456	390	165	147
1982-1993	462	364	154	110
1992-1995	Na	Na	Na	110

Table 9-1 Interannual variability of freshwater flow (yearly average, m^3s^{-1}) from main Cote d'Ivoire rivers reaching the Gulf of Guinea. (Nd: no data; Na: data recorded, but not available)

flow of the 1992-1995 period ($3.5\ 10^9\ m^3\ y^{-1}$) represents only 38% of the 1966-1969 value ($9.1\ 10^9\ m^3\ y^{-1}$). All these changes (decreasing of the freshwater and solid load inputs) have probably limited the continental influence on the neritic area.

Interannual change in upwelling intensity

The coastal upwellings occurring along the Côte d'Ivoire shore are under the influence of several physical, topographical and climatological factors (details in Arfi *et al.*, 1993). They show high interannual variability, and in recent years, the western part of the shoreline was characterized by events that were cooler and of longer duration than in the eastern part (Pézennec and Bard 1992; Pézennec 1994). Until the mid-80s, upwelling intensity was low from Assinie to Fresco and high west of Sassandra (6°W). From 1986, high intensity was also observed along the shoreline east of Fresco. In the last two years, a more classical pattern took over again, with very weak intensity at the extreme east of the coast (Assinie).

Seawater Characteristics

- Temperature and salinity

An obvious annual cycle characterises the upper 30 m of the water column, with low temperature during the main upwelling (from July to September) and high values from November to June (except a few weeks of possible cooling in January-February during a SCE). Fortnightly averaged Sea Surface Temperature (SST) ranged from 21.17°C to 30.50°C, while at -50 m, values were always under 26°C. Sea surface salinity shows also a seasonal cycle with two periods of high values alternating with two periods of low values. T-S plots at the various levels (fortnightly averaged data) show several features:

i.   the deepest level (-75 m) features very low variability;

ii. continental influence is clearly perceptible at a depth of 30m but not perceptible below 50m;

iii. deep waters reach seasonally the intermediate levels, and its influence is still perceptible at 20m during the major cold event (temperature < 21°C from mid-July to end-September). At 10 m, deep water is present between mid-August and mid-September. Near the surface, cooling is induced by mixing of deep water and of superficial water;

iv. the short cold events are not perceptible in the average cycle. It can be very marked (1994, 1997), but its intensity can be very weak in this part of the shoreline. In the continental shelf of Ghana, Mensah (1991) reported a thermocline, which fluctuates between 10 and 40 m depth.

In the surface layer, three steps can be distinguished in the annual T-S cycle (Figure 9-4):

Step 1 : After the flood period, there is a progressive salinity increase induced by mixing with the underlying water during the SCE and by the reduction of continental outputs (low water for the major rivers, local dry season). From January to May, a "coastal" situation can be defined.

Step 2 : From mid-May to June, warm and desalinated waters are observed throughout the West African coastal area (intrusion of "Guinean waters" (GW), a combination of oceanic water and freshwater originated from the coastal area and carried by the GC). From October to December, the main river floods reach the Gulf of Guinea, inducing a warming and a sharp salinity decrease: the hydrological situation is comparable to the GW sequence, but the salinity decrease is less pronounced.

Step 3 : The local rainy season ends when subsurface temperature decreases sharply (July to September). The thermocline reaches the surface, salinity increases rapidly and temperature decreases due to upwelling. This is the main cold period.

- Water transparency

In this coastal area, light attenuation is linked to three components:

i.   non-living seston of continental origin (detritus and particles of various size and shape): the river plumes are flattened eastward along the coast, but small particles can be

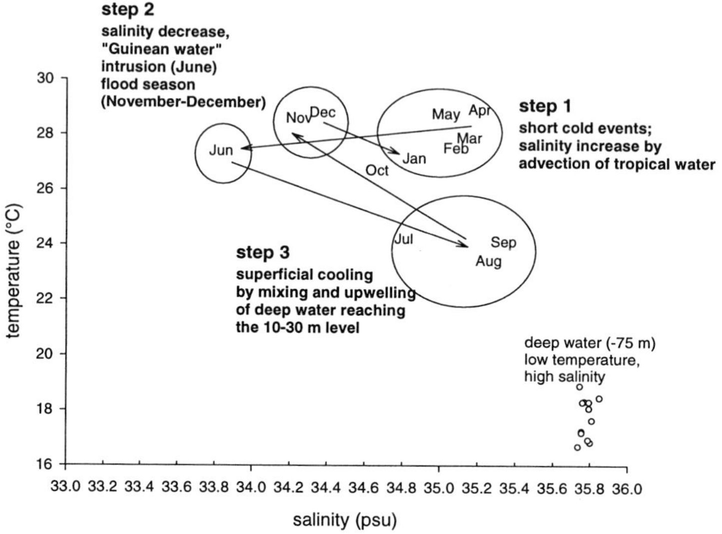

Figure 9-4 Annual T-S cycle in the surface layer.

mobilized by the currents while sedimenting-out, and can thus contribute to the local turbidity.

ii. living seston, essentially phytoplankton and bacterioplankton: this material is abundant just after the enrichment events in the neritic area.

iii. dissolved matter: lagoon and river waters have high concentrations of humic acids, conferring a yellow-brown color. Their effects, rapidly decreasing with the distance from the estuary, would be the most effective during the flood events.

Light attenuation is high from June to October (Secchi disk values: 4-15 m, average 9.3 m) and low from November to May (Secchi disk values: 10-29 m, average 17.6 m). The euphotic layer is 20-25 m thick during the high turbidity period, 35-40 m thick during the low turbidity period (Figure 9-5). Water transparency sharply decreases in June, when the GW invade the coastal area. From July to September, turbidity remains high (slow evacuation of continental particles from the superficial layer, then local phytoplankton production). Transparency slowly increases at the end of the upwelling event, reflecting the limited influence of flood water in the area.

Turbidity has markedly decreased between the 1966-1971 (interannual average Secchi disk value: 10.2 m, euphotic layer: 26.7 m) and the 1992-1997 periods (respective values: 13.9 m and 35.5 m). Turbidity is mainly related to the local rainy season: the decreasing rainfall in the last two decades have reduced the continental influence on the coastal water, so that the consequences of the GW intrusion are now less effective.

- Nutrients

SRP enrichment of the euphotic layer is obvious from mid-June to mid-October when the 0.6 μM contour line reaches the surface. Transient incursion of phosphate-rich water above the $-$ 30 m level is observed from mid-December to mid-January. Working on fortnightly averaged data allows representation of time-depth variations in nutrients over an average year. The study of the whole time series versus the water density permits a discrimination of three homogeneous sequences: continental influence (σ_t at $-$ 50 m < 25.25), coastal conditions (25.26 < σ_t at $-$ 50 m < 25.80) and upwelling influence (σ_t at $-$ 50 m > 25.81). Non parametric ANOVA on grouped data (Table 9-2) illustrates the trophic situation in the water column:

i.   values > 1 μM below 50 m;
ii. values < 0.5 μM at the superficial level;
iii. higher concentrations in upwelling situation than under continental influence;
iv. nitrates show the same annual pattern as the SRP. Average concentration at the bottom level (14.4 μM) is four times higher than at the surface (3.7 μM). At these levels, there is no obvious seasonal cycle, while such a cycle is clear between $-$20 and $-$50 m, with a marked concentration increase from mid-June to mid-October. This increase is also significant in January at $-$50 m. In the three typical situations (Table 9-2), the euphotic layer shows higher concentrations during the upwelling situation than during the sequence under continental influence (respective average: 7.63 and 3.04 μM, p < 0.001).

An opposite scheme is observed for ammonia, with high concentrations from March to May and from October to December. High values reflect active mineralization process in the water column and occur simultaneously with the input of organic matter of continental origin.

Average concentrations of nutrients in the euphotic layer are significantly higher during the upwelling event than during the GW and the coastal phases (Table 9-3). GW clearly have low SRP and nitrate concentrations in the euphotic layer. In most of the cases, this situation is not different from the "coastal" situation. Therefore, upward movements induced by the currents carry nutrient-poor water in the productive layer. When this type water is no longer present (i.e. end of the rainy season), the rich deep water reaches the surface, and therefore, the system could be potentially more productive.

Biological Characteristics

For bacterial abundance and chlorophyll biomass, a typical year can be drawn. An obvious seasonal cycle is observed for each parameter, with several peaks noted during the upwelling periods (Figure 9-6). Isolated peaks are also observed at the surface level, either during the seasonal continental inputs (rain or floods) or during the SCE. The main difference between these processes is their duration. Each peak corresponds to the biological response to an intrusion of cold and nutrient-rich deep water into the euphotic layer. If such a phenomenon is repeated several times at a high frequency, the planktonic enrichment will be significant and sustainable (upwelling situation). If the process has a low frequency (SCE), the enrichment will be less important. In other situations, enrichment is sporadic, even if in some occasions (1994), real fertilization can be detected.

Bacterioplankton abundance fluctuated from 1.5 x 10^5 to 19.0 x 10^5 cells ml^{-1}, with an interannual average of 4.2 x 10^5 cells ml^{-1}. These data confirm the previous results obtained on bacterioplankton biomass in this zone (unpublished data) and are in agreement with those

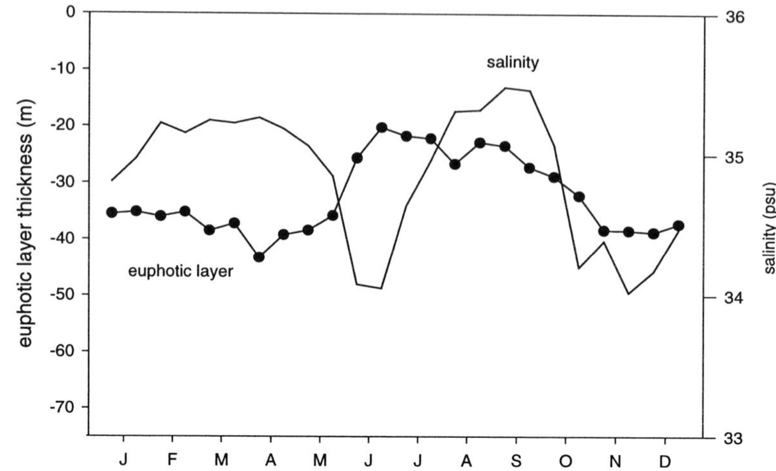

Figure 9-5 Surface salinity and water transparency: annual pattern based on fortnightly grouped data.

PO$_4$-P, µM	0 m	- 10 m	- 20 m	- 30 m	- 50 m	- 75 m
Continental (n=59)	0.47	0.50	0.46	0.49	0.66	1.08
Coastal (n=40)	0.43	0.45	0.58	0.73	1.01	1.19
Upwelling (n=40)	0.45	0.61	0.79	0.94	1.17	1.29
P	H$_o$ nr	0.007	<0.001	<0.001	<0.001	0.002

NO$_3$-N, µM	0 m	- 10 m	- 20 m	- 30 m	- 50 m	- 75 m
Continental (n=59)	3.16	3.61	3.72	4.07	7.03	12.61
Coastal (n=40)	4.18	4.50	5.3	6.94	11.54	15.36
Upwelling (n=40)	4.21	6.30	9.12	10.92	14.50	16.25
P	0.005	< 0.001	<0.001	<0.001	<0.001	<0.001

Chlorophyll, µg l^{-1}	0 m	- 10 m	- 20 m	- 30 m	- 50 m	- 75 m
Continental (n=57)	0.62	0.55	0.63	0.68	0.55	0.27
Coastal (n=39)	1.05	1.02	0.95	0.88	0.59	0.28
Upwelling (n=38)	1.70	1.42	0.99	0.80	0.35	0.30
P	<0.001	<0.001	<0.001	H$_o$ nr	0.003	H$_o$ nr

Table 9-2 Nutrient concentrations and chlorophyll biomass: averages calculated on data grouped by water quality. A sampled day is considered as representative of "continental", "coastal" or "upwelling" situation following of the σ_t at − 50 m. Statistical significance: mean equality rejected (p value provided) or not rejected (H$_o$ nr).

	Continental (n=57)	Coastal (n=39)	Upwelling (n=38)	*p*
SRP (PO_4-P), µM	0.48	0.55	0.70	< 0.001
Nitrate (NO_3-N), µM	3.04	5.23	7.63	< 0.001
Chlorophyll, µg l^{-1}	0.65	0.97	1.22	< 0.001

Table 9-3 Nutrient concentrations and biomass integrated over the euphotic layer: averages calculated on data grouped by water quality. A sampled day is considered as representative of "continental", "coastal" or "upwelling" situation following of the σ_t at − 50 m. Statistical significance: mean equality rejected (p value provided) or not rejected (H_o nr).

encountered in oligotrophic sites of the Atlantic Ocean during the EUMELI cruises (Dufour and Torréton 1994). But they are lower than those observed during seasonal upwelling in the NW Indian Ocean (1.0-2.1 x 10^6 cell ml^{-1}, Wiebinga *et al.* 1997) or those reported in the southern Benguela (10^7 cell ml^{-1}, Painting *et al.* 1993). Knowing the rapid growth potential of bacterial communities, the fact that variations of bacterial densities do not exceed one order of magnitude might suggest a drastic control by grazing or nutrient availability in the sea (Azam *et al.* 1983). Low abundance could also be related to the transport to the surface of deep waters with low bacterial numbers.

Chlorophyll biomasses under the surface ranged from 0.1 to 5.6 µg l^{-1} (average: 1.0 µg l^{-1}) with marked seasonal variations. High values are noted during the main upwelling, with an average of 1.9 μg l^{-1} from July to September at the surface level. From October to June, values are fluctuating around 0.8 μg l^{-1}. On a vertical profile, the highest values are observed at a depth of 10 m to 20 m. Concentrations measured at a depth of 50 m and below are low throughout the year, most of them ranging between 0.2 and 0.5 μg l^{-1}. In the Côte d'Ivoire upwelling, chlorophyll biomass is lower than recorded in other coastal upwelling systems. In Somalia, Wiebinga *et al.* (1997) reported a range of 0.2-2.7 µg l^{-1}, with peaks at 15µg l^{-1}, while in Mauritania, Herbland *et al.* (1973) measured values between 5 and 35µg l^{-1}. In Peru values between 2-5 μg l^{-1} were measured near the coast (Minas *et al.* 1990). Chlorophyll values collected now are higher than those collected near the surface 25 years ago. Sevrin-Reyssac (1993) reported values fluctuating around 0.2 μg l^{-1} from February to May and around 0.6 to 1.0 μg l^{-1} from July to October. Algal biomasses are higher than the historical ones, for the oligotrophic season as well as for the enriched season. This process can be related to the combination of several factors:

a) higher enrichment linked to the SCE, now more intense and lasting longer in this area;
b) a decrease of the duration of the GW period with low nutrient water present in the euphotic layer in a context of enhanced water transparency, under the effect of reduced flow and particulate load from the rivers (drought plus dams);
c) since the system shows a marked gradient from poor (GW) to high production (upwelling situation), the longer the cold event and the shorter the GW intrusion would be, the more productive the euphotic layer would be;
d) short-term variability of physical events affects the distribution of chlorophyll biomass integrated over the euphotic layer;

e) the first peak (January-February) showed a high (1994, 1997) or low (1993) biomass increase, directly linked to the upwelling enrichment and/or to the drift or an advection of enriched superficial waters;

f) the same phenomenon is observed from July to October: the main upwelling allows a more (1994) or less (1992 and 1993) intense algal development, ending with the flood.

Biomass data can be integrated over the vertical; these values (mg m^{-2}) multiplied by time can be integrated over the time series, and expressed in mg d m^{-2}. Such an operation produces an assessment of the relative enrichment of each sequence. The respective average contribution of the various situations is: SCE and upwelling 55%; GW 13%; flood 15%; coastal 17%. The cold events produce more than half of the annual enrichment, while they occur, on average, for 174 days (Figure 9-7).

Conclusions

Hydrological conditions observed at the coastal station studied off Abidjan are strongly influenced by the seasonal variability of three major phenomena: rainfalls, river floods and upwellings. Upwelling enriches the neritic ecosystem, exerting an immediate influence on biological productions, on phytoplankton and consequently, on bacterioplankton. Therefore, during four to five months (main upwelling plus short cold events), the coastal ecosystem can be considered as productive. Freshwater influence can be considered as an impoverishing factor, and oligotrophic conditions are characteristic of this period.

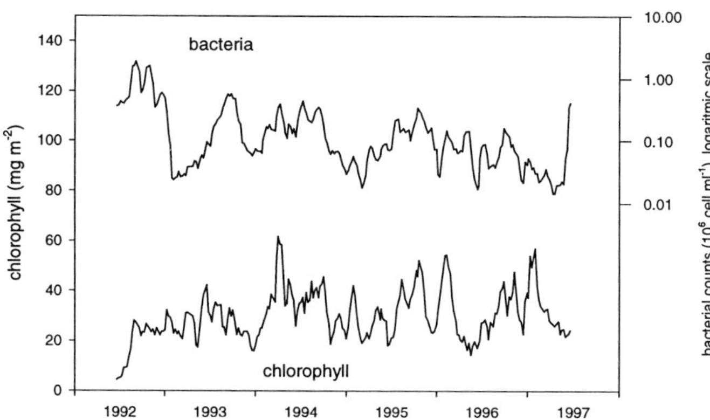

Figure 9-6 Chlorophyll biomass and bacterial abundance(4-term moving average) for the 1992-1997 period.

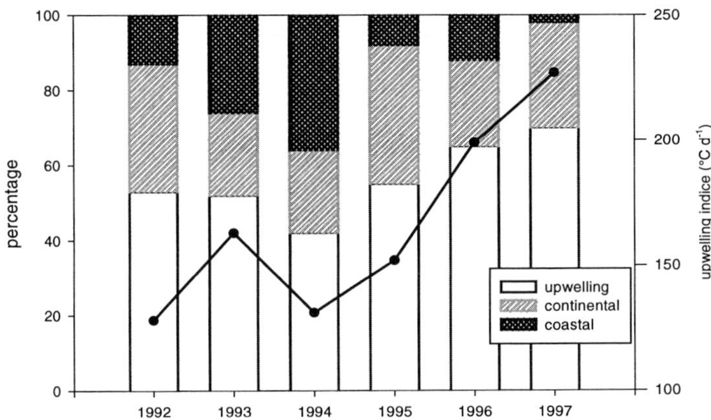

Figure 9-7 Respective percentage of the main hydrological situations for chlorophyll biomass integrated over the euphotic layer and over time (see text) and duration of the cold periods summed annually.

A first consequence of the lasting drought affecting the Sahel zone but also the coastal area is the decrease of continental influence and therefore, of the oligotrophic situation. Increased upwelling duration, and the recent major importance of SCE, enhances primary productivity and, in turn, the bacterial compartment. The neritic area along the eastern Côte d'Ivoire coastline can be presently considered as more productive than a few decades ago with the nutrient poor situation lasting less time, and the nutrient rich situation lasting longer. This change could explain the recent outburst of small pelagic fishes (such as *Sardinella aurita*) in this part of the Gulf of Guinea.

Acknowledgments

The authors gratefully acknowledge F. Sanséo and O. Sin for the sea sampling. We also thank R. Leborgne and M. Rodier for helpful comments and discussions.

References

Aminot, A., and M. Chaussepied 1983 *Manuel des analyses chimiques en milieu marin.* BNDO Documentation, Centre CNEXO de Brest. 395 p.

Arfi, R., O. Pézennec, S. Cissoko and M.A. Mensah 1991 Variations spatiales et temporelles de la résurgence ivoiro-ghanéenne. pp162-172. *In*: P. Cury and C. Roy (eds). *Pêcheries ouest-africaines: variabilité, instabilité et changements.* ORSTOM, Paris.

Arfi, R., O. Pézennec, S. Cissoko and M.A. Mensah 1993 Evolution spatio-temporelle d'un indice caractérisant l'intensité de la résurgence ivoiro-ghanéenne. pp 111-122. *In*: . P. LeLoeuff, E. Marchal and J-B. Amon-Kothias (eds) *Environnement et ressources aquatiques de Côte d'Ivoire. I: Le milieu marin.* ORSTOM, Paris.

Azam, F., T. Fenchel, J.G. Field, L.A. Meyer-Reil and F. Thingstad 1983 The ecological role of water-column microbes in the sea. Marine Ecological Progress Series 10: 257-263.

Bakayoko, S. 1990 Stations côtières et station Vridi. Exploitation des données des années 1984 et 1985. Arch. Sci., Centr. Rech. Océanogr. Abidjan 13 (2): 33-79.

Binet, D. 1983 Phytoplancton et production primaire des régions côtières à upwellings saisonniers dans le Golfe de Guinée. Oceanogr. trop. 18: 331-355.

Binet, D. 1993 Zooplancton néritique de Côte d'Ivoire. pp167-193. *In*: P. Le Loeuff, E. Marchal and J-B. Amon-Kothias (eds). *Environnement et ressources aquatiques de Côte d'Ivoire. I: Le milieu marin.* ORSTOM, Paris.

Binet, D. and J. Servain 1993 Have the recent hydrological changes in the Northern Gulf of Guinea induced the *Sardinella aurita* outburst. Oceanologica acta 16: 247-260.

Binet, D., E. Marchal and O. Pézennec 1991 *Sardinella aurita* de Côte d'Ivoire et du Ghana: fluctuations halieutiques et changements climatiques. pp320-342. *In*: P. Cury and C. Roy (eds). *Pêcheries ouest-africaines. Variabilité, instabilité et changement.* ORSTOM, Paris.

Cissoko, S., S. Bakayoko, J. Abe and S.B. Bamba 1995 Station hydrologique de Vridi et stations côtières de 1989 à 1991. Arch. Sci. Centr. Rech. Océanogr. Abidjan 14(3).

Cissoko, S., S. Bakayoko, J. Abe and S.B. Bamba 1996 Station hydrologique de Vridi et stations côtières de 1992 à 1995. Arch. Sci. Centr. Rech. Océanogr. Abidjan 15(1).

Colin, C. 1988 Coastal upwelling events in front of Ivory coast during the FOCAL program. Oceanologica acta 11: 125-138.

Colin, C. and S.L. Garzoli 1988 High frequency variability of the in situ wind, temperature and current measurements in the equatorial Atlantic during the FOCAL/SEQUAL experiment. Oceanologica acta 11: 139-148.

Colin, C., Y. Gallardo, R. Chuchla and S. Cissoko 1993 Environnements climatique et océanographique sur le plateau continental de Côte d'Ivoire. pp75-110. *In*: P. Le Loeuff, E. Marchal and J-B. Amon-Kothias (eds). *Environnement et ressources aquatiques de Côte d'Ivoire. I: Le milieu marin.* ORSTOM, Paris.

Dandonneau, Y. 1973 Etude du phytoplancton sur le plateau continental de Côte d'Ivoire. III Facteurs dynamiques et variations spatio-temporelles. Cah. ORSTOM ser. Oceanogr. 11: 431-454.

Dufour, Ph. and J.P. Torréton 1994 Biomasses et activités bactériennes de la colonne d'eau à EUMELI 4, Leg 1. Premiers résultats des campagnes EUMELI. A. Morel (ed.). 33 p.

Durand, J.R. and Guiral D. 1994 Hydroclimat et hydrochimie. pp59-90. *In*: J.R. Durand, Ph. Dufour, D. Guiral and G.S. Zabi (eds). *Environnement et ressources aquatiques de Côte d'Ivoire. I: Les milieux lagunaires.* ORSTOM, Paris.

Fiala, M. and D. Delille 1992 Variability and interactions of phytoplankton and bacterioplankton in the Antarctic neritic area. Marine Ecology Progress Series 89: 135-146.

Hanson, R.B., M.T. Alvarez-Ossorio, R. Cal, M.J. Campos, M. Roman, G. Santiago, M. Varela and J.A. Yolder 1986 Plankton response following a spring upwelling event in the Ria de Arosa, Spain. Marine Ecology Progress Series 32: 101-113.

Herbland, A., R. Le Borgne and B. Voituriez 1973 Production primaire, secondaire et régénération des sels nutritifs dans l'upwelling de mauritanie. Doc. Scient. Centre Rech. Océanogr. Abidjan 6: 1-75.

Hisard, P. 1973 Variations saisonnières à l'équateur dans le Golfe de Guinée. Cah. ORSTOM, ser. Oceanogr. 11: 349-358.

Leborgne R. and D. Binet 1979 Dix ans de mesures de biomasses de zooplancton à la station côtière d'Abidjan 1969-1979. Doc. Sci. Cent. Rech. Océanogr. Abidjan 10(2): 165-176.

Mahé, G. 1991 La variabilité des apports fluviaux au golfe de Guinée utilisée comme indice climatique. pp147-161. *In*: P. Cury and C. Roy (eds). *Pêcheries ouest-africaines. Variabilité, instabilité et changement.* ORSTOM, Paris.

Mensah, M.A. 1991 The influence of climatic changes on the coastal oceanography of Ghana. pp67-79. *In*: P. Cury and C. Roy (eds) *Pêcheries ouest-africaines. Variabilité, instabilité et changement.* ORSTOM, Paris.

Minas, H.J., B. Coste, M. Minas and P. Raimbault 1990 Conditions hydrologiques, chimiques et production primaire dans les upwellings du pérou et des îles Galapagos en régime d'hiver austral (campagne Paciprod). Oceanologica Acta 10: 383-390.

Morlière, A. 1970 Les saisons marines devant Abidjan. Doc. scient. Cent. Rech. océanogr., Abidjan 1 1-15.

Morlière, A. and J.P. Rebert 1972 Etude hydrologique du plateau continental ivoirien. Doc. Sci. CRO Abidjan, III(2): 1-30.

Painting, S.J., M.I. Lucas, W.T. Peterson, P.C. Brown, L. Hutchings and B.A. Mitchell-Innes. 1993 Dynamics of bacterioplankton, phytoplankton and mesozooplankton communities during the development of an upwelling plume in the southern Benguela. Marine Ecology Progress Series 100: 35-53.

Pézennec, O. 1994 *Instabilité et changements dans l'écosystème pélagique côtier ivoiro-ghanéen, variabilité de la ressource en sardinelles: faits, hypothèses et théories.* Thèse de Doctorat en Océanologie Biologique, Université de Bretagne Occidentale.

Pézennec, O. and F.-X. Bard 1992 Importance écologique de la petit saison d'upwelling ivoiro-ghanéenne et changements dans la pêcherie de *Sardinella aurita.* Aquatic living Resources 5: 249-259.

Philander, S.G.H. 1986 Unusual conditions in the tropical Atlantic Ocean in 1984. Nature 322: 236-238.

Picaut, J. 1983 Propagation of the seasonal upwelling in the eastern equatorial Atlantic. Journal of Physical Oceanography 13: 18-37.

Porter, K.G. and Y.S. Feig 1980 The use of DAPI for identifying and counting aquatic microflora. Limnology and Oceanography 25: 943-948.

Roy, C. 1995 The Côte d'Ivoire and Ghana coastal upwellings dynamics and changes. pp346-361. *In*: F.X. Bard and K.A. Koranteng (eds). *Dynamics and use of Sardinella resources from upwelling off Ghana and Ivory Coast.* ORSTOM, Paris.

Sevrin-Reyssac, J. 1993 Phytoplancton et production primaire dans les eaux marines ivoiriennes. pp151-166. *In*: P. Le Loeuff, E. Marchal and J-B. Amon-Kothias (eds) *Environnement et ressources aquatiques de Côte d'Ivoire. I: Le milieu marin.* ORSTOM.

Servat, E., J.E. Paturel and H. Lubes 1996 La sécheresse gagne l'Afrique Tropicale. La Recherche 290: 24 –25.

Shannon, L.V., A.J. Boyd, G.B. Brundrit and J. Taunton-Clark 1986 On the existence of an El-Nino-type phenomenon in the Benguela system. Journal of Marine Research 44: 495-520.

Voituriez, B. 1981 *The equatorial upwelling in the eastern Atlantic Ocean. Recent progress in equatorial Oceanography.* Nova University Press 229-247.

Wiebinga, C.J., M.J.W. Veldhuis and H.J.W. De Baar 1997 Abundance and productivity of bacterioplankton in relation to seasonal upwelling in the northwest India Ocean. Deep Sea Research 44: 451-476.

Yentsch, C.S. and D.W. Menzel 1963 A method for the determination of phytoplankton, chlorophyll and phaeophytin by fluorescence. Deep Sea Research 10: 221-231.

II

Environmental Forcing & Productivity

Productivity

The Gulf of Guinea Large Marine Ecosystem
J.M. McGlade, P. Cury, K.A. Koranteng and N.J. Hardman-Mountford (Editors)

10

Environmental and Resource Variability off Northwest Africa and in the Gulf of Guinea: A Review

C. Roy, P. Cury, P. Fréon and H. Demarcq

Abstract

Environmental monitoring off West Africa relies mainly on a set of coastal stations, on the COADS data base and on satellite imagery. This provides useful information on a limited set of variables such as SST and wind. These variables can be related to fish population dynamics at different scales of observation including short-term changes in fish availability, year to year abundance or lower frequency regime shifts. Tools such as multiple time series analysis, GAM (General Additive Models) and IBM (Individual Based Models) can help to track time lags, non-linear relationships and discontinuities that exist between environmental variables and fish populations. These methods can help to further understand ecological processes in relation to environmental variability. Relatively few oceanographic surveys have been done off West Africa and the existing oceanographic data are difficult to access. As new information on environment, resources, fisheries and their interaction are needed for management purposes and research, particular attention should be devoted to process oriented studies. Given constraints on human and financial resources, the challenge is to achieve an appropriate balance between monitoring and process-oriented studies.

Introduction

Scales of observation and scales of study are a subject of interest in marine as well as in terrestrial ecology (Steele 1995). Variability may affect ecological processes on very different scales ranging from micro-turbulence that affects encountering rates between plankton and fish larvae, to global climatic changes which might strengthen upwelling intensity and therefore global ecosystem productivity (Bakun1990). Scales of observation and of a particular studies might be different (disconnected). For example, a particular process might act on a very local scale yet be observable on a more global scale (e.g. recruitment may be due to a very small fraction of the parental spawning success in a particular retention area but observable at the scale of the whole fishery). West African examples of short-term, medium-term and long-term changes and their impact on fisheries are presented in this paper. When referring to short-term we consider changes in fish availability from one fortnight/month to the next. Seasonal or inter-annual variability affects fish abundance and longer time periods may play a significant role in changing reproductive patterns and decadal variations in fish abundance. A review of previous work done in West-Africa is presented, emphasizing the importance of addressing proper scales when studying climate and fisheries.

Monitoring Environmental Variability

In most West-African countries, oceanography is the responsibility of the national fisheries research institutes and consequently limited financial and human resources have been allocated to support environmental monitoring and research. Oceanographic research vessels do exist in many West-African countries, but due to lack of modern scientific equipment, trained technicians and financial support, alternative ways to gather environmental information have to be considered.

A brief review of the available environmental data

In the late 1950s, ORSTOM set up a network of coastal stations in Senegal, Côte d'Ivoire and Congo. At each station, Sea Surface Temperature (SST) is measured daily from the beach or a wharf using a bucket and a thermometer. Some of these stations are still operating and continuous records of SST are now available for the past 30 or 40 years. Salinity and nutrient data were also collected at some stations, but data density is very low. In Côte d'Ivoire and Congo, time series of plankton displacement volume are also available from 1969 to 1981 (Binet 1983). In the late 1960s, the MFRD of Ghana also implemented a set of coastal stations along the coast. Since 1968, MFRD has also maintained a weekly transect, consisting of four stations from the coast to the 1000m isobath. Salinity, temperature, oxygen and zooplankton samples are collected at each station (Mensah 1995; Koranteng 1998).

Another source of information is provided by meteorological data collected at several airports along the coast of West Africa. It has been shown that the Dakar-Yoff (Senegal) and Nouadhibou (Mauritania) airports, each being located on a cape, provide useful data for assessing wind variability over the coastal domain. These wind data are often used to calculate an index of the upwelling intensity (Arfi 1985; Roy 1989). This index called 'Coastal Upwelling Index' (CUI) is the offshore component of the wind induced Ekman transport (Bakun 1973). CUI has been used to characterize the upwelling strength on both seasonal and interannual scales.

The Comprehensive Ocean Atmosphere Data Set (COADS) provides an alternative when limited information is available. COADS summarizes over 100 million surface meteorological observations collected by ships of opportunity and other platforms over the world oceans which have been quality controlled and put into a consistent format (Woodruff *et al.* 1987). This dataset is the most complete record of surface marine climate to date. It has world-wide coverage and the earliest data dates back to 1854. The five CD-ROMs and the software developed by the CEOS project (Roy and Mendelssohn 1998) provide an easy way to extract time series of SST, wind, air temperature and atmospheric pressure along the ship tracks located off the west African coast.

The COADS dataset appears to be a useful surrogate when oceanographic data are not available. It is also well suited for comparative analyses between different areas, as one can expect the data to be homogeneous over large areas. Despite some biases such as a spurious trend in the wind and a shift in SST in the early 1940s, useful information on the long-term variability of the environment can be developed using COADS (Roy and Mendelssohn 1998). Satellite data such as METEOSAT IR or NOAA/AVHRR images are also of special interest for analysing the coastal dynamics of the upwelling system. Several long-term databases, going back to the early 1980s, exist for the Atlantic ocean : the CORSA from JRC (Nykjaer and Van Camp 1994) or the IR METEOSAT images from UTIS/ISRA (Demarcq and Citeau 1995). NOAA MCSST or PATHFINDER dataset also provides useful information for large-scale studies.

Oceanographic data were collected over North West Africa during the CINECA and JOINT experiments which took place in the late 1970s (Hempel1982). The data collected during these international projects remain the most comprehensive oceanographic dataset collected off Northwest Africa. Some oceanographic data are collected during fishery surveys (national or co-operative) but most often these are difficult to access. In some countries, oceanographic surveys are performed under bilateral agreement or fishing counterpart. Off Senegal, the CIRSEN project lasted for three years (1985-1988) and was designed to provide oceanographic data over the whole continental shelf.

In summary, continuous environmental monitoring off West Africa has been limited. However time series are available for a couple of parameters (wind, SST) for several decades. These time series provide useful basic information on the local marine environment and its variability.

Environmental variability off West Africa

Off West-Africa, the density of data collected by ships of opportunity is high as main ship tracks connecting the northern and the southern hemispheres are located within a few miles off the coast. From Morocco to Ghana, continuous monthly SST and wind time series can be constructed using the COADS dataset. Using COADS, a comparison of the seasonal dynamics of the West-African coastal ecosystems was performed using the mean seasonal cycle of SST and wind-derived variables such as upwelling index and wind mixing index (Roy 1991). It provided useful insight into the latitudinal and seasonal variability of the main physical forcing factors of the West-African coastal upwelling (Table 10-1). The resulting data have been used in several studies, which related small pelagic fish reproductive strategy and the environment (Roy *et al.* 1989, 1992; Fréon *et al.* 1997; Shin *et al.* 1998).

Along the West African coast, COADS SST anomalies during the upwelling season highlight the main patterns of upwelling activity over the last 45 years (Figure 10-1, colour plate). South of 20°N, interannual variability is well defined and one of the major features is a pronounced cooling that occurred from 1971 to 1977, followed by a warming that lasted for several years, until 1984. These alternations of cold/warm periods have been well documented using other data sources (Binet 1988; Roy 1989; Koranteng 1998; Koranteng and McGlade, this volume) and they have had important ecological consequences. North of 25°N, the intensity of SST interannual variability during the upwelling season appears to be weaker than further south.

Upwelling dynamics off Senegal

Upwelling off Senegal starts to develop in late November as a consequence of the onset of the trade-winds which blow quite steadily from November through May. Along the coast, SST drops within a month from 28°C to less than 18°C. SST can be as low as 14°C in upwelling centres such as the tip of the Cape-Vert peninsula or the northern coast of Senegal. The negative correlation between CUI and SST suggests that local wind is the dominant forcing

Location	Upwelling duration	SST (°C)			Wind Speed (m.s^{-1})			CUI (m^3.s^{-1}.m^{-1})		
		min.	max.	mean	min.	max.	mean	min.	max.	mean
34°N-36°N	April-August	16.21	22.37	19.03	5.18	6.44	5.77	0.04	0.35	0.19
32°N-34°N	April-August	16.80	22.25	19.36	4.96	6.31	5.75	0.11	0.71	0.37
30°N-32°N	March-August	17.14	21.30	19.12	5.18	7.91	6.73	0.41	1.39	0.87
28°N-30°N	March-August	17.63	21.65	19.47	5.29	7.92	6.47	0.31	1.05	0.70
26°N-28°N	Permanent	18.48	22.40	20.27	5.47	8.59	6.79	0.54	1.53	0.92
24°N-26°N	Permanent	18.57	22.03	20.15	5.50	7.87	6.68	0.60	1.45	0.98
22°N-24°N	Permanent	18.28	21.38	19.66	5.58	8.16	7.00	0.78	1.73	1.28
20°N-22°N	Permanent	18.22	22.62	19.88	5.73	8.52	7.10	0.92	2.18	1.52
18°N-20°N	October-June	18.92	26.38	21.89	4.54	7.41	6.21	0.92	2.15	1.48
16°N-18°N	October-June	19.54	27.76	23.22	4.15	6.66	5.51	0.52	2.11	1.37
14°N-16°N	December-May	19.68	28.09	24.05	3.62	5.71	4.86	0.21	1.96	1.16
12°N-14°N	December-May	20.25	28.37	24.83	2.83	5.06	4.16	-0.14	1.69	0.91
10°N-12°N	January-April	22.18	28.25	25.91	2.12	5.41	3.40	0.07	1.04	0.54

Table 10-1 Main environmental characteristics of the Canary current upwelling ecosystem (derived from COADS dataset).

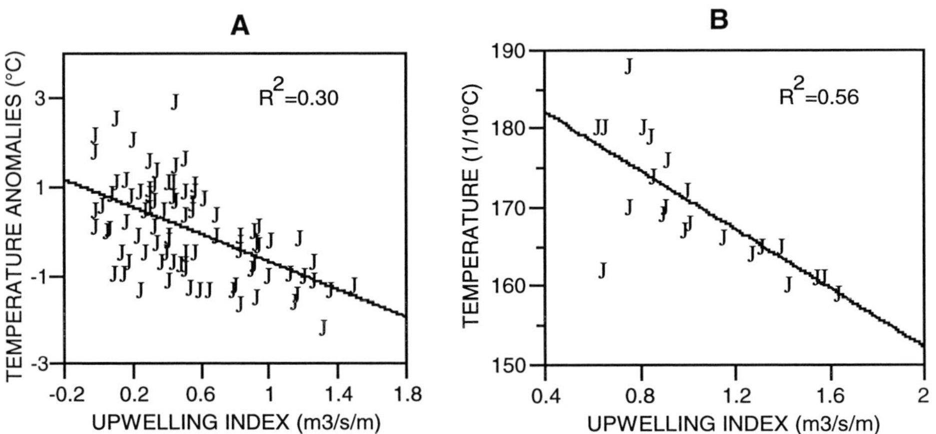

Figure 10-2 Relationship between SST anomalies and upwelling index (A) and SST and upwelling index (B) at Dakar Thiaroye, 1966-1987.

factor of the upwelling (Roy 1989). Stronger than average trade winds lead to enhanced coastal upwelling and cold SST (Figure 10-2). This simple model of the response of upwelling to wind shows that data collected at coastal stations can be used to characterise the interannual variability of the Senegalese upwelling. Both coastal SST and CUI data have been widely used to relate fish population dynamics to the environment (Fréon 1983; Cury and Roy 1988, 1989; Roy 1998).

Spatial dynamics of the upwelling from METEOSAT data

Sea Surface Temperature (SST) is routinely computed using METEOSAT satellite infrared data. METEOSAT's half-hourly observation frequency allows production of daily images with low cloud contamination. Despite the limitation of METEOSAT IR data (low spatial resolution, low accuracy), a continuous time-series of images with a time step of several days can be constructed; these time-series are extremely useful for studies of the dynamics of the coastal upwelling off West Africa.

The satellite data are initially processed on a daily basis; the initial spatial resolution (5 km) as well as the thermal resolution (0.5 °C) is preserved. As a compromise between the variability of the upwelling process and the quantity of data to be analysed, daily images are averaged over 5 days time periods without altering the 5 km spatial resolution (Demarcq and Citeau 1995). These 5 day composites are then used to derive structural parameters such as the value and the position of the near-coastal upwelling maximum. These data are also used to derive a satellite based upwelling index (Figure 10-3, colour plate) and to study SST gradient over the shelf in order to identify areas where retention may be favoured.

Short-term Environmental Changes and Pelagic Fish Availability

Availability of sardinellas off Côte d'Ivoire and associated environmental conditions

Availability of pelagic fish might be associated with instant or short-term environmental processes as fish act to seek food in their proximate environment. The fishery off Côte d'Ivoire occurs in a region with two upwelling seasons, a smaller one around January and a larger one from June to September. Mendelssohn and Cury (1987) primary goal was to model the relative impact of the environment on the availability of pelagic species off Côte d'Ivoire and the manner in which the environment affects the dynamics of catch per unit effort for these species on time scales as short as two weeks. Using multivariate time series methods, a fortnightly abundance index of the Ivorian coastal pelagic species was related to SST in the areas in which the fishery operates. This multivariate autoregressive model is able to explain 43% of the observed variance in catch per unit effort (cpue, in tonnes per unit of search time) from 1966 to 1982. The following AR model was used:

$$\text{Lncpue (t)} = \begin{aligned}&0.248\ \text{Lncpue (t-1)} + 0.254\ \text{Lncpue (t-2)} + 0.162\ \text{Lncpue (t-4)} + \\ &0.143\ \text{Lncpue (t-24)} - 0.143\ \text{SST (t-1)} + 0.112\ \text{SST (t-2)} + \\ &0.144\ \text{SST (t-17)}\end{aligned} \qquad \textbf{Eq. 10-1}$$

where Lncpue (t) = LOG (cpue (t) + 0.05).

The model shows that, all things being equal, cpue will show some persistence on its own; at low levels it will tend to remain low and conversely so for high levels remain high. From the model estimates it appears that a drop in SST from two to one fortnight previously provides an increase in cpue. The sharper the drop, the greater the effect on cpue. Binet (1976) presented results that show a correlation between a drop in SST and an increase in zooplankton biomass a fortnight later (Figure 10-4). Binet (1983) also suggested that cold

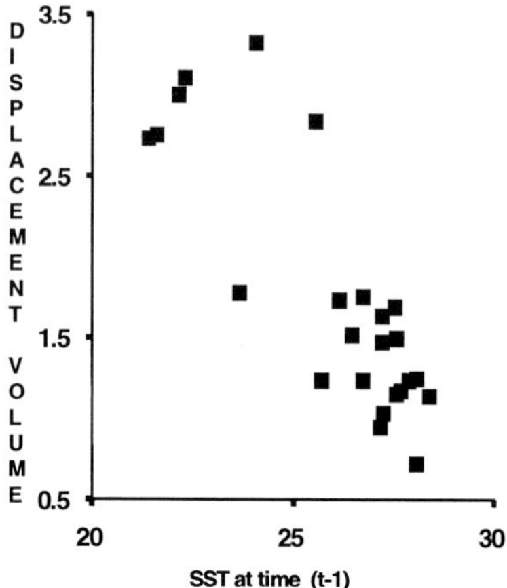

Figure 10-4 Mean plankton displacement volume (ml/m3) per fortnight as a function of SST off Cote d'Ivoire from 1966 to 1990; (plankton data from D.Binet).

waters tend to increase the aggregation of zooplankton at the surface and, hence to aggregate fish as catch per set tends to increase with plankton volume (Figure 10-5). Availability of fish depends on food availability as fish tend to form larger schools when the environmental conditions are favourable. Pelagic species probably come to the surface more and school more when there is abundant zooplankton biomass around which to aggregate. Variables such as SST are not sufficient to explain the evolution of cpue particularly when measured at the same time period as the catch. SST acts as a surrogate variable for oceanographic and biological processes that create favourable conditions for the pelagic fish species. Therefore it is the dynamics of the variables between periods that are important, not a static value in one time period. Similar results were found in tuna in the Gulf of Guinea (Mendelssohn and Roy 1986; Stretta 1991), illustrating the importance of considering short-term process when studying fish availability.

Catch per set of pelagic fish in Senegal and Côte d'Ivoire and associated environmental conditions

Fish school size appears to be highly variable on extremely short time periods (from hour to hour) or on larger time scales (seasonal or inter-annual). This has significant consequences for the availability of fish to the fishers. Fish school size can be measured directly using acoustics or indirectly using fisheries data. Many factors can affect the catch per set of a purse-seiner: the size of the net, the boat-loading capacity, the fishers strategy, the size of the schools. However it appears that the catch per set can provide a reliable index of mean school size and abundance. The mean catches per set of Senegalese (Fréon 1991) and Ivorian seiners were calculated per month from 1969 to 1987 and per fortnight from 1966 to 1990. As noted previously there are two upwelling seasons off Côte d'Ivoire, however small pelagic fish are caught all year round. Catch per set for the pelagic fisheries have been calculated in Senegal and Côte d'Ivoire on different scales of interest for the fisheries (from a fortnightly mean to an annual mean). When comparing the two data sets it appears that larger schools are fished in Senegal compared to those fished in Côte d'Ivoire: the average catch per set is around 9.7 tonnes in Senegal and only 4.6 tonnes in Côte d'Ivoire. In both countries the catch per set exhibits large seasonal fluctuations which can be related to the SST patterns (Figure 10-6). Thus it appears that larger catches per set are observed during the upwelling seasons. When calculating annual mean indices, it appears that there exists a strong correlation between cpue (in tonnes per unit of search time) and catch per set (in tonnes) (Figure 10-7). Catch per set roughly reflects fish school size and is strongly related to environmental fluctuations and probably to food availability. Changes in fish school size would depend primarily on the productivity of the upwelling area; a more productive area like Senegal would be able to sustain larger pelagic fish school sizes compared to Côte d'Ivoire. Mean fish school size appears to be related to the seasonal environmental patterns in the way that during the more productive periods, probably corresponding to higher food density observable during the upwelling seasons, fish tend to form larger schools (Figure 10-6) (Blaxter and Hunter 1982). On the whole, this has an effect on fisheries as catch per set and cpue appear to be strongly correlated on an annual scale (Fréon 1991).

Seasonal or Inter-annual Environmental Changes and Pelagic Fish Abundance

Pelagic fish abundance and surplus production models

Conventional surplus production models for stock assessment use only one input variable: the fishing effort. These models are not suitable for most of the pelagic stocks because fishing effort (E) variations only explain a small part of the total variability of annual production and catch per unit of effort (CPUE). In such a so called "unstable stock", residual variability often originates from the influence of environmental phenomena, which affects either the abundance (surplus production) or catchability of a stock. Fréon (1986, 1989) gave a theoretical basis for production models incorporating an environmental variable as a second independent variable in addition to fishing effort.

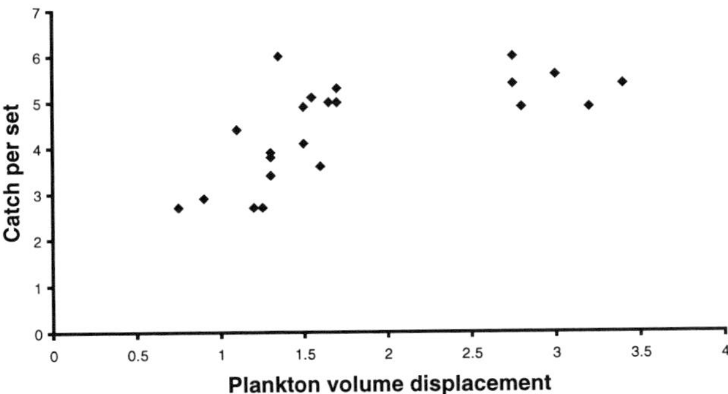

Figure 10-5 Catch per set as a function of plankton volume displacement per fortnight from 1966 to 1990; (data from CROA and D. Binet).

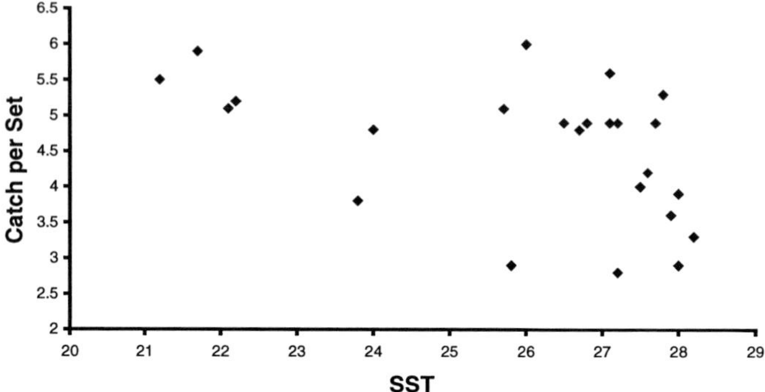

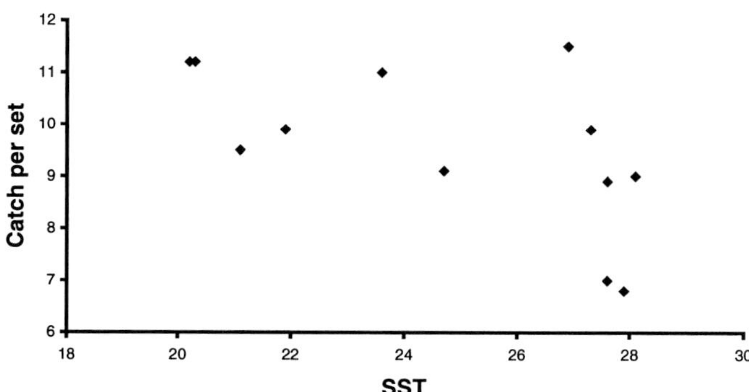

Figure 10-6 Mean total pelagic catch per set (tons/searching day) off Côte d'Ivoire and Senegal per fortnight and per month from 1966 to 1990 and 1969 to 1987.

The influence of this variable has been considered at two levels: on stock abundance and on stock catchability. For each case the conventional linear and exponential models (and sometimes the generalised model) are considered. The CLIMPROD software is an expert-system which first helps the user to select a corresponding model according to objective criteria. Then CLIMPROD fits the selected model using a non-linear regression routine and tries to assess the fit with parametric and non-parametric tests before presenting tables and graphs of the results in EXCEL format (Fréon *et al.* 1993).

This approach has been successfully applied to several pelagic stocks in Senegal, Côte d'Ivoire and Morocco (Fréon 1986) and also in Chile and Peru. In the case of Senegal, the interannual variations of the upwelling index are empirically related to the abundance of the stock (in addition to CPUE), while in the case of the *Sardina pilchardus* stock off the northern part of Morocco they are related to the catchability of this species. In the case of northern Morocco, the environmental effect is mainly on the catchability and the conventional Fox's exponential model was modified in order to let the coefficient of catchability, q, vary according to an upwelling index.

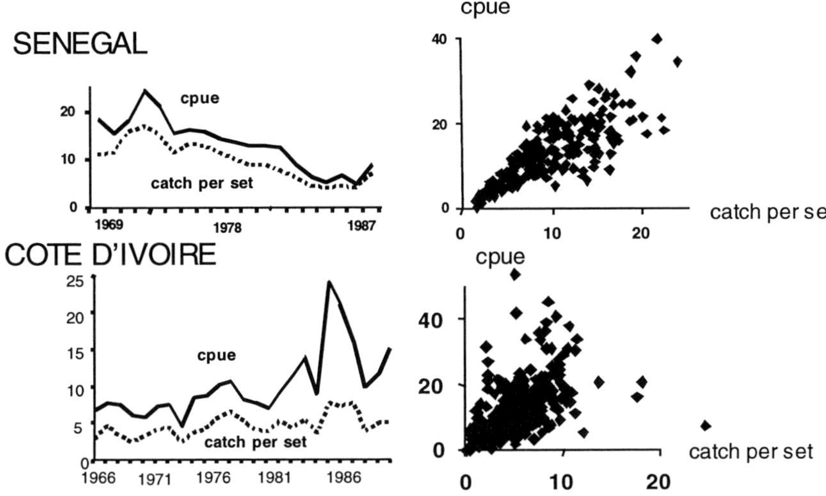

Figure 10-7 Annual cpue and catch per set in Senegal and Côte d'Ivoire from 1969 to 1987 and 1966 to 1990 (left). Cpue versus catch per set in Senegal and Côte d'Ivoire per month and per fortnight (right).

Using the same approach Cury and Roy (1987) analysed the catch and CPUE off Côte d'Ivoire by introducing SST into a Fox production model, but here the environmental effect was found to be on the abundance of the stock rather than on its catchability.

Optimal environmental window and pelagic fish recruitment

A dome shape relationship was found between annual recruitment indices and upwelling intensity using GAM (Generalized Additive Models) for different pelagic species such as sardines and anchovies in different upwelling systems (Cury and Roy 1989; Cury *et al.* 1995, Serra *et al.* 1998) (Figure 10-8). This model suggests that very weak winds may have a negative effect on recruitment by decreasing the production of food for larvae, while strong turbulence generated by high wind speeds has a negative effect on larval survival by desegregating food and larvae patches. This dome shaped relationship is known as the OEW (Optimal Environmental Window) (Cury *et al.* 1995; Bakun 1996). The value of 5-6 m.s^{-1} was found to be a threshold common to many ecosystems and species. The OEW explains approximately 20-25% of the recruitment variance in upwelling areas. For a non-Ekman type upwelling such as occurs in Côte d'Ivoire, the relationships between recruitment and upwelling intensity appears to be linear as the winds appear to be lower than the threshold value of 5-6 m s^{-1} (Figure 10-11).

Seasonal fish migration off Senegal and Mauritania

Many pelagic and demersal species migrate seasonally between Senegal and Mauritania. The appearance of a migrant population of thiof (*Epinephelus aeneus*) along the north coast of

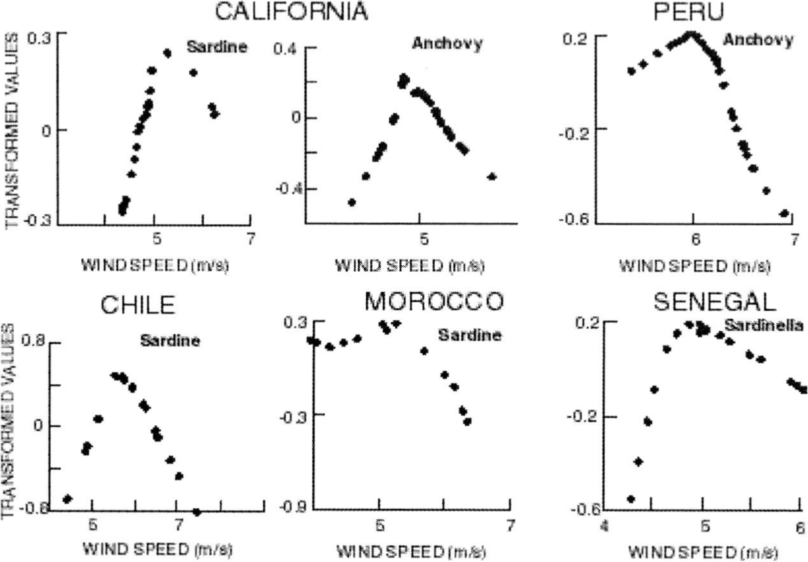

Figure 10-8 The relationship between recruitment and wind speed in several upwelling ecosystems.

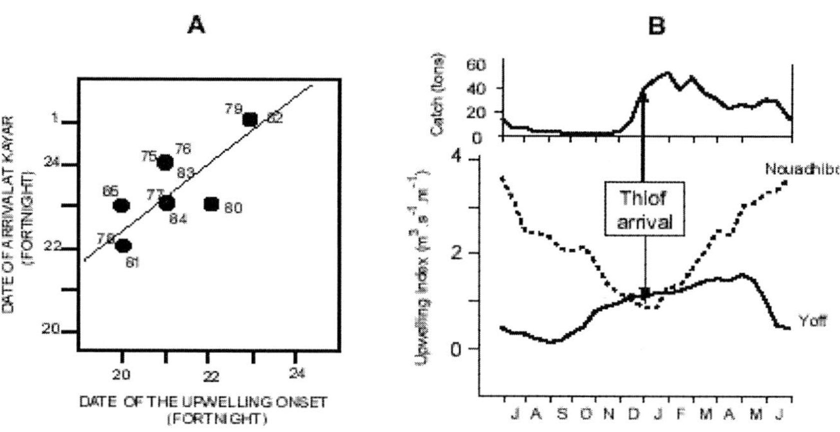

Figure 10-9 a-the relationship between the seasonal arrival of Thiof at Kayar and the onset of the upwelling (1975-1985). **b**-Mean monthly upwelling (m3/s/m) indices at Yoff (Senegal) and Nouadhibou (Mauritania) and mean monthly thiof catches (tons) at Kayar (adapted from Cury and Roy 1988).

Senegal is related to the onset of the Senegalese upwelling which takes place from November to May (Cury and Roy 1988).

Anomalies of SST data collected at coastal stations were used to characterise the upwelling intensity. Using cpue in the two main landing ports (Saint-Louis and Kayar), a mean lag of about one month was found between the occurrence of the upwelling and the arrival of the thiof off Kayar (Figure 10-9a). The migration of the thiof from Mauritania to Senegal appears to be not only related to the onset of the Senegalese upwelling but is also linked to the relaxation of the upwelling off northern Mauritania (Figure 10-9b). Fish species also seem to migrate seasonally in order to take advantage of the most productive areas (Fréon 1986; Pauly 1994).

Long-term Changes and Pelagic Fish Dynamics

Long-term changes in pelagic fish abundance off Côte d'Ivoire and Ghana and associated environmental changes

The analysis of several environmental time series off Côte d'Ivoire and Ghana, Koranteng and Pézennec (1998) showed that the mean annual wind speed exhibits a significant positive trend from 1966 to 1990. However when taking into consideration the annual mean SST, the minor and the major upwelling seasons show opposite trends during the same time period. As a result the difference between the characteristic minimum temperatures of the two upwelling seasons decreased by about one degree between 1970 and 1990 in both countries (Koranteng and Pézennec 1998). Meanwhile one of the two species of sardinella, *Sardinella aurita*, showed a transition from a depleted to a prosperous state as cpue increased from 0.8 to 7.2 t/day before and after 1980 (Pézennec and Bard 1992). The minor upwelling occured during an environmentally unfavourable period for the global productivity of the pelagic ecosystem. The minor upwelling, which started to intensify in the late 1970s, looks to have played a major role in driving the abundance of *Sardinella aurita* (Figure 10-10). It seems to act as a 'bottleneck', with the duration and strength of the minor upwelling appearing to significantly affect recruitment of the *Sardinella* population. This example illustrates quite well the importance of considering not only the main environmental events but also the minor events that can affect fish populations.

Long-term changes in fish distribution

In the southern part of the Canary current a large sardine fishery has developed off the Western Sahara which did not exist before 1965. Two southward expansions of the sardine (*Sardina pilchardus*) occurred 23 years apart and are correlated to multi-year periods of trade wind strengthening which occurred in 1972-1975, in 1982 and from 1991 onward (Binet et al., 1998). The dynamics of the upwelling, computed from satellite data (Meteosat), for the period 1984-1994 also reveals spatial changes in recent years (Figure 10-3) that are associated with sardine availability to the small scale fisheries off Senegal (Demarcq 1998).

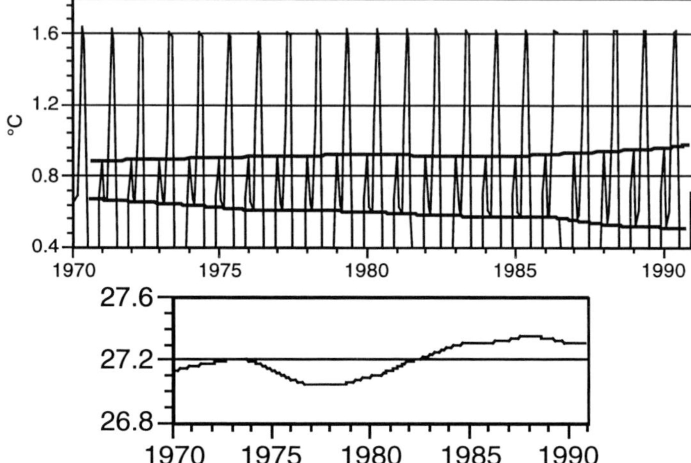

Figure 10-10 Seasonal component (upper) and long term trend (lower) of a decomposition of the SST signal off western Côte d'Ivoire (from Roy 1995).

Fish reproductive strategy

When comparing small pelagic fish reproductive strategies in several areas within the Canary and Guinea Current ecosystems, differences in timing are observed (Roy *et al.* 1989, 1992). In some areas spawning occurs during the upwelling season (Senegal, Côte d'Ivoire) and in other areas outside the upwelling season (Morocco) or when upwelling activity reaches a minimum (Sahara). The mean environmental conditions (wind speed and CUI) during the spawning season in the main spawning grounds were calculated using data extracted from COADS. It appears that there is no apparent relationship between the upwelling intensity and reproduction (Figure 10-11) but rather a striking correspondence between the timing of reproduction and the occurrence of a composite-average wind speed of about 5-6 m s^{-1} (Roy *et al.* 1992). This range in wind speed corresponds to the optimal wind conditions as defined by the OEW. The correspondence between spawning peaks and the 5-6 m.s-1 wind speed illustrates the long-term adaptation of small pelagic fish reproductive strategies to the environment.

Upwelling and retention

Off Senegal, there is a surprising correlation between spawning of several major fish species and the upwelling process. Intense spawning is concentrated in late spring south of Cape Vert. Using satellite data and surface oceanographic measurements, it is shown that the surface distribution of several environmental parameters in this location is quite unusual for an upwelling area. Sea surface temperature is minimum over the shelf and increases toward both offshore and coastal directions; the distribution of chlorophyll is also contrasted with a

being the result of a 'double cell' structure of the upwelling vertical circulation (Figure 10-12) (Roy 1998):

- a first cell is located on the shelf-break; it is the main upwelling cell that brings cold and nutrient rich subsurface water to the surface.

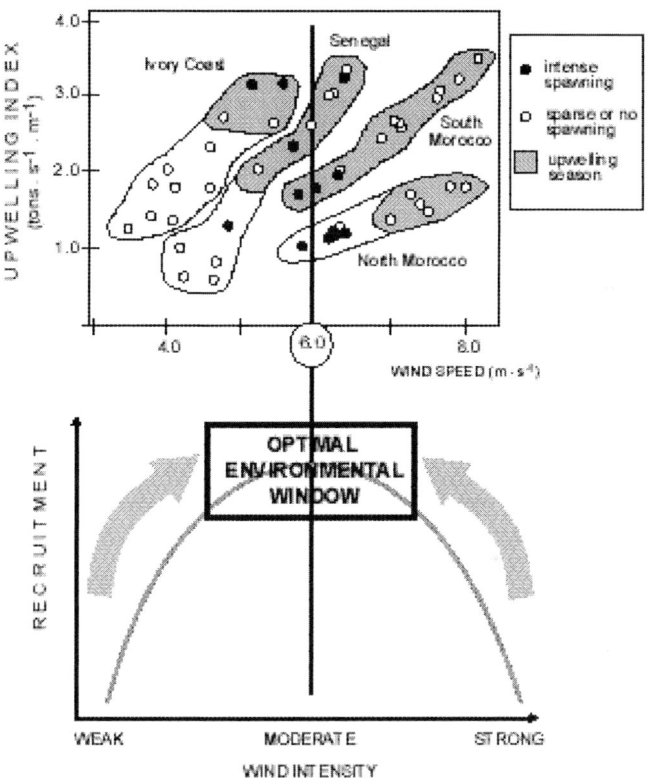

Figure 10-11 The spawning activity of sardine and sardinellas in relation to wind speed and upwelling off West-Africa : north Morocco (30°N to 27°N), south Morocco (26°N to 22°N), Senegal (15°N to 9°N) and Côte d'Ivoire (5°N, 2°W to 1°E). Mean monthly upwelling indices (tons.s⁻¹.m⁻¹) are plotted against mean monthly wind speed (m.s⁻¹); wind speed and upwelling indices were calculated using COADS. The upwelling seasons are shaded and months corresponding to intense spawning are indicated by a black dot. Note that black dots (intense spawning) for all regions are With this type of configuration, upwelling and retention can occur simultaneously and act together to provide a favourable reproductive habitat. In other areas (e.g. Côte d'Ivoire and Ghana, the southern coast of Morocco, South-Africa) there is also a co-occurrence between spawning and upwelling. In these ecosystems, similar physical processes must allow a positive coupling between upwelling and retention. clustered around 6 m.s⁻¹ (upper figure); this value corresponds to the average wind intensity of the OEW (lower figure).

- A second cell is located on the coastal side of the shelf-break. A convergence occurs in the nearshore area. In this cell, phytoplankton and other biological components tend to become trapped and retained along the coastal side of the shelf.

In an upwelling with a double-cell circulation structure, the coastal cell represents a very favourable environment for fish to reproduce: eggs and larvae are not spread in the offshore environment but can be retained in the productive and relatively stable coastal environment.

Fish reproductive behaviour and environmental changes

The way we consider group of biological entities and utilise them is a controversial subject in ecology, particularly when referring to relevant scales of observation. In ecology we consider demographic exchangeability (i.e., individuals have the same dynamics when they are confronted by any perturbation) but we know that every individual is different from another. Thus, according to Cury (1994), individuals tend, when they reproduce, to try to find in their surroundings the environmental conditions with which they were imprinted at an early stage. This hypothesis was developed using arguments developed within comparative ecology (Cury 1994) and its implications for fish population dynamics was investigated using an IBM (Individual Based Model) (Lepage and Cury 1997; McGlade 1999). This represents an ecology based on individual dynamics and consequently the impact of environmental changes can be explored using spatially explicit models (Lepage and Cury 1996). It is also possible to test the effect of long term and global climatic changes on fish population dynamics such as a slow increase in temperature over the years.

Discussion: the Right Scale for the Right Process, the Right Tool for the Right Process

Relatively few oceanographic data have been collected off West Africa, however these data have been collected over long time periods. Thus, the spatial structure of the environment has been poorly documented whereas the temporal aspects have been extensively addressed. The quality of these data appears to be adequate to explore patterns that relate fish population dynamics to the environment, as drastic changes have been observed at different scales (month, seasons, year, decades). The research done in West Africa over the past forty years stresses the importance of collecting environmental time series of relevant variables such as SST or wind but it is crucial, as spatial dynamics is now of prime importance in fisheries related problems, that particular attention be devoted to the collection of more oceanographic data through surveys related to well-defined related fishery or ecological questions.

Table 10- 2 presents some of the ecological processes that are addressed in this paper and tries to relate the scales at which each process could be studied. Scale of observations and scale of studies are sometimes disconnected in time and space and difficult to identify. Moreover time lags often exist between environmental processes and the response of the ecological process. There is thus a requirement for tools that allow us to deal with time and space. Usually, an ecological process responds in a non-linear fashion as threshold effects and discontinuities are observed when related to environmental factors (Cury and Roy 1991). Methods and tools are now available that address these features in ways that can help to better understand the coupled dynamics between the environment and the ecological processes in the Canary Current and Guinea Current ecosystem.

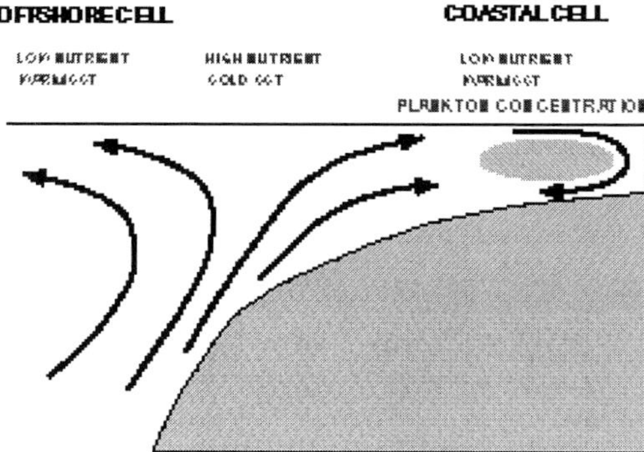

Figure 10-12 Sketch view of a double-cell vertical circulation structure in an upwelling located on a wide shelf, down-wind of a cape.

Ecological process	Scale of observation	Ecological data	Environmental data	Method	Results	Countries
Availability	fortnight	cpue	SST (coastal, COADS)	Multivariate time series analysis	Depend on enrichment process	Côte d'Ivoire
School size	fortnight, month	catch per set	SST (coastal, COADS)	regression	Depend on food availability	Senegal Côte d'Ivoire
Seasonal Migration	month	catch	CUI SST coastal and COADS	Comparative dynamics (CUI)	Depend on differential food production	Senegal Morocco
Changes in Migration	Month, annual	Catch	SST coastal, Satellite (Meteosat)	Spatial upwelling index	Depend on yearly strength of the upwelling	North-west Africa
Inter- annual abundance	annual	Catch cpue	SST, wind	Climprod (production models) GAM	Depend on availability/ Abundance OEW (optimal environmental window)	Côte d'Ivoire Senegal Morocco
Long term abundance	decadal	Catch cpue	SST (coastal, COADS)	GAM, STL (generalized additive models)	Change in the seasonal pattern and in the long term environment	Côte d'Ivoire
Retention area	decadal	Eggs and larvae	SST (COADS), satellite	Models (3D, IBM)	Double cell circulation	Morocco, Senegal,
Reproductive behavior	microscale	Individual fish dynamics in space	Global change	Comparative Evolutionary ecology, IBM	Ecology of individuals	

Table 10-2 Ecological processes and related scales of observation for the ecological and environmental data. Methods used and main results obtained in West-Africa.

References

Arfi, R. 1985 Variabilité inter-annuelle d'un indice d'intensité des remontées d'eaux dans le secteur du Cap-Blanc (Mauritanie). Canadian Journal of Fisheries and Aquatic Sciences 42: 1969-1978.

Bakun, A. 1973 *Daily and weekly upwelling indices, West Coast of North America 1946-71.* U.S. Dep. Comm., NOAA Tech. Rep. NMFS SSRF-671, 103p.

Bakun, A. 1990 Global climate change and intensification of coastal ocean upwelling. Science 247: 198-201.

Bakun, A. 1996 *Patterns in the Ocean: Ocean Processes and Marine Population Dynamics.* University of California Sea Grant, San Diego, California, USA, in co-operation with Centro de Investigaciones Biolgicas de Noroeste, La Paz, Baja California Sur, Mexico. 323 p.

Binet, D., Samb B., Sidi M.T., Levenez J.J. and J. Servain 1998 Sardine and other pelagic fisheries changes associated with multi-year trade wind increases in the southern Canary current. pp211-233. *In* Durand M.H., Cury P., Mendelssohn R., Roy C., Bakun A. and D. Pauly (eds.) *Global versus local changes in upwelling systems.* Orstom, Paris.

Binet, D. 1976 Biovolumes et poids secs zooplanctoniques en relation avec le milieu pélagique au dessus du plateau ivoirien. Cah ORSTOM Ser. Oceanogr. 14: 301-326.

Binet, D. 1983 Zooplancton des régions cotières à upwelling saisonniers du golfe de Guinée. Cah ORSTOM Ser. Oceanogr. 18: 357-380.

Binet, D. 1988 Rôle possible d'une intensification des alizés sur le changement de répartition des sardines et sardinelles le long de la côte ouest-africaine. Aquatic Living Resources 1:115-132.

Blaxter, J.H.S. and J.R. Hunter 1982 The biology of the clupeoid fishes. Academic Press. Advances in Marine Biology 20: 1-223.

Cury, P. 1994 Obstinate Nature: an ecology of individuals. Thoughts on reproductive behavior and biodiversity. Canadian Journal of Fisheries and Aquatic Sciences 51: 1664-1673.

Cury, P. and C. Roy 1987 Upwelling et pêche des espèces pélagiques cotières de Côte d'Ivoire: une approche globale. Oceanologica Acta 10: 347-357.

Cury, P. and C. Roy 1988 Migration saisonnière du thiof (*Epinephelus aeneus*) au Sénégal: influence des upwelling sénégalais et mauritanien. Oceanologica Acta 11: 25-36.

Cury, P. and C. Roy 1989 Optimal environmental window and pelagic fish recruitment success in upwelling areas. Canadian Journal of Fisheries and Aquatic Sciences 46: 670-680.

Cury, P. and C. Roy 1991 *Pêcheries ouest-africaines: variabilité, instabilité et changement.* Orstom, Paris. 525p.

Cury, P., C. Roy, R.Mendelssohn, A. Bakun and D. M. Husby 1995 Moderate is better: exploring nonlinear climatic effect on the Californian northern anchovy. pp417-424. *In*: R.J. Beamish (ed.) *Climate change and northern fish populations.* Canadian Special Publications Fisheries and Aquatic Sciences 121.

Demarcq, H. 1998 Spatial and temporal dynamics of the upwelling off Senegal and Mauritania: local change and trend. pp149-165 *In*: Durand M.H., Cury P.,

Mendelssohn R., Roy C., Bakun A. and D. Pauly (eds) *Global versus local changes in upwelling systems*. Orstom, Paris.

Demarcq, H. and J. Citeau 1995 Sea surface temperature retrieval in tropical area with Meteosat: the case of the Senegalese coastal upwelling. International Journal of Remote Sensing 16: 1371-1395.

Fréon, P. 1983 Production models as applied to sub-stocks depending on upwelling fluctuations. pp1047-1064 *In:* G.D. Sharp and J. Csirke (eds). *Proceedings of the expert consultation to examine changes in abundance and species composition of neritic fish resources*. FAO, Fish. Rep. 291.

Fréon, P. 1986 *Réponses et adaptation des stocks de clupéidés d'Afrique de l'Ouest à la variabilité du milieu et de l'exploitation: analyse et réflexion à partir de l'exemple du Sénégal.* Thèse Doct. Univ. Aix-Marseille. 287p.

Fréon, P. 1989 Introduction of climatic variables into global production models. pp 481-528. *In:* Larrañeta, M.G. and Wyatt, T. (eds). *International Symposium on Long Term Changes in Marine Fish Populations, Vigo (Espange)* Consejo Superior de Investigaciones Cientificas.

Fréon, P. 1991 Seasonal and interannual variations of mean catch per set in the senegalese sardine fisheries: fish behavior or fishing strategy? pp135-145 *In*: T. Kawasaki (ed.) *Long-term variability of pelagic fish populations and their environment*. Pergamon Press.

Fréon, P. Mullon, C. and G. Pichon 1993 CLIMPROD: Experimental interactive software for choosing and fitting surplus production models including environmental variables. FAO Computerized Information Fisheries 5. 76 p.

Fréon, P., El Khattabi, M., Mendoza, J. and R. Guzmán 1997 Unexpected reproductive strategy of *Sardinella aurita* off the coast of Venezuela. Marine Biology 128: 363-372.

Hempel, G., 1982 The Canary current: studies of an upwelling system. Introduction. Rapp. P.-v. Réun. Cons. int. Explor. Mer 180: 7-8.

Koranteng, K.A. 1998 *The impact s of environmental forcing on the dynamics of demersal fishery resources of Ghana*. Ph.D. thesis University of Warwick, U.K. 377p.

Koranteng, K.A. and O. Pézennec 1998 Variability and trends in some environmental time series along the Ivoirian and the Ghanaian coasts. pp167-177. *In*: Durand M.H., Cury P., Mendelssohn R., Roy C., Bakun A. and D. Pauly (eds.) *Global versus local changes in upwelling systems*. Orstom, Paris.

Le Page, C. and P. Cury 1996 How spatial heterogeneity influences population dynamics: Simulations in SEALAB. Adaptive Behavior 4: 249-274.

Le Page, C. and P. Cury 1997 Population viability and spatial fish reproductive strategies in constant and changing environments: an individual-based modelling approach. Canadian Journal of Fisheries and Aquatic Sciences 54: 2235-2246.

McGlade, J. (ed.) 1999 *Advanced Ecological Theory. Principles and Applications.* Blackwell Science, Oxford.

Mendelssohn, R. and C. Roy 1986 Environmental influences of the French, Ivory-coast, Senegalese and Moroccan tuna catches in the Gulf of Guinea. pp 170-88. *In: Proceedings of the ICCAT Conference on the International Skipjack Program. Comptes-rendus de la Conference ICCAT sur le programme de l'année internationale du listao. Actes de la Conferencia ICCAT sobre el programa de Ano International del listao.* Madrid, ICCAT.

Mendelssohn, R. and P. Cury 1987 Fluctuations of a fortnightly abundance index of the Ivoirian coastal pelagic species and associated environmental conditions. Canadian Journal of Fisheries and Aquatic Sciences 44: 408-421.

Mensah, M.A. 1995 The occurrence of zooplankton off Tema during the period 1969-1992. . pp279-289. *In*: F.X. Bard et K.A. Koranteng (eds) *Dynamique et usage des ressources en sardinelles de l'upwelling côtier du Ghana et de la Côte d'Ivoire.* ORSTOM éditions, Paris.

Nykjaer, L. and L.V. Van Camp 1994 Seasonal and interannual variability of coastal upwelling along northwest Africa and Portugal from 1981 to 1991. Journal of Geophysical Research 99 (C7): 14197-14207.

Pauly, D. 1994 Un mécanisme explicatif des migrations des poissons le long des côtes du Nord-Ouest africain. pp p235-244 *In*: Barry-Gérard M., Diouf T. et A. Fonteneau (eds). *L'évaluation des ressources exploitables par la pêche artisanale sénégalaise.* ORSTOM, Paris.

Pézennec, O. and F.X. Bard 1992 Importance écologique de la petite saison d'upwelling ivoiro-ghanénne et changements dans la pêcherie de Sardinella aurita. Aquatic Living Resources 5: 249-259.

Pézennec, O. and K.A. Koranteng 1998 Changes in the dynamics and biology of small pelagic fisheries off Côte d'Ivoire and Ghana: the ecological puzzle. pp, 329-343 *In*: Durand M.H., Cury P., Mendelssohn R., Roy C., Bakun A. and D. Pauly (eds) *Global versus local changes in upwelling systems.* Orstom, Paris.

Roy, C. 1989 Fluctuations des vents et variabilité de l'upwelling devant les côtes du Sénégal. Oceanologica Acta 12: 361-369.

Roy, C. 1991 Les upwellings : le cadre physique des pêcheries côtières ouest-africaines. pp38-66. *In* : P. Cury et C. Roy (eds) *Variabilité, instabilité et changement dans les pêcheries ouest africaines.* ORSTOM, Paris.

Roy,C. 1995 The Cote d'Ivoire and Ghana coastal upwellings : dynamics and Change. pp346-361. *In*: F.X. Bard et K.A. Koranteng (eds.) *Dynamique et usage des ressources en sardinelles de l'upwelling côtier du Ghana et de la Côte d'Ivoire.* ORSTOM éditions, Paris.

Roy, C. 1998 Upwelling-induced retention area off Senegal: a mechanism to link upwelling and retention processes. In: Benguela Dynamics. Pillar, S. C., Moloney, C., Payne, A. I. L. and F. A. Shillington (Eds). South African Journal of Marine Science 19: 89-98.

Roy, C. Cury P., Fontana A. and H. Belvèze 1989 Stratégies spatio-temporelles de la reproduction des clupéidés des zones d'upwelling d'Afrique de l'Ouest. Aquat. Living Resourc., 2:21-29.

Roy, C. P. Cury and S. Kifani 1992 Pelagic fish recruitment success and reproductive strategy in upwelling areas : environmental compromises. *In*: A.I.L. Payne, K.H. Brink, K.H.Mann and R. Hilborn (eds). *Benguela Trophic Functioning.*South African Journal of Marine Science 12: 135-146.

Roy, C. and Mendelsohn R. 1998 The development and the use of a climatic database for CEOS using the COADS dataset. pp27-44. *In* : Durand M.H., P. Cury, R. Mendelssohn, C. Roy, A. Bakun et D. Pauly (eds). *Global versus local changes in upwelling systems.* Editions ORSTOM Paris.

Serra, R., P. Cury and C. Roy 1998 The recruitment of the Chilean sardine (Sardinops sagax) and the optimal environmental window. pp267-274. *In* Durand M.H., P. Cury, R. Mendelssohn, C. Roy, A. Bakun et D. Pauly (eds). *Global versus local changes in upwelling systems*. Editions Orstom.

Shin, Y.J., C. Roy and P. Cury 1998 Clupeoids reproductive strategies in upwelling areas : a tentative generalization. pp409-422. *In:* Durand M.H., Cury P., Mendelssohn R., Roy C., Bakun A. and D. Pauly (eds.) *Global versus local changes in upwelling systems*. Orstom, Paris,

Steele, J.H. 1995 Can ecological concepts span the land and ocean domains? pp5-19. *In:* Ecological time series. Powell T.M. and J.H. Steele (eds). *Ecological time series*. Chapman & Hall, New York.

Stretta, J.-M. 1991 Contribution de la télédétection aérospatiale à l'élaboration des bases de l'halieutique opérationnelle: l'exemple des pêcheries thonières tropicales de surface (aspect prédictif) doctoral dissertation, Paris VI, Grenoble, France.

Woodruff, S.D., R.J. Slutz, R. L. Jenne and P. M. Steurer 1987 A Comprehensive Ocean-Atmosphere Data-Set. Bulletin of the American Meteorological Society 68: 1239-1250.

The Gulf of Guinea Large Marine Ecosystem
J.M. McGlade, P. Cury, K.A. Koranteng and N.J. Hardman-Mountford (Editors)
© 2002 Elsevier Science B.V. All rights reserved.

11

Monitoring Levels of 'Phytoplankton Colour' in the Gulf of Guinea Using Ships of Opportunity

A.W.G. John, P.C. Reid, S.D. Batten and E.R. Anang

Abstract

Continuous Plankton Recorders were used in a survey of the plankton of the Gulf of Guinea from December 1995 to April 1999. 'Phytoplankton Colour', a measure of phytoplankton biomass, was estimated visually for all samples and showed a maximum in the upwelling period in all years. Significant inter-annual variability was evident.

Introduction

Monitoring the plankton of the Gulf of Guinea forms a component of the Productivity Module of the Gulf of Guinea Large Marine Ecosystem (LME) Project. The aim of this module is to establish the carrying capacity of the fishery and ecosystem variability that may result from climate change or anthropogenic impacts. To this end, the Sir Alister Hardy Foundation for Ocean Science (SAHFOS) commenced towing Continuous Plankton Recorders (CPRs) in December 1995 in the northern Gulf of Guinea.

Data from the CPR survey have been used in the past to illustrate, amongst other things, spatial distribution (Oceanographic Laboratory 1973), seasonal changes (Robinson 1970), vertical migration (Hays *et al.* 1994), isolated events (Edwards *et al.* 1999), harmful algal blooms (Robinson 1968), and introduced species (Robinson *et al.* 1980). In recent years, it has been used successfully in the North Atlantic, the North Sea, the Mediterranean, the Baltic, the Southern Ocean and the Pacific Ocean. In 1999, the CPR survey was incorporated into the Initial Observing System of the Global Ocean Observing System (GOOS).

Measurements of Phytoplankton Colour, a coarse estimate of phytoplankton biomass, were made on CPR samples taken between December 1995 and April 1999 in the Gulf of Guinea. Colour estimates are made immediately on return of the CPRs to the SAHFOS laboratory in Plymouth. Long-term changes of CPR Phytoplankton Colour in the North Atlantic and North Sea have been the subject of a number of studies (Colebrook 1979; Reid *et al.* 1987). It has recently been shown (Reid *et al.* 1998) that levels of Phytoplankton Colour in three areas of the north-east Atlantic and North Sea increased during 1948-1995, with a step-wise increase after the mid 1980s. This change has been attributed to meteorological forcing, possibly linked to global warming. Edwards *et al.* (submitted and in prep.) have shown good correlation between phytoplankton colour in the north-east Atlantic and the North Atlantic Oscillation.

141

As part of the Gulf of Guinea LME project, three two-week Plankton Training Workshops were held in the Marine Fisheries Research Division (formerly the Fisheries Research and Utilisation Branch) Laboratory in Tema, Ghana in 1996 and 1997. These were organised and run by SAHFOS and attended by six scientists from Côte d'Ivoire, Ghana (two), Benin, Nigeria and Cameroon. During these workshops participants were trained in phytoplankton and zooplankton taxonomy and briefed on the logistics of establishing and running a routine CPR monitoring survey.

This paper describes preliminary results from visual estimates of chlorophyll *a* determined from CPR samples taken in the Gulf of Guinea.

Methods

CPRs were towed at approximately six-week intervals by merchant ships along standard routes in the Gulf of Guinea at a depth of approximately 7-10 metres (Hays & Warner 1993). Water passes through a sampling aperture of 1.24 cm x 1.24 cm at the front of the towed body, with plankton retained on a continuously moving band of bolting silk with a mesh size of 270 µm (Warner & Hays 1994). A visual assessment of the colour of the CPR silks was made immediately on return of the CPRs to the SAHFOS laboratory in Plymouth. Following division of the silk by ruled lines into 10 nautical mile (18 km) 'blocks', the silk is passed over a white background and the green colour (chlorophyll *a*) of each 'block' is visually assessed as Nil, Very Pale Green, Pale Green, or Green by reference to a standard colour chart (Colebrook 1960). Subsequently, the silk is cut into 10 nautical mile blocks prior to identification and counting of phytoplankton and zooplankton under a microscope. With the exception of samples taken in December 1995, most samples have been stored at SAHFOS prior to detailed taxonomic analysis at the Marine Fisheries Research Division Laboratory in Tema, Ghana.

Between December 1995 and April 1999, CPRs were towed on 114 occasions along four routes in the Gulf of Guinea. Three of these routes were 'coastal': the 'GH' route from Lagos, Nigeria to Tema/Takoradi, Ghana; the 'GG' route from Tema/Takoradi to Abidjan, Côte d'Ivoire; and the 'GJ' route from Abidjan to San Pedro/Cape Palmas. The fourth route, the 'GI', runs over deeper water across the southern Bight of Benin from Douala, Cameroon to Cape Three Points, Ghana. The location of these routes and of places mentioned in the text is shown in Figure 11-1.

Results

The first three CPR tows in December 1995 were analysed by microscopy at SAHFOS and a total of 76 phytoplankton taxa were recorded. In the eastern Gulf of Guinea, south of the Niger Delta, the phytoplankton was dominated by *Rhizosolenia calcar-avis*, *R. alata* forma *alata*, *Chaetoceros* spp and *Navicula* spp. From east of Takoradi to Cape Three Points *Chaetoceros* spp, *Bacteriastrum* spp and the cyanophycean *Trichodesmium* spp were the dominant taxa. *Trichodesmium* spp. was abundant south and east of San Pedro.

Figure 11-2 (colour plate) shows Phytoplankton Colour on all CPR tows in the Gulf of Guinea from December 1995 to April 1999 plotted by longitude. No sampling took place in the eastern Gulf of Guinea between March 1996 and July 1997 due to problems with shipping access at Douala. For the western Gulf of Guinea, between Cape Palmas and Lagos, there is a clear distinction in levels of Phytoplankton Colour in the period before July 1997,

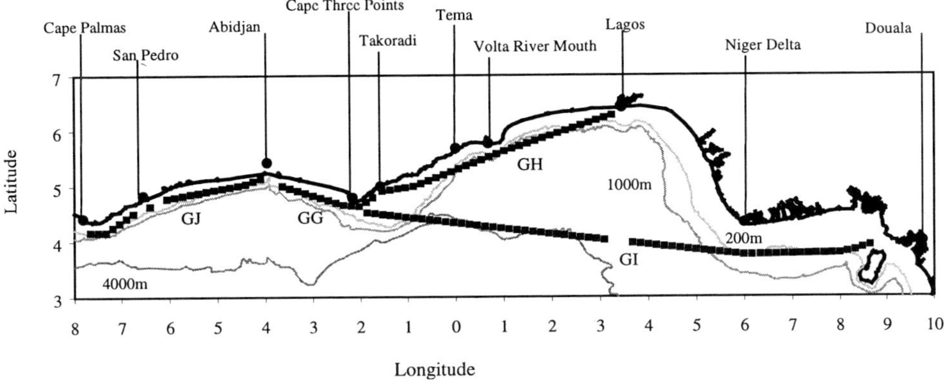

Figure 11-1 The Gulf of Guinea coast, showing location of the CPR routes and places mentioned in the text.

when levels were higher, and subsequently.

The mean Phytoplankton Colour for all CPR samples taken in each degree of longitude averaged for three-month periods is shown in Figure 11-3. In the first year (July 1996 to June 1997), the level of Phytoplankton Colour was high during the upwelling period (July-September), and remained high for the next two quarters. In contrast, in the second and third years (July 1997 to June 1998, and July 1998 to March 1999) Phytoplankton Colour was at a maximum only during the upwelling period. There appears to be a downward trend in the colour in the upwelling period (July-September) between the three years. Even with only a few years of data, significant inter-annual variability is evident.

Discussion & Conclusions

There have been few detailed studies of offshore phytoplankton in the Gulf of Guinea. Binet (1983) summarised work done by Reyssac (1970), Reyssac & Roux (1972), and Dandonneau (1971, 1973) off Côte d'Ivoire in the 1970s and early 1980s, and by Dessier off the coasts of the Congo, Gabon and Angola. Anang (1979) investigated seasonal changes in the phytoplankton off Tema, Ghana, from September 1973 to November 1974. She found high cell counts during the upwelling period (July to October) when the phytoplankton was dominated by diatoms. Anang *et al.* (1976) reported a bloom of *Mesodinium rubrum* off Tema in August-September 1974. Binet (1983) studied the neritic phytoplankton and primary production in the upwelling areas of the Gulf of Guinea; he listed eight groups of phytoplankton species associated with different ecological parameters.

The 'phytoplankton colour' recorded by the CPR survey in the North Sea was shown by Hays & Lindley (1994) to be a quantitative index of chlorophyll *a* concentration, but only when the numbers of phytoplankton cells on the CPR silks are not high. However, as this study was based only on a limited number of samples from February, April and May 1991, further research is needed to demonstrate a clear relationship. Edwards *et al.* (submitted) have shown good correlation in the north-east Atlantic between phytoplankton colour and

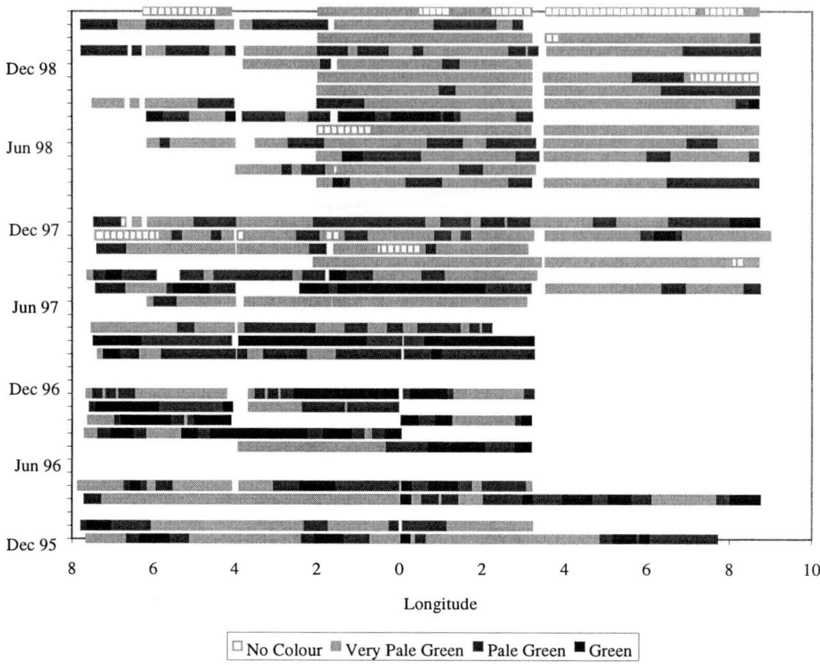

Figure 11-3 Seasonal and year-to-year differences in mean Phytoplankton Colour from August 1996 to March 1999 plotted by longitude. A 'x' means no sampling in that period.

spatial patterns of chlorophyll determined by other sampling methodologies, including satellite remote sensing.

Despite its limitations, the CPR is a proven sampling technology that has established patterns of spatial and temporal change in the plankton of the North Atlantic over many decades. This approach has now also provided valuable information on a little studied region, the Gulf of Guinea, providing data required for developing management strategies for this ecosystem.

The samples taken in the Gulf of Guinea provide a unique snapshot of the marked inter-annual and spatial variability of plankton over a very large area. Analysis of the remaining CPR samples will be undertaken at the Marine Fisheries Research Division Laboratory in Tema, Ghana in due course.

Acknowledgments

The CPR survey in the Gulf of Guinea is supported by UNIDO. The survey depended on the voluntary cooperation of the shipping companies, masters and crews of the merchant vessels that towed CPRs. The following ships were involved in the Gulf of Guinea survey: the 'Repubblica di Amalfi' (Grimaldi Line), the 'Kumasi' (OT Africa Line Ltd) and the 'NDS Prominence' (Nile Dutch Africa Line). We wish to thank Nick Hardman-Mountford for constructive comments.

References

Anang, E.R. 1979 The seasonal cycle of the phytoplankton in the coastal waters of Ghana. Hydrobiologia 2: 33-45.

Anang, E.R., E.K. Obeng-Asamoa and D.M. John 1976 Observations on the red-water ciliate *Mesodinium rubrum* Lohmann in Ghanaian coastal water. Bulletin de l'Institut Fondamental d'Afrique Noire 38A: 235-240.

Binet, D. 1983 Phytoplancton et production primaire des régions côtières à upwellings saisonniers dans le Golfe de Guinée. Océanographie Tropicale 18: 331-355.

Colebrook, J.M. 1960 Continuous plankton records: methods of analysis, 1950-1959. Bulletins of Marine Ecology 5: 51-64.

Colebrook, J.M. 1979 Continuous plankton records: seasonal cycles of phytoplankton and copepods in the North Atlantic Ocean and the North Sea. Marine Biology 51: 23-32.

Dandonneau, Y. 1971 *Étude du phytoplancton sur le plateau continental de Côte d'Ivoire. I. Groupes d'espèces associées.* Cahiers ORSTOM, série océanographie 9: 247-265.

Dandonneau, Y. 1973 *Étude du phytoplancton sur le plateau continental de Côte d'Ivoire. III. Facteurs dynamiques et variations spatio-temporelles.* Cahiers ORSTOM, série océanographie 11: 431-454.

Edwards, M., P.C. Reid and B. Planque (submitted). Long-term changes in phytoplankton biomass in the north-east Atlantic (1960-1995). ICES Journal of Marine Science.

Oceanographic Laboratory, Edinburgh. 1973 Continuous Plankton Records: a plankton atlas of the North Atlantic and the North Sea. Bulletins of Marine Ecology 7: 1-174.

Edwards, M., A.W.G. John, H.G. Hunt and J.A Lindley 1999 Exceptional influx of oceanic species into the North Sea late 1997. Journal of the Marine Biological Association of the U.K. 79: 737-739.

Hays, G.C. and J.A. Lindley 1994 Estimating chlorophyll *a* abundance from the 'phytoplankton colour' recorded by the Continuous Plankton Recorder survey: validation with simultaneous fluorometry. Journal of Plankton Research 16: 23-24.

Hays, G.C. and A.J. Warner 1993 Consistency of towing speed and sampling depth for the Continuous Plankton Recorder. Journal of the Marine Biological Association of the U.K. 73: 967-970.

Hays, G.C., C.A. Proctor, A.W.G. John and A.J. Warner 1994 Interspecific differences in the diel vertical migration of marine copepods: the implications of size, color and morphology. Limnology and Oceanography 39:1621-1629.

Reid, P.C., G.A. Robinson and H.G. Hunt 1987 Spatial and temporal patterns of marine blooms in the Northeastern Atlantic and North Sea from the continuous plankton recorder survey. Rapports et Proces-verbaux des Réunions. Conseil International pour l'Exploration de la Mer 187: 27-37.

Reid, P.C., M. Edwards, H.G. Hunt and A.J.Warner 1998 Phytoplankton change in the North Atlantic. Nature 391: 546.

Reyssac, J.1970 Phytoplancton et production primaire au large de la Côte d'Ivoire. Bulletin de l'Institut Fondamental d'Afrique Noire 32A: 869-981.

Reyssac, J. and M.Roux 1972 Communautés phytoplanctoniques dans les eaux de Côte d'Ivoire. Groupes d'espèces associées. Marine Biology 13:14-33.

Robinson, G.A. 1968 Distribution of *Gonyaulax tamarensis* Lebour in the western North Sea in April, May and June 1968. Nature 220: 22-23.

Robinson, G.A. 1970 Continuous Plankton Records: variations in the seasonal cycle of phytoplankton in the North Atlantic. Bulletins of Marine Ecology 4: 333-345.

Robinson, G.A. and A.R. Hiby 1978 The Continuous Plankton Recorder survey. pp 59-63 *In*: A. Sournia (ed). Phytoplankton Manual,UNESCO, Paris.

Robinson, G.A., T.D. Budd, A.W.G. John and P.C. Reid 1980 *Coscinodiscus nobilis* (Grunow) in Continuous Plankton Records, 1977-78. Journal of the Marine Biological Association of the U.K. 60: 675-680.

Warner, A.J. and G.C. Hays 1994 Sampling by the Continuous Plankton Recorder survey. Progress in Oceanography 34: 237-256.

The Gulf of Guinea Large Marine Ecosystem
J.M. McGlade, P. Cury, K.A. Koranteng and N.J. Hardman-Mountford (Editors)

12

Spatial and Temporal Variations in Benthic Fauna and Communities of the Tropical Atlantic Coast of Africa

P. Le Loeuff and G.S.F. Zabi

Abstract

This paper provides an account of different models of faunal and bionomic variation encountered in the West African marine benthic ecosystems. Spatial variations at local, regional and continental scales are linked to environmental factors such as hydrology, climate and sedimentation. Vertical zonation of the benthic fauna is mainly dependent on the hydrological structure. The nature of the communities depends closely on the bottom sedimentary dynamics. On a regional scale, differences in environmental conditions considerably influence the faunal composition and structure. On the continental scale, the faunal biodiversity over the whole of the tropical Atlantic coast of Africa depends on the regional hydroclimatic typology.

Nyctitropic and seasonal cycles in the benthic communities of Côte-d'Ivoire provide good examples for temporal variations. Typical nyctitropic cycles are seen under normal conditions, however, perturbed cycles exist when light penetration down to the bottom is affected by the passing of turbid waters. The faunal composition of the communities is also dependent on seasonal biological cycles specific to each species.

Introduction

This paper is intended to demonstrate, with the help of some examples, the major types of faunal and bionomic variations encountered at different spatial and temporal scales in the benthic ecosystems of the tropical Atlantic coast of Africa. Spatial scales include:
- bathymetric, in relation with the vertical hydrological structure;
- sedimentary, in relation with the nature of the bottom;
- regional, for comparison of oceanic sub-systems that differ as a function of hydroclimate and, hence, in their functional organization;
- continental, i.e. the African coast of the African continent, and the tropical zone in particular), which integrates the other three scales.

Time scales include:
- diel (day/night) cycles,
- seasonal (annual) cycles.

Spatial Variations

• Variations as a function of bathymetry (Figure 12-1)

The communities under consideration here were located along a transect off Grand-Bassam, on a more or less silted bottom. Data were collected over a period of one year and subsequently analysed (Le Loeuff and Intès 1998b). The well-recognised zonation pattern, ranging from the *"infralittoral"* through the *"circalittoral côtier"* to the *"circalittoral du large"* zones (Chardy and Glémarec 1974; Le Loeuff and Intès 1993), can be seen distinctly. These zones correspond, from a hydrological point of view, to the water masses of the mixed layer, the thermocline and the infrathermocline, respectively. Therefore, the bathymetric distribution of the fauna is highly dependent on hydrological structure.

• Variations as a function of sediment dynamics (Figure 12-2)

Two major types of sediment series occur on the continental shelf of Côte-d'Ivoire. The first is found off the mouths of large rivers, where there is active sedimentation of fine particles. The second is characterised by weak, fine sedimentation, with coastal sandy substrates giving way further offshore to a more or less coarse and silted *"détritique organogène"* deposit (Martin 1977; Tastet *et al.* 1993).

Samples were obtained during exploratory trawling for shrimps on the Côte-d'Ivoire continental shelf from similar depths at two stations: one off the mouth of the Bandama river and the other on a transect a few miles to the west of Abidjan. The results of the correspondence analysis and analysis of gradient (Whittaker 1967; Intès and Le Loeuff 1984),

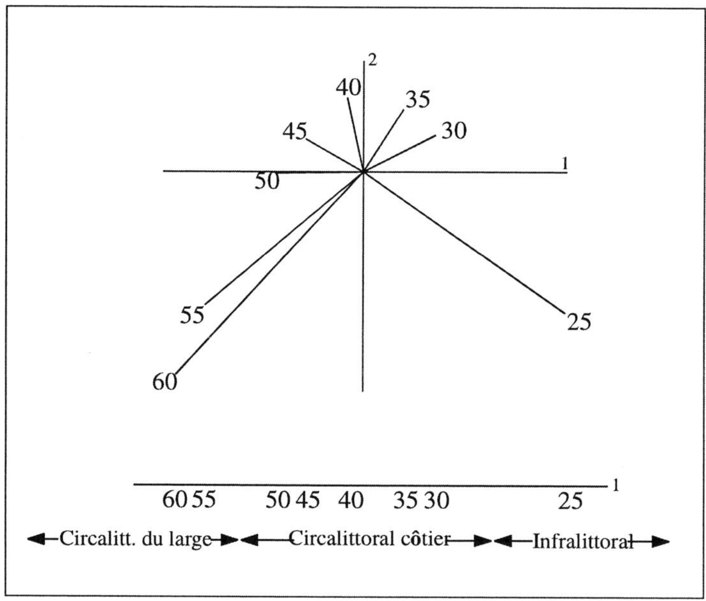

Figure 12-1 Bathymetric zonation of the soft bottom communities as seen from the results of a correspondence analysis using data from samples obtained from shrimp trawling along a transect (25-60 m) off Grand-Bassam (Côte-d'Ivoire), running from the *"infralittoral"* to the *"circalittoral du large"*.

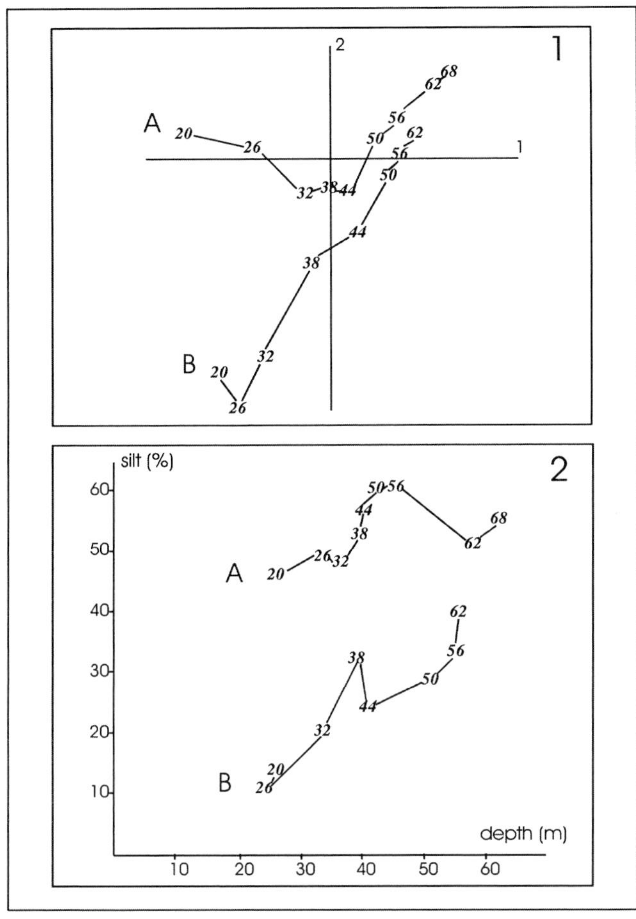

Figure 12-2 Benthic fauna (obtained using a shrimp trawl) on two transects across the continental shelf of the Côte-d'Ivoire. A: samples from silty bottoms near the Bandama river estuary (strong muddy sedimentation); B: communities on sandy (coastal) or "*détritique organogène*" (further offshore) bottoms (weak muddy sedimentation). Results of a correspondence analysis (1) and a gradient analysis (2). Depth of sampling stations are shown on the figure.

show that there is a distinct difference between the communities found on these two types of sedimentary series (Le Loeuff and Intès 1993, 1998a).

• Comparative study of the Guinea and Côte-d'Ivoire communities (Figure 12-3)

A correspondence analysis was used to compare data on benthic fauna from the continental shelf off Guinea and the Grand-Bassam transect (Le Loeuff and Intès 1968; Le Loeuff 1993). Data were obtained under analogous sampling conditions (fish trawl) and at the same period of the year (November).

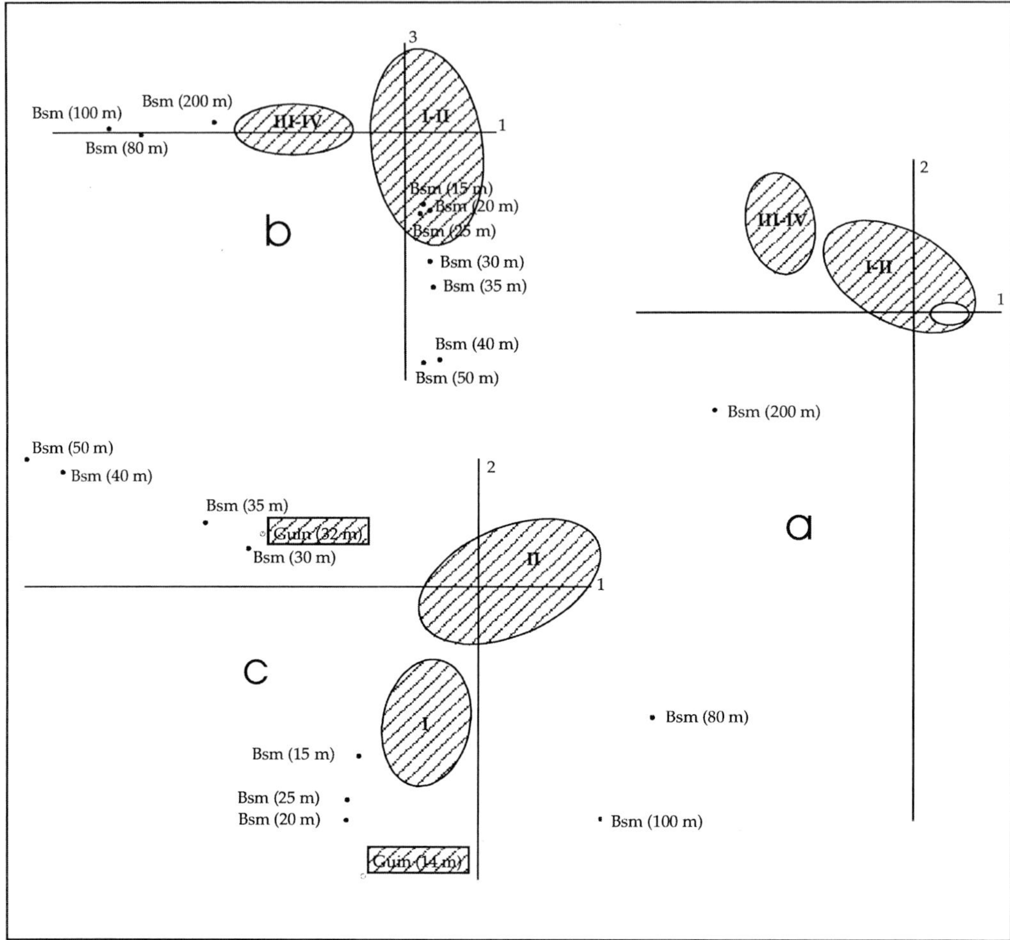

Figure 12-3 Benthic fauna (fish trawl samples) off Guinea (survey on the continental shelf, November 1992) and Côte-d'Ivoire (transect off Grand-Bassam, November 1966); results of an analysis of correspondence (**a** and **b**, all the samples were taken in account; **c**, only the samples of the *"infralittoral"* and *"circalittoral côtier"* zones were considered). Côte-d'Ivoire samples are denoted Bsm or inside grey envelope; most of the Guinean samples are inside ginger dotted envelopes or are denoted Guin. Roman numerals: I, samples in the *"infralittoral"* zone; II, samples in the *"circalittoral côtier"* zone; III, samples in the *"circalittoral du large"* zone; IV, samples in the continental edge zone.

The first analysis was made on the whole data set (Figure 12-3a, b). Axis 1 on these figures separates the deep and littoral samples, axis 2 isolates the fauna of the deep bottom-areas of Grand-Bassam (numerous species found only in these samples) and axis 3 contrasts the samples from the silty substrates of Côte-d'Ivoire with those of the sandy substrates of Guinea.

The second analysis (Figure 12-3c) used only the data from collections made in the near-littoral (< 60 m) zone. Axis 1 refers to the nature of the bottom deposit (sand to silt gradient) and axis 2 to the depth (coast to offshore gradient). Here again a distinct difference between the Guinea and the Côte-d'Ivoire communities can be seen.

The environmental conditions and functional dynamics of the littoral ecosystems of Guinea are characterised by vast stretches of sandy sediments beyond 20 m depth (Domain and Bah 1993) and are under the influence of continental waters (Domain 1989). In contrast those of Côte-d'Ivoire are characterised by more or less silted bottoms (Martin 1973) and are under the simultaneous influence of upwelling and continental waters (Morlière 1970; Colin *et al.* 1993). The strong dissimilarities seen between these areas are reflected on the faunal composition and structural organization of the benthic communities.

- Faunal biodiversity along the tropical Atlantic coast of Africa (Figures 12-4 – 12-8)

In the eastern Atlantic, the tropical zone extends from Cape Blanc in the north to Cape Frio in the south, with permanently cold waters beyond both (Figure 12-4).

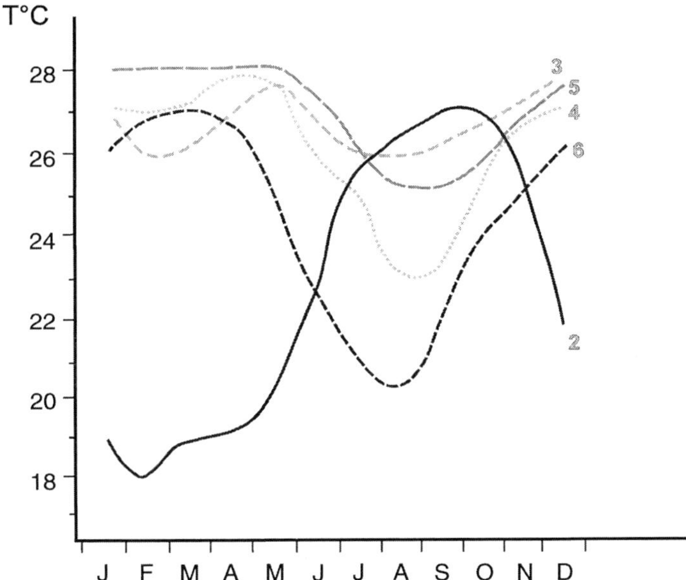

Figure 12-4 Hydroclimatic regions on the tropical West African coast (2, 3, 4, 5, 6) as given by their monthly temperature profiles. External regions: 1, Atlantic coast of Northwest Africa and Europe, Mediterranean; 7, Namibia and Atlantic coast of South Africa. Zones 3,4,5,6 and 2 represent permanent and periodic upwelling respectively.

This tropical zone comprises (Longhurst 1962; Berrit 1973; Intès and Le Loeuff 1984):

i. the *"alternance"* regions, with alternating hydrological conditions characterized by the presence, over a large part of the year, of cold waters whose extension is limited respectively by the Guinea and Lopez Cape fronts;

ii. the typical tropical regions, without upwelling, the dynamics of which are subject only to riverine advection of fresh water; and

iii. the central atypical tropical region, where the upwelling occurs during boreal summer. These different regions can be distinguished (Figure 12-5) by their monthly mean temperature curves (Hastenrath and Lamb 1977).

The distribution of 1449 benthic species, belonging to well known zoological groups (gorgonians, polychaetes, brachyurans, pagurids, bivalve molluscs, echinoderms), was studied in this environmental context. Their presence to the north of Cape Blanc and to the south of Cape Frio was also seen (Le Loeuff and Cosel 1998).

A number of species show specific distribution patterns, termed here as "discontinuous". About 30 % of them avoid the waters of the typical tropical regions, which remain permanently warm and often become less saline, and of the atypical tropical region, which remain warm over a large part of the year and occasionally become less saline (Figure 12-6). About 16 % avoid the typical tropical regions without upwelling (Figure 12-7), where only 2 % of the species can be described as endemic to one or the other of these two regions.

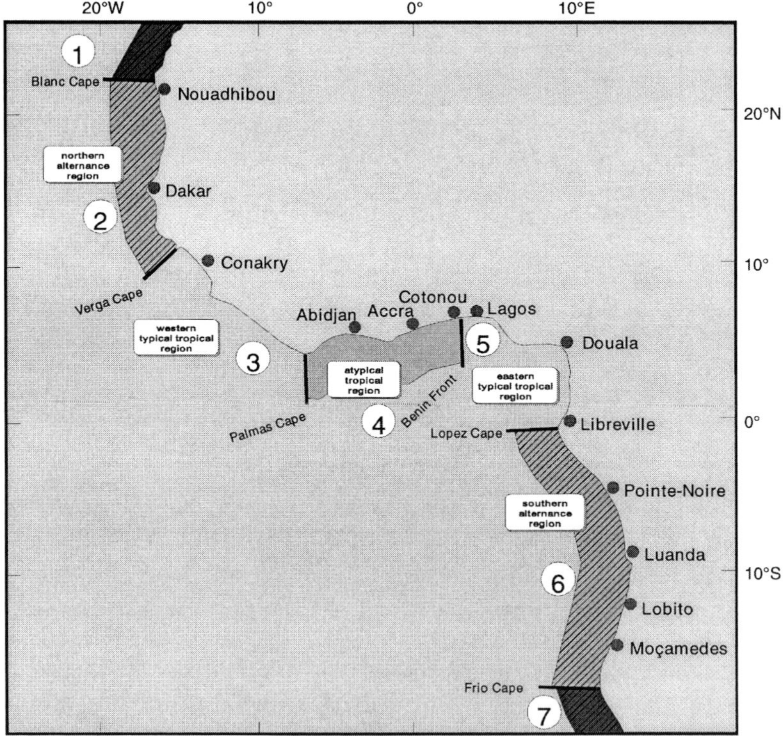

Figure 12-5 Mean monthly surface water temperatures of the hydroclimatic regions of the tropical Atlantic coast of Africa (see Figure 12-4 for details).

Figure 12-6 Distribution pattern of species avoiding the atypical tropical region (4) and the two typical tropical regions (3, 5). Strip width proportional to the number of species indicated on the right (the distribution of 1449 species was examined, with the following break-down: gorgonians, 44; polychaetes, 606; brachyurids, 170; pagurids, 59; bivalves, 380; echinoderms, 190).

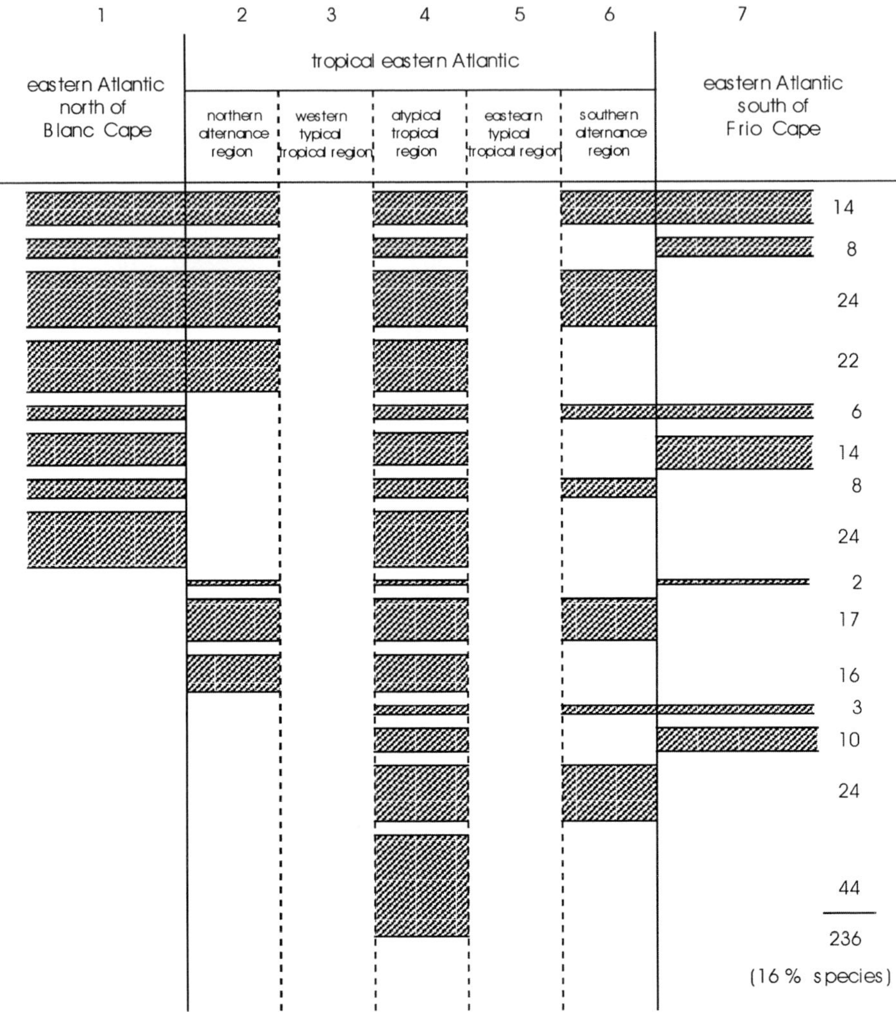

Figure 12-7 Distribution pattern of species that are always present in the atypical region (4) and avoiding the two typical tropical regions (3, 5). Strip width proportional to the number of species indicated on the right.

The species richness index shows that the upwelling regions sustain the highest number of species (Figure 12-8a). Furthermore, when the faunal composition of each of these regions was subjected to a correspondence analysis, the upwelling regions were distinctly separate from those where warm waters occur permanently (Figure 12-8b).

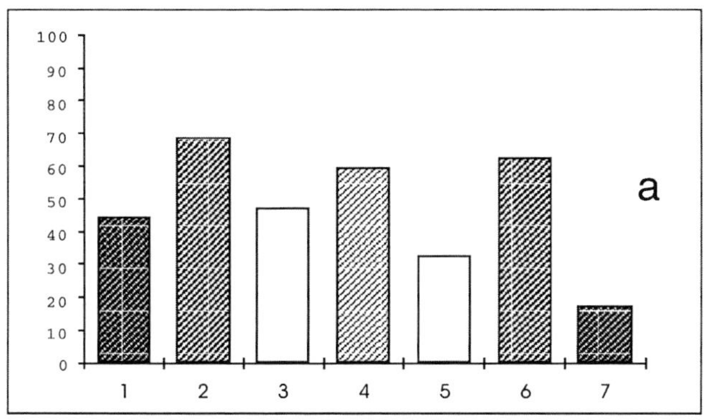

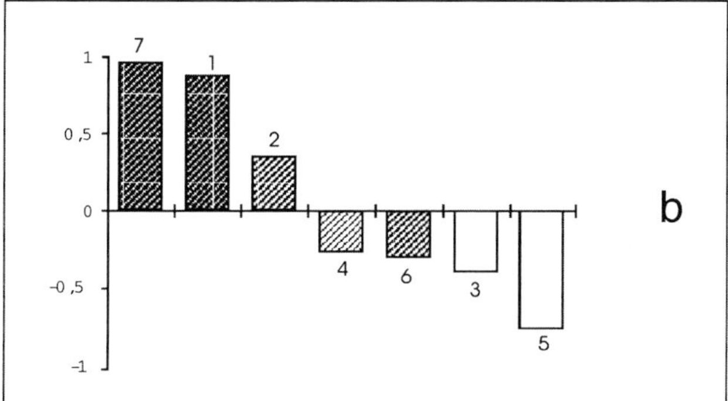

Figure 12-8 Biodiversity patterns along the Atlantic coast of tropical West Africa. **a** species richness, percentage of the number of tropical West African species present in each region; **b** results of a correspondence analysis based on presence-absence of the species in each region. To demonstrate the degree of separation of their fauna; the regions are arranged in order of their values along axis 1 (27.4 % of total variance).

These results tend to suggest that the marine benthic fauna of West Africa is not truly tropical. Numerous species are of European Atlantic and Mediterranean origin and they succeed the ancient circumtropical fauna of the Tethys Sea. Paleogeographic studies show that there is in fact a strong similarity of the present West African fauna and the Pliocene fauna of southern Europe (Cosel 1993).

During the Pliocene, and later during the Pleistocene, the effects of glacial and interglacial periods were pronounced on the Atlantic front of the African Continent (McMaster *et al.* 1970; Martin 1972; Tastet 1989). These cooling and warming cycles of the climate would have strongly perturbed the faunal diversity in the tropical West African oceanic environment, which is not higher than that observed in European seas. The frequent presence of cold waters over a large part of this zone, at the present time, is conducive to the maintenance of this fauna that can perhaps be qualified as warm temperate.

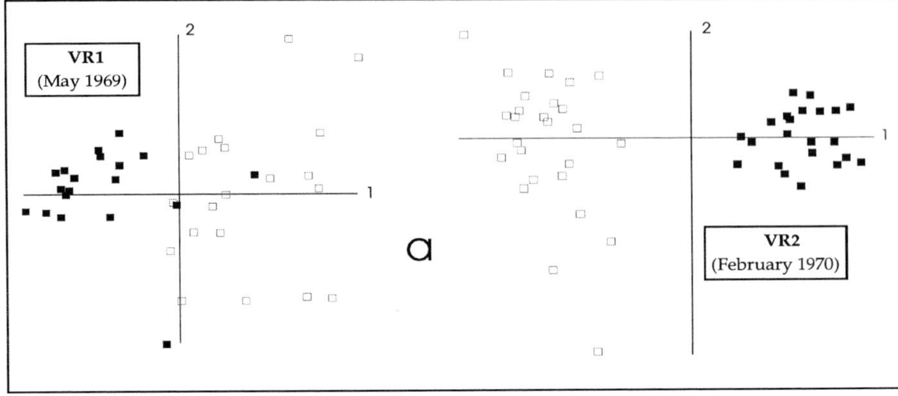

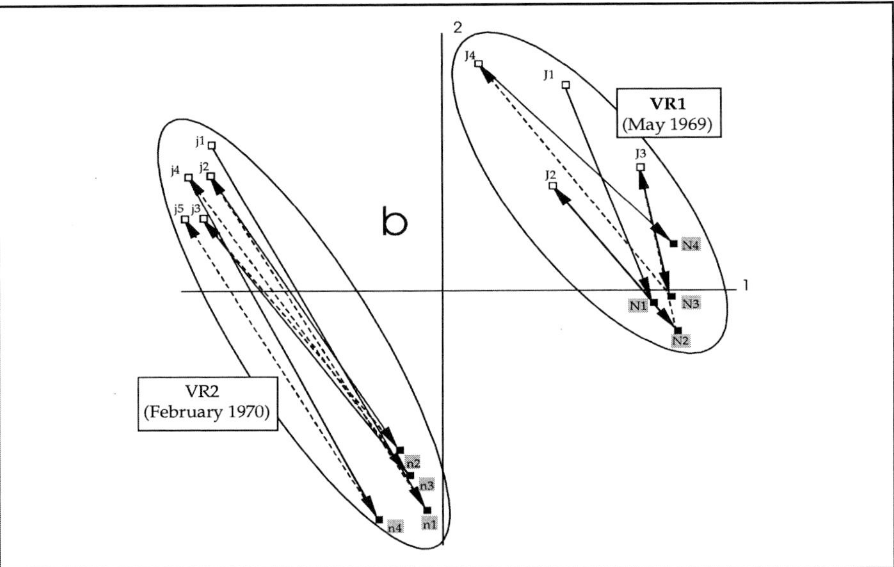

Figure 12-9 Nyctitropic variations in the benthic community (shrimp trawl samples from 45 m depth) off Grand-Bassam, Côte-d'Ivoire, in May 1969 (VR1 cruise) and February 1970 (VR2 cruise). Results of a correspondence analysis on: **a** all samples from each data set; **b** pooled data from both surveys (analysis based on the margin faunal distribution in the successive day and night phases).

Temporal Variations

- Diel cycle (Figure 12-9)

Repetitive sampling with shrimp trawls at 45 m depth off Grand-Bassam (Côte-d'Ivoire) in May 1969 (VR1 cruise) and February 1970 (VR2 cruise) showed that there were large variations between night and day catches of the benthic species belonging to the community associated with the exploited populations of the shrimp *Penaeus notialis*. A typical nyctitropic cycle and a perturbed cycle could also be seen (Le Loeuff and Intès 1998c).

The separation of data points by the correspondence analysis, representing the day and night collections, is much less distinct in May 1969 than in February 1970. Precipitation, heralding the arrival of the rainy season, usually occurs in June but occurred in May 1969. Consequentially, turbid waters from the Ébrié lagoon drifted over the area of sea-bottom under study, thereby reducing the light penetration and affecting the nyctitropic rhythm of a number of species, particularly those that are normally nocturnal (stomatopods, some brachyurids). Under these conditions, the activity of these species would not be substantially different between periods of the circadian cycle. Hence, the faunal composition of the day and night hauls tend to be more similar to each other.

- Seasonal cycle (Figure 12-10)

Benthic fauna were collected with a shrimp trawl along a transect (25-60 m) off Grand-Bassam from January 1969 to January 1970 (14 cruises). A correspondence analysis of the data from these collections showed a clear seasonal cycle: the data points for the brief cold season (January–February 1969) are followed successively by those of the warm season (March-June), the main cold season (July-October) and the following warm season (November-January 1970). A principal components analysis of the hydrological data (temperature, salinity and dissolved oxygen at the bottom) obtained at the same time as the benthos was collected, shows only the contrast between the warm and cold seasons; the departure of the data from the November cruise away from others is due to a strong reduction in salinity, associated with an enhanced river discharge during the flood season of 1969 (Le Loeuff and Intès 1998b).

If the results show that the biological cycle of the benthos is obviously dependent on the hydroclimatic cycle, then the variations in the faunal composition and the structure of the communities is, in fact, closely related to seasonal cycles specific to each of the species, as already noticed by Le Loeuff and Intès (1968, 1993). A majority of these are short-lived and are recruited at different periods of the year, mainly in the warm season but sometimes in the cold seasons. After recruitment, they steadily decrease in abundance and may disappear altogether from the collections. The community structure is, therefore, constantly modified throughout the annual cycle, each seasonal period being characterised by a typical fauna.

Further Work

We hope that this study, carried out on the biodiversity of the benthic marine fauna of the West African coast, will be complemented by an analysis of data collected from Senegal and Guinea (spatial variations). In addition, shrimp trawl collections have been programmed for areas off Grand-Bassam, one in the cold season (August), and the other in the warm season (December), using sampling conditions as close as possible to those that prevailed in 1969. The objective is to compare the present state of the benthic ecosystem with that of 1969, after a gap of about 30 years during which the hydroclimate has undergone vast changes, in particular a long period of drought, while the benthic areas have continued to be intensively trawled for fish and shrimp.

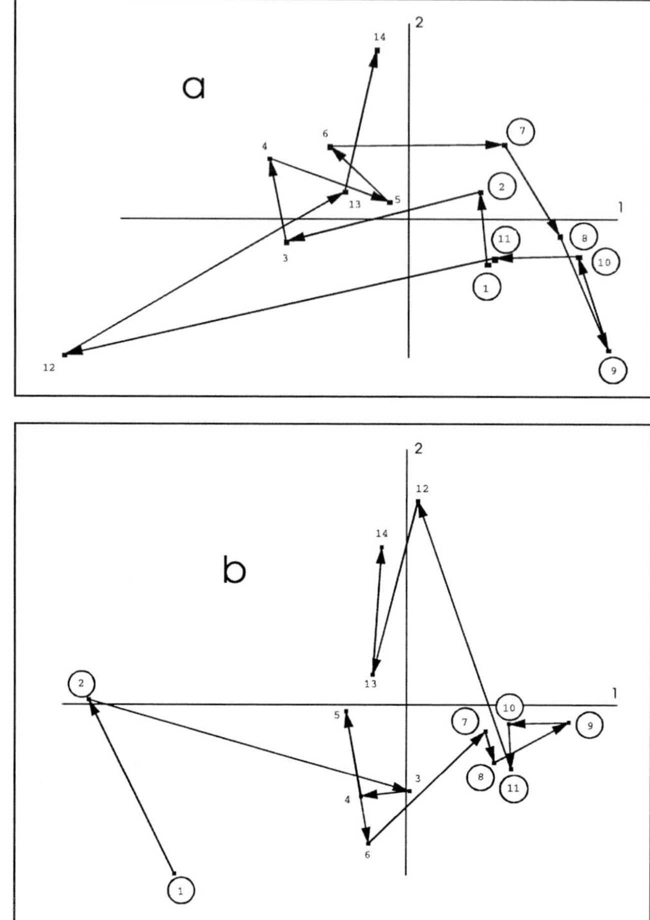

Figure 12-10 a Results from the correspondence analysis of benthic data (1969 to January 1970; 14 cruises); **b** principal components of the hydrologic data (temperature, salinity and dissolved oxygen at the bottom) obtained at the same time.

References

Berrit, G.R. 1973   Recherches hydroclimatiques dans les régions côtières de l'Atlantique tropical. État des connaissances et perspectives. Bulletin Mus. Nat. Histoire Natur. Paris, 3e sér., Écologie Génerale 4: 85-99.

Chardy, P. and M. Glémarec 1974 Contribution au problème de l'étagement des communautés benthiques du plateau continental Nord-Gascogne. C. R. Academie Sci. Paris, 278, série D: 313-316.

Colin, C., Y. Gallardo, R. Chuchla and S. Cissoko 1993 Environnement climatique et océanographique sur le plateau continental de Côte-d'Ivoire. pp76-110. *In*: P. Le Loeuff, É. Marchal and J.B. Amon Kothias (eds.). *Environnement et ressources aquatiques de Côte-d'Ivoire*, tome I- *Le milieu marin,* ORSTOM, Paris.

Cosel, R. von 1993 Aspects zoogéographiques des mollusques bivalves de l'Atlantique tropical vus à partir de l'Afrique occidentale. Thèse d'habilitation, Univ. Bretagne Occidentale, Brest. 71p.

Domain, F. 1989 Rapport des campagnes de chalutage du N/O "André Nizery" dans les eaux de la Guinée de 1985 à 1988. Doc. Sci. Centre Rech. Hal. Boussoura, 5. 80 p.

Domain, F. and M.O. Bah 1993 Carte sédimentologique du plateau continental guinéen. Notice explicative108 ORSTOM, Paris. 15p.

Hastenrath, S. and P.J. Lamb 1977 *Heat budget atlas of the tropical Atlantic and eastern Pacific ocean.* University of Wisconsin Press. 90p.

Intès, A. and P. Le Loeuff 1984 Les annélides polychètes de Côte-d'Ivoire. III- Relations faune-conditions climatiques. Unités régionales faunistico-climatiques dans le golfe de Guinée. Océanographie tropicale 19 (1): 3-24.

Le Loeuff, P. 1993 La faune benthique des fonds chalutables du plateau continental de la Guinée. Premiers résultats en référence à la faune de la Côte-d'Ivoire. Revue Hydrobiologie tropicale 26 (3): 229-252.

Le Loeuff, P. and R. von Cosel 1998 Biodiversity patterns of the marine benthic fauna on the Atlantic coast of tropical Africa in relation to hydroclimatic conditions and paleogeographic events. Acta Oecologica 19 (3): 309-321.

Le Loeuff, P. and A. Intès 1968 La faune benthique du plateau continental de Côte-d'Ivoire. Récoltes au chalut: abondance, dominance, répartition, variations saisonnières. Doc. Scienifiques Prov. Centre Recherche Océanographiques Abidjan, 25. 110p.

Le Loeuff, P. and A. Intès 1993 La faune benthique du plateau continental de Côte-d'Ivoire. pp76-110 *In*: P. Le Loeuff, É. Marchal and J.B. Amon Kothias (eds). *Environnement et ressources aquatiques de Côte-d'Ivoire*, tome I- *Le milieu marin,* ORSTOM, Paris.

Le Loeuff, P. and A. Intès 1998a Étude des peuplements benthiques du plateau continental de la Côte-d'Ivoire, de Grand-Bassam au cap des Palmes; mise en évidence de la faune associée aux fonds à crevettes (*Penaeus notialis*). Doc. Scient. Tech. Centre ORSTOM Brest 82: 5-38.

Le Loeuff, P. and A. Intès 1998b Les peuplements benthiques au large de Grand-Bassam (Côte-d'Ivoire); étude du cycle saisonnier. Doc. Scient. Tech. Centre ORSTOM Brest 82: 39-79.

Le Loeuff, P. and A. Intès 1998c Le peuplement associé aux fonds à crevettes (*Penaeus notialis*) de Grand-Bassam (Côte-d'Ivoire); étude détaillée des rythmes nycthéméraux. Doc. Scient. Tech. Centre ORSTOM Brest 82: 111-150.

Longhurst, A. R. 1962 A review of the oceanography of the Gulf of Guinea. Bull. IFAN, sér. A 14 (3): 633-663.

Martin, L. 1972 Variations du niveau de la mer et du climat en Côte-d'Ivoire depuis 25 000 ans. Cah. ORSTOM, sér. Géol. 4 (2): 105-120.

Martin, L. 1973 Carte sédimentologique du plateau continental de Côte-d'Ivoire. Notice explicative 48, ORSTOM, Paris. 19p.

Martin, L. 1977 Morphologie, sédimentologie et paléogéographie au quaternaire récent du plateau continental ivoirien. Trav. Doc. ORSTOM, 61. 266p.

McMaster, R.L., T.P. Lachance and A. Ashraf 1970 Continental shelf geomorphology features of Portuguese Guinea, Guinea and Sierra Leone, West Africa. Mar. Géol. 9 (3): 203-213.

Morlière, A. 1970 Les saisons marines devant Abidjan. Doc. Scient. Centre Rech. Océanogr. Abidjan 2 (3): 1-30.

Tastet, J.P. 1989 Continental shelf and sea level evolution during the last 120 000 years on the north-west African margin: Ivory Coast example. pp. 213-215. *In: International Symposium on global changes in South Africa during the Quaternary, past-present-future.* May, Sao Paulo.

Tastet, J.P., L. Martin and K. Aka 1993 Géologie et environnements sédimentaires de la marge continentale de Côte-d'Ivoire. pp21-61. *In:* P. Le Loeuff, É. Marchal and J.B. Amon Kothias (eds). *Environnement et ressources aquatiques de Côte-d'Ivoire,* tome I- *Le milieu marin,* ORSTOM, Paris.

Whittaker, R.H. 1967 Gradient analysis of vegetation. Biological Reviews 42: 207-264.

The Gulf of Guinea Large Marine Ecosystem
J.M. McGlade, P. Cury, K.A. Koranteng and N.J. Hardman-Mountford (Editors)

13

The Contribution of Coastal Lagoons to the Continental Shelf Ecosystem of Ghana

M. Entsua-Mensah

Abstract

Coastal lagoons are a resource within the coastal area. They are simple systems that are easily disturbed both by natural processes and by pollution from adjacent urban and industrial development. There is increasing pressure on coastal resources as a result of anthropogenic activities. The contribution of coastal lagoons to the coastal ecosystem in Ghana and some basic management measures are discussed. A number of issues impacting on lagoon resources are identified.

Introduction

Coastal lagoons form part of the interface between the land and the sea. The Coastal Zone Indicative Management Plan of Ghana has defined the landward extension of the coastal zone in Ghana as "the line joining the landward limits of the lagoons, lagoonal depressions, marshes and estuarine swamps together with the intervening interfleuve areas". Lagoons are simple systems in a delicate natural balance and are therefore easily susceptible to any detrimental environmental impact from pollution or other human activities which, if uncontrolled, would exceed their capacities and their ability to sustain their levels of use.

The coastline of Ghana, which forms part of the Gulf of Guinea, is between latitudes 4° 30' N at Cape Three Points and 6° N in the east and stretches for a distance of 550 km between 3° W and about 1° E. The shoreline is made up of a series of rocky headlands linked by sandy beaches behind which lie many brackish-water lagoons.

The geology of the coast of Ghana has been described by Kitson (1916) and Dickson and Benneh (1970). There is the Western section, between Newtown in the Western Region and Axim, which has stretches of beach sand behind which lies swamp. The Central section stretches from Axim to Apam and is mostly rocky cliffs, interrupted where rivers emerge, and fringes of sandbars that separate off coastal lagoons. The eastern section is from Apam to Keta where there is an alternating succession of lagoons and coastal plain, with or without low cliffs. Most of the rivers in southern Ghana (e.g. Tano, Ankobra, Pra, Ayensu and Densu) enter the Gulf of Guinea through coastal lagoons.

Bernard (1937) and de Rouville (1946) recognized two types of lagoons that fringe the West African Coast corresponding to the two types of estuaries described by Boughey (1957). These are the 'open' and 'closed' lagoons.The 'open' lagoon (often referred to as an estuary) has sufficient volume of water at all seasons to maintain a permanent outflow from its mouth into the sea. In Ghana, such lagoons occur more in the western part of the coastline where rainfall is heavy (mean of about 1250 mm per annum) and the lagoons are continuously fed by rivers.

The closed lagoons in Ghana are fed by seasonal rivers and streams. They usually lie behind a sand barrier which separates them from the sea and are normally opened for one or two months a year during the rainy season (May and September). Most of these lagoons are located on the eastern coastal region where rainfall is low.

This paper discusses the contribution of lagoon resources to the continental shelf ecosystem of Ghana and suggests strategies for managing the use of these resources.

Lagoon Resources

Lagoons and their wetlands present unique environments and habitats providing valuable products and services which include the support of fisheries, flood assimilation, regulation and supply of water and protection of biodiversity. Wetlands also provide important roosting, nesting and feeding sites for many species of birds, especially migratory ones. Associated with all lagoons in tropical areas are coastal wetland mangrove swamps. In Ghana, *Avicennia* spp. and *Rhizophora* spp. are the most common plants found in and around wetlands. *Paspalum vaginatum* and *Sporobolus* spp. are some of the grasses which occur in wetland areas.

The grasses are used for thatched mats. The mangroves, which support fisheries, act as windbreaks and prevent soil erosion, are under constant threat from anthropogenic activities. These include salt winning and indiscriminate harvesting of mangroves as fuel wood for cooking and smoking of fish. In Ghana, exploitation of mangrove wood is very significant because of the high density of rural populations along the coast, and the dwindling nature of the country's wood resources.Fish are a major component of the nektonic community in lagoons. In Ghana lagoons have a great influence on the socio-economic well-being and health of the communities that live close to and beyond them. This is because the usually small-scale, labour-intensive fisheries in lagoons provide income, employment and protein for coastal people.   Shells and shellfish obtained from the lagoon environment are an invaluable resource in the ceramic, animal feed and building industries.

The mobility of fish enables them to avoid unfavourable conditions, exploit a wide variety of habitats and food sources, evade predatory attacks and move from the freshwater to the brackish environment (Whitfield 1994). Fish also reside in the lagoons as marine or freshwater "visitors", euryhaline "permanent residents" and migratory catadromous or anadromous "transients".

Following Pauly (1975) the fish in Ghanaian lagoons can be grouped as follows:

a) Fresh water fishes which swim into the lagoon through permanent or temporary rivers: *Tilapia zillii, Clarias anguillaris, Heterobranchus bidorsalis, Hemichromis fasciatus, Hemichromis bimaculatus.*

b) Those that spend most of their lifetime in the lagoon: *Sarotherodon melanotheron, Porogobius schlegeli, Gobioides ansorgii, Periophthalmus papilio.*

c) Those that have their juvenile forms washed into the lagoon from the sea after the rainy season: *Ethmalosa fimbriata, Syacium microrum, Gerres melanopterus, Mugil cephalus, Liza falcippinis, Elops lacerta.*

d) Marine species which make short incursions into the lagoon: *Lutjanus fulgens, Caranx hippos, Epinephelus aenus, Trachinatus ovatus, Pegusa lascaris, Pseudotolithes senegalensis.*

Fish in lagoons face complex environmental factors, each of which may vary widely.

Contribution of Coastal Lagoons to the Continental Shelf Ecosystem

The coastal wetland habitat forms an integral part of the marine fishing industry. This is because the lagoon environment provides important spawning and nursery grounds for many marine fish species.

Mangroves located in the wetlands lie between two mature ecosystems, the rainforest and the sea. As a result, fish from these areas migrate to the 'immature' mangrove system to exploit it for food, to make use of the under utilised energy of the mangrove system and to take shelter and avoid predation. Mangrove swamps can also absorb pollutants from coastal and upland sources, trap sediments from rivers and accumulate washed silt to raise the ground level. They also protect the coast against erosion, stabilize shorelines and improve water quality.

Of the 20 fish species recorded in a study in three coastal lagoons in Ghana (Koranteng *et al.* 1998), eight (i.e. 40%) were typical marine species. These included economically important species like flatfish (*Syacium micrurum*, Bothidae) and grouper (*Epinephelus aeneus*, Seranidae). Others were snappers (*Lutjanus fulgens*, Lutjanidae and *Lethrinus atlanticus*, Lethrinidae), jack mackerel (*Caranx hippos*, Carangidae) and bonga shad (*Ethmalosa fimbriata*, Clupeidae). These species either have their juvenile forms washed into the lagoon from the sea or make short incursions from the sea into the lagoon. Penaeid shrimps (mainly *Penaeus notialis* and *Parapenaeopsis atlanticus*) were also encountered in the three lagoons in the study. Many shell and finfish need the lagoons either as nursery habitat or spawning areas in order to complete their life cycles. Thus the lagoon ecosystem is an important component of the continental shelf ecosystem.

The increasing rates of urban population growth along the Gulf of Guinea (an average 4-7% growth), agricultural run-off from irrigation, domestic and industrial pollution have created negative synergies in terms of human and environmental impact along lagoons and their wetlands in the Gulf of Guinea. Observation has shown that the mangroves, which filter inland water as it flows into the sea, are being rapidly destroyed (Entsua-Mensah 1996). As a result of the high population density, there is considerable exploitation of mangroves for fuel wood for cooking and smoking of fish. The loss of mangrove cover in the coastal wetlands leads to loss of biological diversity.

Factors Mitigating the Important Role of Lagoons to the Coastal Ecosystem

The issues identified as factors mitigating the important role of lagoons to the coastal ecosystem can be categorised under six major headings: water quality, habitat degradation, overfishing, education, population pressure and poverty, and a lack of management policy.

Water Quality

The quality of coastal wetland waters is controlled by the transfer of solutes from the sea and the influx of fresh water. It is also affected by human interference in the carbon flow by the release of sewage, organic matter, nutrients and sediments into them. Lagoons receive pollutants from agricultural pesticides and fertilizers from their immediate catchment area and from flood waters of larger streams and rivers.

In most lagoons in Ghana, drainage from the surrounding areas occurs during the rainy season. Domestic sewage from the villages is discharged into the lagoon at some parts of the lower reaches. Biney (1982) has identified pollution from domestic activities as the major cause of deterioration in water quality in Ghanaian lagoons. This is due to inadequate provision of basic sanitary facilities in most coastal settlements. Thus sewage and garbage are either deposited on lagoon banks or beaches. Pollution can affect lagoons directly, by increasing the amount of toxic chemicals, or indirectly. The decrease of water quality reduces the suitability of the lagoon habitat for fish (Alabaster and Lloyd 1982).

In an assessment of the state of pollution of some lagoons and estuaries in Ghana Biney (1982) classified 12 as polluted to various extents. Such pollutants may reach toxic levels and chronic exposure of the fish may affect their biology and population parameters, such as fecundity, recruitment, age at first maturity and mortality. Increases in biological oxygen demand (BOD) and the nutrient and ammonia levels, caused by domestic, agricultural and some industrial effluent into the lagoons, impact negatively on the fish community. Two of the lagoons, Korle and Chemu, were found to be grossly polluted.

Habitat Degradation

As a result of their high levels of biological productivity, lagoons and their wetlands provide feeding and breeding habitats for a number of commercially important fish species. Degradation caused by human settlements and industrial development causes loss of lagoon resources that prevent marine fish and shellfish from benefiting from the productivity of the lagoons.

- Exploitation of mangroves

Mangrove vegetation, which provide breeding and nursery grounds for commercially important fish species, are under threat from clearing to make way for construction of salt pans and also cutting for fuel wood.

In the Densu delta and areas around the Keta and Songhor lagoon, mangroves and other trees are being cut for "Acadja" fishing and fuel wood and also to make way for salt production. The Acadja fisher leaves a pile of branches for two to three weeks for the environment to stabilize.

In harvesting, the area is surrounded with a net and the twigs and branches are then removed and put at a different but nearby area of the lagoon. Cichlids, mullets and crabs are caught. Salt mining activities tend not only to desiccate the soil but also make it too compact for seeds to regenerate. Clear cutting of mangrove is practised and the harvested areas are left to regenerate from coppice shoots. Unfortunately, the harvesting of mangroves occurs at rates which far outstrip their regenerating capacity.

Entsua-Mensah (1996) has described the extent of degradation of mangroves in some lagoon ecosystems in Ghana as follows:

i.   Areas of Critical Degradation

Djange, Laloi, Sakumo 2, Chemu, Korle, Nyanya, Muni, Fosu, Benya lagoons, parts of the Amisa Estuary and the Densu delta. These areas are located near urban settlements and industrial activities. The mangroves are endangered by high levels of pollution or overexploitation.

ii.   Areas of Moderate Degradation

Songhor, Kpeshie, Brenu, Kako, Iture, Nakwa and parts of Amisa Estuary. These mangroves are exploited by local communities and do not have a lot of industrial pollution.

iii.   Important Mangrove Areas

Wetlands, most of the areas along the Western Region lagoons and estuaries, parts of the Volta Estuary and Oyibi lagoon. Mangroves here are relatively intact.

- Cape St Paul's Wilt

In recent years there has been a spread of this disease which affects coastal-zone trees in epidemic proportions in the Western and Central Regions of Ghana. The disease, discovered in Ghana nearly 70 years ago, is caused by a mycoplasma-like organism, believed to be transmitted by the insect *Myndus adiopodoumensis.* Coconut trees, which bind the sand on the sand bar separating the sea from closed lagoons, and help to check erosion of the beach, encourage erosion when they become dead and hollow, which in turn can have adverse consequences on nearby villages.

Overfishing

The consequences of heavy fishing pressure on fish populations include changes in the size structure and species composition of catches. Large predatory fishes and intermediate fishes are replaced by short-lived species. This is clear in the predominance of *Sarotherodon melanotheron* in most lagoons in Ghana. There is intense fishing pressure in the lagoons in spite of traditional management practices which prohibit fishing on certain days.

Education

A study conducted in coastal wetland communities in Ghana (FOE 1994) showed that two out of every three adults in those communities are illiterate. The study also revealed that in communities where the literacy rate was relatively high, direct dependence of the inhabitants on the resources of the wetlands is relatively low. Most fishers in the wetland communities usually lack any formal education that can secure them occupation other than fishing.

Generally, there is lack of awareness of the benefits of conservation amongst the majority of the fishers, even though some of them have good indigenous knowledge of the ecology of the wetlands. This could account for the degradation of the wetland environment and poor management of its resources.

Population pressure and poverty

Population growth in coastal areas in Ghana is generally higher than for inland areas. This is due to migration from inland areas. Obviously, such large numbers of people exert enormous fisheries as well as harvesting of mangroves and other types of vegetation is seriously threatened as a result of the massive influx of people.

Poverty and environmental degradation have been characterised as being part of a vicious cycle. This vicious cycle of absolute poverty and environmental degradation has been well documented in coastal communities by Westing (1989) and Lonergen (1993). The poverty arising from a lack of basic health facilities, low income and declining natural ecosystem productivity forces fishers to place a greater priority on their immediate subsistent needs. This also leads to further exploitation and degradation of lagoon resources. Pauly *et al.* (1989) and Pauly (1994) describe such a situation as *Malthusian overfishing*, which "occurs when poor fishers, faced with declining catches and lacking any other alternative, initiate wholesale resource destruction in their effort to maintain their incomes". In Ghana, this situation is made worse by the fact that there is a market for all sizes of fish, and therefore there appears to be some benefit in catching juvenile fish (Koranteng 1995).

Lack of Management Policy

In the past, management of lagoon resources was in accordance with traditional rules in the form of taboos. Tradition holds that every lagoon has a god or goddess protecting it with its unique set of rules and regulations and a traditional priest acting as a custodian of the lagoon. The priest performs the traditional rites and is responsible for enforcing the laws and the regulations governing the use of the lagoon. Unfortunately, these management rules are not written and are not backed by any form of legislative instruments. With modernisation and Christianity these rules are usually not obeyed and offenders cannot be prosecuted due to lack of legislation.

A recent decentralization of political and state power, in order to enhance participatory democracy through local (community and district) level institutions, with the District Assembly as the pivot (World Bank 1996), should be able to forge a closer link between modern and traditional systems. There are challenges when it comes to environmental management that need to be addressed using traditional knowledge in order to succeed. There is the need to focus on local and culturally accepted procedures for conflict management and resolution over issues connected with the management of environmental resources.

Management Options to maintain the integrity of the Lagoon Ecosystem

It is important to note that the management of fisheries in lagoons depends on understanding the socio-economic nature of fisheries as well as the biological knowledge of the resources and capture characteristics of the fishery. Some management options are thus proposed.

Community-based activities to improve water quality can be designed in such a way as to ensure proper garbage disposal. It is essential to monitor the levels of phosphates, nitrates and silicates in the lagoons because these are indicators of the amount of organic matter which influence phyto-plankton growth and hence primary productivity.

The recently established District Wetland Management Units in some parts of the country is a step in the right direction, in spite of the fact that most are not very effective as it is difficult to get the communities to accept new forms of environmental hygiene.

Environmental education activities, especially involving children, should be conducted to enable the wetland communities to maintain a healthy environment. Any activity which will affect primary productivity (e.g. siltation, dumping of sewage and industrial effluent into the lagoons), leading to loss of fish diversity, should be curtailed. Failure to do so could result in low phytoplankton diversity and primary productivity, thus affecting fish diversity.

The loss of mangrove cover is one of the most drastic catastrophes experienced by lagoons and could have contributed to the apparent loss of the shellfish fishery in some lagoons (Entsua-Mensah 1996). To provide greater habitat for fish in the lagoon, especially for juveniles, re-planting of mangroves along the lagoons is highly recommended. This is also likely to improve oyster production because the prop roots of some of the mangroves serve as substrate for the oysters. There has been a programme for planting mangroves in other West African countries, e.g. Benin and Sierra Leone (EEC 1987). The potential of shellfish (oysters and cockles) aquaculture in the mangroves should be assessed.

The complexity of fishing gears and increasing population pressure is leading to even higher exploitation levels of fish in the lagoons in the near future. The lack of viable livelihood options for fishers makes it difficult to implement the limitation of fishing effort as a management strategy. However, some efforts must be made to strengthen and improve legislative backing. It would be better to manage the fish stocks and the fishers. This can be done by:

a) gear and mesh size control (certain types of gears are not permitted in certain lagoons);
b) closed areas and seasons (in many lagoons fishing is prohibited in some areas or on certain days and periods).

In many wetland areas there are some fishers who are quite knowledgeable about the laws of nature and have invaluable indigenous knowledge of lagoon resources. However, it is essential that all fishers and school children, as well as all adults, be educated on the effects of irresponsible fishing and environmental degradation. Education will resolve the issue of low awareness as well as poor understanding of environmental issues amongst coastal communities. The local communities should be involved in the planning, implementation and enforcement of policies and programmes.

The educational efforts of the Environmental Protection Agency, the Wildlife Society of Ghana, and other non-governmental organizations, such as Friends of the Earth and Green Earth, are a step in the right direction. However, a lot of education is still needed.

The use of land around lagoons and wetland areas for settlement should be properly planned. Urban encroachment on wetlands is occurring at a very fast rate. The Family Planning Programme should be intensified around these areas. The issue of poverty among wetland communities needs to be addressed. Other trades, for example carpentry and masonry, can be introduced to the coastal communities. Policies and programmes that enable women to participate and contribute to the effective sustainable management of the lagoons should be put in place.

The principal traditional management practices of conservation value employed in all the lagoons need to be strengthened. The virtually open access and apparently uncontrolled fishing in areas where no closed seasons exist (e.g. the Densu delta and Keta lagoon) does not augur well for the sustainability of the fishery resources.

For the lagoon fisheries, the following measures could be put in place:

a) The District Assembles should be assisted by the Fisheries Department and traditional authorities to determine total allowable catches for each lagoon in their area, and the probable blend of gear and effort needed to achieve that.
b) The traditional authorities should be supported financially by the District Assembly to register traditional fishers and their gear, and to license them to fish each year for an appropriate fee.
c) Either the traditional authorities or the District Assembly could undertake legally backed enforcement of a system of appropriate controls and fines.

Conclusions

The degradation of coastal lagoon resources in Ghana is occurring at an alarmingly fast rate. Environmental degradation, overfishing and pollution are some of the most important causes of coastal resource variability. The major underlying factor is the rapid population growth that is taking place along the coast and urbanisation. It is therefore critical that management strategies include a blend of traditional and modern methods.

References

Alabaster, J.S. and R. Lloyd 1982 *Water quality criteria for freshwater fish*. (2nd edition). Butterworth, London. 316p.

Bernard, A. 1937 Geographic Universalle XI Afrique. Septemtrionalo et occidentale, 1-529, Paris.

Biney, C.A. 1982 Prelimnary survey of the state of pollution of the coastal environment of Ghana. Oceanology Acta 4 (54): 39-42.

Boughey, A.S. 1957 Ecological studies of tropical coastlines. The Gold Coast, West Africa. Journal of Ecology 45: 665-687.

de Rouville, M.A. 1946 *Le regimes de cotes*. Paris

Dickson, K.B. and G. Benneh 1970 *A New Geography of Ghana*. Longman Group Ltd., London.

European Economic Commission 1987 Mangroves of Africa and Madagascar. Conservation and Reclamation. Societé d'ecomanagement (SECA). Centre for Environmental Studies, University of Leyden (CML).

Entsua-Mensah, M. 1996 The importance of mangroves as transitional ecosystems in Ghana. p105-110. *In*: S.M. Evans, C.J. Vanderpuye and A.K. Armah (eds). *The coastal zone of West Africa problems and management*. Accra.

FOE (Friends of the Earth, Ghana) 1994 Wetlands management in Ghana. A Research Project Sponsored by the International Development Research centre. Canada (Summary Report).

Kitson, A.E. 1916 The Gold Coast. Some consideration of its structure, people and natural history. Geographical Journal 47: 369.

Koranteng, K.A. 1995 Fish and fisheries of three coastal lagoons in Ghana. Global Environmental facilty. Ghana Coastal Wetlands Management Programme. GW/A. 285/SF2/3.

Koranteng, K.A., P.K. Ofori-Danson and M. Entsua-Mensah 1998 Comparative study of the fish and fisheries of three coastal lagoons in West Africa. International Journal of Ecology and Environmental Science 24:371-382.

Lonergen, S.L. 1993 Impoverishment, population and environmental degradation: the case for equity. Environmental Conservation 20 (4): 328 -334.

Pauly, D. 1975 On the ecology of a small West African lagoon. Ber.dt.wiss.Komm. Meerest. 24:46-52.

Pauly, D. 1994 *On the sex of fish and the gender of scientists.* Chapman and Hall, London. 250p.

Pauly, D., G. Silvestre and I.R. Smith 1989 On development, fisheries and dynamite: a brief review of tropical fisheries management. Natural Resource Modeling 3: 307 - 329.

Westing, A. 1989 The environmental component of component of comprehensive security. Bulletin of Peace Proposals 20: 129 -134.

Whitfield, A.K. 1994 Fish species diversity in South African estuarine systems: an evolutionary perspective. Environmental Biology of fishes 40: 37-48.

World Bank 1996 *Towards an Integrated Coastal Management Strategy for Ghana.* World Bank, Washington DC. 138p.

III
Fish & Fisheries

Fish Ecology

14

Fish Species Assemblages on the Continental Shelf and Upper Slope off Ghana

K. A. Koranteng

Abstract

Using two-way indicator species analysis and detrended correspondence analysis, species on the continental shelf and upper slope of Ghana are assigned to six assemblages. The assemblages are comparable to those described thirty years ago from recurrent species analysis and for the whole Gulf of Guinea. The structure of the assemblages is determined primarily by depth and type of sediment on the seabed, the latter of which is more important when considering a compressed depth gradient. There are clear faunal discontinuities at 30 m, 100 m and 200 m deep. The dynamics of the assemblages are influenced by the physico-chemical parameters of regional and local water masses, primarily temperature, salinity and dissolved oxygen which are periodically modified by the seasonal coastal upwelling that occurs in the area.

Increased fishing activity, environmental forcing and the proliferation of *Balistes capriscus* singly or conjointly influenced the observed changes in the composition and relative importance of species in the assemblages.

Introduction

The objectives of this paper are (i) to establish assemblage units of demersal fish species on the continental shelf and upper slope off Ghana, (ii) to establish the factors that determine the structure of these assemblages, and (iii) to assess the impact of environmental forcing on the structure and dynamics of the assemblages.

Over the past three decades, the structure of species assemblages of exploited fish stocks around the world has been established. Examples are in the Narragansett Bay, USA (Oviatt and Nixon 1973), Scotian shelf, Canada's Atlantic coast (Mahon 1985), southern Gulf of Mexico (Yáñez-Arancibia *et al.* 1985), Gulf of Carpentaria, Australia (Blaber *et al.* 1994), Congo, Gabon and Angola in south-west Africa (Bianchi 1992) and north-western Indian Ocean and East Africa (Bianchi 1992). In the Gulf of Guinea, Fager and Longhurst (1968) and Longhurst (1969) worked out the assemblage structure of demersal species on the continental shelf using data from the Guinean Trawling Survey (Williams 1968). Caddy and Sharp (1986) pointed out that such a study is a necessary step towards understanding the dynamics of multispecies stocks and their management.

In this paper, the structure of fish assemblages on the continental shelf and upper slope off Ghana in West Africa are determined from trawl surveys conducted in the area between

1963 and 1990. From these, it is possible to examine how the fish assemblages have changed during the last three decades since Longhurst's (1969) work, thus contributing to the global debate on ecosystem effects of fishing, especially the effect of exploitation on fish assemblages. Spatial and temporal changes in the assemblages are also investigated to shed light on the dynamics of the assemblages over this period.

Materials and Methods

Between 1956 and 1992, a number of stock assessment surveys of the demersal fishery resources on the continental shelf and upper slope off Ghana were conducted (Koranteng 1998). The first comprehensive and most extensive survey in Ghanaian waters, and indeed, in the waters of West Africa (the Guinean Trawling Survey, GTS) was conducted in 1963-64 under the auspices of the Scientific Committee of the Organization of African Unity (Williams 1968). Trawls hauls were taken according to a systematic sampling scheme that covered depths between 15 and 600 m in the waters between Guinea Bissau and the Congo (*ibid.*). Following the establishment of the Fishery Research Unit (now the Marine Fisheries Research Division) in Ghana in 1962, a number of bottom trawl surveys were carried out in Ghana's shelf waters. For each survey, there were cruises undertaken in the upwelling and non-upwelling (or thermocline) seasons.

Systematic sampling designs were used in the surveys conducted between 1969 and 1980 and stratified random designs were used from 1980 (Rijavec 1980; Koranteng 1984, 1998). The sampling procedures were almost identical in all the surveys. When the vessel arrives at a sampling station, the net is shot at a pre-determined depth (in the case of the systematic surveys) or the depth and trawling direction are chosen at random (in the case of the stratified random surveys). The average between the depth at start and end of the haul is taken as the depth of the station sampled. The duration of tow was one hour in the surveys before 1980 and thirty minutes in the surveys since then. This is the time between engagement of the brakes on the trawl winch drum when the net and bridles have been paid out, and subsequent release of the brakes to commence hauling in of the net.

After the survey net has been hauled in, the catch is sorted according to species. Each species is weighed separately and the number of individuals counted. When the catch is too large, sub-samples are taken and weighed and the total catch extrapolated from the weighed samples. Names of fishes follow Blache *et al.* (1970), Fischer *et al.* (1981) and Schneider (1990) and for this work, these names were cross-checked, in this work, with entries in FishBase (FishBase 1996).

During the trawl surveys water temperature is obtained from thermometers mounted on Nansen reversing bottles. Salinity and dissolved oxygen are determined from the water samples collected with the Nansen bottles.

The following survey data sets were used in the study of community structure:

i. subsets of the Guinean Trawling Survey fish data pertaining to the transects worked off Ghana;

ii. subsets of data from the Marine Fisheries Research Division (MFRD) surveys of 1981-82 and 1989. Only trips that covered the entire survey area were used;

iii. water temperature, salinity, and dissolved oxygen measured at depths trawled; and

iv. depth sampled.

All catch data were entered into the NAN-SIS computer program for survey data logging and analysis (Stromme 1992). All catches were extrapolated to catch-per-hour of trawling.

Bottom type information was obtained from sediment maps produced during the Guinean Trawling Survey (Williams 1968) and also from Ramos *et al.* (1990). The continental shelf off Ghana is characterised by a belt of soft, muddy substrate in shallow waters down to about 30 m deep, followed by a wide area of mixed to hard bottom type. The descriptions of sediment types as used by Williams (1968) and Ramos *et al.* (1990) are as follows:

- Hard: predominantly sand, shell, rock, gravel, grit or coral
- Soft: predominantly mud
- Mixed: combination of hard and soft (sand, mud, grit).

Using Two-way Indicator Analysis (TWIA) implemented by TWINSPAN (Hill 1979), the groupings of species caught in the trawl surveys were obtained. In the TWIA method, a classification of the samples is constructed which is then used to obtain a classification of the species according to their ecological preferences. 'The two classifications are then used together to obtain an ordered two-way table that expresses the species' synecological relations as succinctly as possible' (Hill 1979). Stations and species are then arranged along the major gradients in the data. The number of sub-divisions of the data is determined, *inter alia*, by the length of the gradient, the size of the eigen numbers and presence of suitable indicator species which are species that are representative of the groups.

TWINSPAN uses a divisive cluster analysis algorithm to classify the samples and correspondence analysis (CA) to perform the ordination. CA is an ordination method which does not assume that species respond linearly to environmental changes but rather assumes a unimodal Gaussian curve response (Digby and Kempton 1987; ter Braak 1991; Jongman *et al.* 1995). From a suite of environmental variables, CA constructs 'theoretical variables' (ordination axes) that best explain the variation in the species data. The ordination axes are referred to as eigen vectors and the importance of each ordination axis is determined by its eigen value which is a measure of the dispersion of the species scores on the ordination axis. Thus the eigen value measures the percentage of total variation in the data that the axis explains.

The importance values (weights) are converted to a scale based on lower class limits before being used in the analysis. For this work, the class limits were 0, 0.5, 5, 50 and 500 kg.

A further ordination of the data was performed using Detrended Correspondence Analysis (DCA, Hill and Gauch 1980). DCA is a modification of the method of correspondence analysis (CA) and is intended to remove the two defects of CA, namely the 'arch effect' and compression of the ends of the first ordination axis (Hill 1974; Gauch 1994). The DCA routine contained in CANOCO (ter Braak 1991), a community ecology computer program, was used in this work. Weights of the catches were used for the analysis (Bianchi and Hoisæter 1992). Each weight (x) was converted to log (x + 1) to stabilize the variance as a Gaussian relationship between species abundance and each environmental variable was assumed. The environmental variables included in the analysis were bottom temperature, salinity and dissolved oxygen. Also included were depth sampled and type of sediment. The CANOCO program package also correlates the ordination axes with the environmental variables. The significance of each correlation was assessed with a student t-statistic.

The GTS cruises were first analysed and the resultant groups of stations plotted on trawl survey charts (Figure 14-1). Hauls in the same group were denoted by the same symbol. Summaries of salient properties of the groups identified in the GTS data are presented in Table 14-1. Using the results of TWINSPAN to label the sites/hauls, the DCA scores (from CANOCO) were plotted using the drawing tools in the computer program CANODRAW (Smilauer 1992).

The data collected during 1981-82 and 1989 Marine Fisheries Research Division surveys (MFRD3 and MFRD5 respectively) were similarly analysed. These surveys covered depths between 10 and 75 m deep and occasionally hauls were made between 75 and 100 m. For the GTS results to be comparable to those of the MFRD, the former were re-analysed using only hauls made between 10 and 75 m deep.

To obtain the most important species in each assemblage, an index of relative importance (IRI) was used. The index used, modified from that of Pinkas et $al.$ (1971), was defined as:

$$IRI = \%W \times \%F$$

Eq. 14-1

where %W is the percentage contribution by weight of each species in the assemblage and %F is the percentage of the number of times that the species occurred in hauls from the assemblage. Species with IRI values of 50 or more in each assemblage were included in a short list of the most important species of the assemblage.

Seasonal (upwelling and non-upwelling (or thermocline)) and long-term changes in the assemblages were investigated. For each survey and for each assemblage (the first three assemblages in the GTS data), the list of the most important species for the upwelling and thermocline seasonal cruises were compared. This was to determine species that were present mainly during the upwelling period, those present in the thermocline period and those that were regularly present in the assemblage ('residents').

Similarities in the composition of the various assemblages were assessed using the Jaccard Index, Sj (Southwood 1988; Magurran 1988) and the Similarity Ratio, Sr (Ball 1966). For two sampled sites (1 and 2), Sj is calculated as:

$$Sj = \frac{c}{A + B - c}$$

Eq. 14-2

where c is the number of species common to both sites, and A and B are the total number of species at the first and second sites respectively. To be adapted for use in this work, all stations in each assemblage were grouped and the assemblage treated as a 'site'. The similarity ratio is basically, a quantitative equivalent of the Jaccard Index (van Tongeren 1995). Following the notation of van Tongeren (1995), the Similarity Ratio for the comparison of two sites (i and j) is calculated from:

$$Sr_{ij} = \frac{\sum_{k} y_{ki} y_{kj}}{\sum_{k} y_{ki}^{2} + \sum_{k} y_{kj}^{2} - \sum_{k} y_{ki} y_{kj}}$$

Eq. 14-3

where y_{ki} is the abundance of the kth species at site i, y_{kj} is its abundance at site j, and $y_{ki} y_{kj}$ is the product of the abundance of the kth species occurring at both sites.

Results

Figure 14-1 shows the groupings of the stations as indicated on the survey charts. The position of the points represents approximate locations of the trawls hauls. Figure 14-2 is a dendrogram showing the relationship between the various groups and Figure 14-3 (for the GTS data) is an example of the CANODRAW bi-plots of sites and environmental parameters in the DCA axis 1 against DCA axis 2 plane. Hauls in the same group (assemblage) are indicated by the same symbol and enclosed in an ellipse. To be able to compare the plots for the various surveys, the axes are scaled to lie between -1 and +1.

Summaries of salient properties of the groups identified in the GTS data are presented in Table 14-1. The table gives average values of each environmental parameter, the indicator species and some of the other important species of each assemblage. Each level of the nominal variable (bottom type) is scaled from 0 to 1 where 0 denotes non-existence and 1 is a predominance of the type of bottom. Table 14-2 gives the correlation coefficient of the first two axes with the environmental variables, depth and bottom type. The list of species with IRI value of 50 or more in each of the six groups identified from the GTS data is given in Table 14-3. In this table, W is total weight caught in the survey, %W is percentage of total weight contributed by a particular species and F is the number of hauls in which the species was caught throughout the survey.

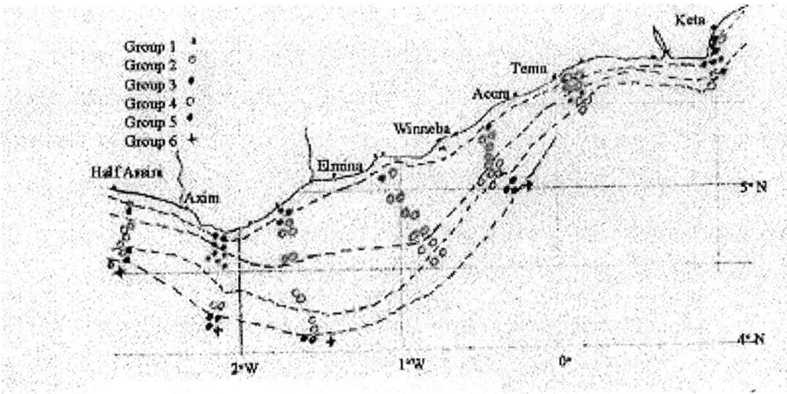

Figure 14-1 Position of trawl hauls made off Ghana in the Guinea Trawling Survey, 1963-64.

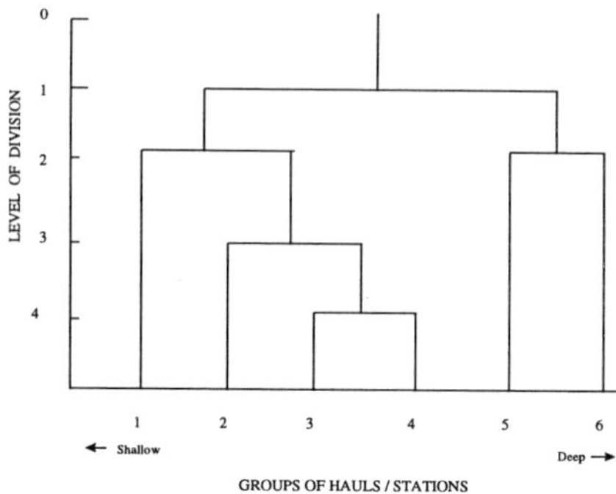

Figure 14-2 Dendrogram showing clustering order of groups of stations for the GTS data analysed with TWINSPAN.

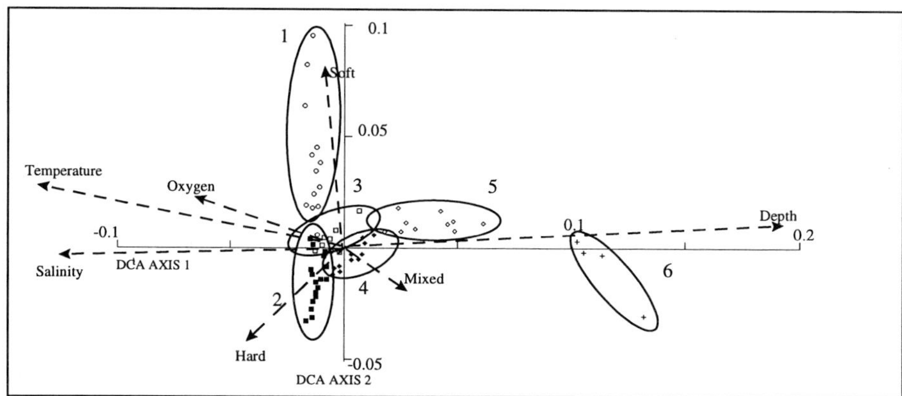

Figure 14-3 Bi-plot of trawl hauls and environmental parameters in DCA axis 1 versus axis 2 for the GTS data.

Tables 14-4.1 – 14-4.3 give the list of most important species in each of the two main seasons (upwelling and non-upwelling). In these tables, the names in bold type face are those of the 'resident' species. The species above the 'resident' species are present mainly during the upwelling period and those below are present mainly during the thermocline period.

Calculated values of the Jaccard Index and Similarity Ratio are presented in Tables 14-5 and 14-6 respectively. These were computed for both upwelling and thermocline seasons.

Discussion

The results from the GTS data, using the TWINSPAN and CANOCO analyses identified the principal fish assemblages similar to those described by Fager and Longhurst (1968) and Longhurst (1969) for the entire Gulf of Guinea.

Survey	Variable	Group 1	Group 2	Group 3
GTS	Hard		0.4	
	Mixed	0.4	0.6	0.4
	Soft	0.6		0.6
	Depth (m)	28 (8)	37 (10)	49 (10)
	Temperature (°C)	20.8 (1.0)	19.8 (1.3)	19.4 (1.2)
	Salinity (‰)	35.85 (0.09)	35.82 (0.12)	35.88 (0.09)
	Oxygen (ml l^{-1})	3.13 (0.74)	2.98 (0.62)	2.44 (0.70)
	Indicator species	*S. dorsalis*	*B. forcipatus*	*P. notialis*
		G. decadactylus	*A. punctatus*	*B. auritus*
			A. guineensis	
	Other important	*B. auritus*	*P. bellottii*	*P. bellottii*
	species	*I. africana*	*B. auritus*	*Trachurus* sp.
		P. senegalensis	*Trachurus* sp.	*R. miraletus*

Survey	Variable	Group 4	Group 5	Group 6
GTS	Hard	0.3		
	Mixed	0.6	0.7	0.8
	Soft	0.1	0.3	0.2
	Depth (m)	87 (32)	217 (43)	411 (23)
	Temperature (°C)	17.5 (1.3)	14.0 (1.4)	11.0 (3.0)
	Salinity (‰)	35.74 (0.09)	35.41 (0.15)	35.37 (0.46)
	Oxygen (ml l^{-1})	2.74 (0.54)	2.23 (0.54)	2.07 (0.65)
	Indicator species	*D. congoensis*		*H. italicus*
		B. boops		
		S. oculata		
	Other important	*Trachurus* sp.	*P. ledanoisi*	*S. fernandus*
	species	*S. japonicus*	*Trachurus* sp.	*C. uyato*
		I. africana	*Loligo* sp.	*H. bella*

Table 14-1 Summary (mean) of community environmental parameters and indicator species. For full nomenclature see Table 14-4).

The occurrence of *Brachydeuterus auritus* and *Trachurus* sp. (*Trachurus trachurus* and *Trachurus trecae)* in many of the groups is notable. The six groups identified in this work correspond to the following assemblages (as named by Longhurst):

Group 1: Sciaenid community   Group 2: Lutjanid community
Group 3: Sparid community (shallow part)   Group 4: Sparid community (deep part)
Group 5: Deep shelf community   Group 6: Upper slope community.

The group of species referred to as eurybathic or thermocline species by Longhurst (1969) is not isolated in this work; these are generally included in the second assemblage. Also the estuarine sciaenid community is not represented in these surveys.

From Figure 14-2 the six communities appear to group in pairs; namely a pair in waters shallower than 40 m, a second pair in 40-100 m and the last pair in waters deeper than 100 m. Off Ghana, the average thermocline depth is about 41 m (Koranteng 1998). This means that the sciaenid and lutjanid assemblages are generally above the thermocline (Table 14-1) whereas the two sparid communities originate from within the thermocline depth and stretch seaward into deeper waters. Obviously, the deep shelf and upper slope assemblages lie well below the thermocline layer.

The significantly high correlation between depth, bottom temperature, bottom salinity and bottom dissolved oxygen and the first DCA axis and between sediment type and the second axis, for the GTS data (Table 14-2), show the importance of these parameters in the determination of the structure of the demersal species assemblages in the study area.   With this data set, depth appears to be the most important variable in the ordination. It is followed by bottom temperature, salinity and dissolved oxygen. These physico-chemical parameters are themselves closely related to depth in the oceans and usually change by seasons. Thus the upwelling, which appears to change the properties of water masses may also have effect on the structure of demersal species assemblages.

Sediment type then follows in importance being highly significant (1 % level) on the second DCA axis (Table 14-2). This shows the importance of this feature, which like depth, is invariant with time (at least within the time frame under consideration).

In the MFRD surveys, sediment type appears to be the most important factor determining assemblage structure. All levels of this variable are highly significant on the first DCA axis (Table 14-2). The second axis is dominated by depth and then temperature, dissolved oxygen and salinity. These results are rather different from the GTS results and which seem to imply that with a long depth gradient, as was the case in the GTS, sediment type becomes secondary to depth as the principal factor influencing assemblage structure.

Survey	Variable	Species Axis 1	Species Axis 2	Survey	Variable	Species Axis 1	Species Axis 2
GTS	Hard	-0.13	-0.29**	MFRD3	Hard	-0.49***	0.09
(all hauls)	Mixed	0.13	-0.14		Mixed	0.30**	0.01
	Soft	-0.02	0.43**		Soft	0.30**	0.13
	Depth	0.95**	0.01		Depth	-0.29**	0.79***
	Temperature	-0.81**	0.10		Temperature	0.23*	-0.65***
	Salinity	-0.64**	-0.01		Salinity	-0.02	0.28
	Oxygen	-0.34**	0.11		Oxygen	0.13	-0.54***
	Eigenvalue	0.68	0.50		Eigenvalue	0.40	0.34
GTS	Hard	-0.31**	-0.02	MFRD5	Hard	-0.29**	-0.30**
(hauls at	Mixed	-0.11	-0.29*		Mixed	0.16	0.12
≤ 75 m)	Soft	0.44**	0.37**		Soft	0.20*	0.26**
	Depth	-0.51***	-0.71***		Depth	-0.42***	0.60***
	Temperature	0.47***	-0.47***		Temperature	0.18	-0.54***
	Salinity	0.18*	-0.07		Salinity	0.08	-0.07
	Oxygen	0.21*	-0.19		Oxygen	-0.02	-0.47***
	Eigenvalue	0.54	0.40		Eigenvalue	0.46	0.36

Table 14-2 Pearson's product-moment correlation coefficient r, of species axes 1 - 4 with bottom environmental variables: (r significant at 5% level; ** at 1% level; *** at 0.1 % level).

4.1 GTS

Group 1	Group 2	Group 3

Group 1:
Drepane africana
Epenephelus aeneus
Ilisha africana
Loligo sp.
Pomadasys jubelini
Pseudotolithus brachygnathus
Pseudotolithus typus
Sparus caeruleostictus

Brachydeuterus auritus
Galeoides decadactilus
Pagellus bellottii
Raja miraletus
Trichiurus lepturus
Pseudotolithus senegalensis
Pteroscion peli

Sphyraena sp.

Group 2:
Decapterus spp.
Fistularia villosa
Lutjanus dentatus
Pomadasys incisus
Lutjanus dentatus
Trachinocephalus myops
Acanthurus monroviae
Balistes forcipatus
Loligo sp.
Lutjanus agennes
Lutjanus fulgens
Raja miraletus
Scomber japonicus
Turtles

Brachydeuterus auritus
Dactylopterus volitans
Dentex canariensis
Epenephelus aeneus
Pagellus bellottii
Priacanthus arenatus
Pseudupeneus prayensis
Sparus caeruleostictus
Trachurus sp.
Sardinella aurita

Group 3:
Dentex gibbosus
Raja miraletus
Sardinella aurita
Boops boops
Scomber japonicus

Sparus caeruleostictus
Priacanthus arenatus
Epenephelus aeneus
Pagellus bellottii
Dentex congoensis
Trachurus sp.
Pseudupeneus prayensis
Dentex angolensis

Brachydeuterus auritus
Sphyraena sp.
Pentheroscion mbizi
Sardinella maderensis

Table 14-4.1 GTS **4.2** MFRD3 1981-82 and **4.3** MFRD5 1989. Table indicating seasonal membership of species in the various assemblages; names in bold type face are for species found in both seasons (resident) in the indicated assemblages, those at the top were found in the upwelling season only and those below were found in the thermocline season only.

From these datasets, it appears that when the depth range is not wide, then the most important factor affecting community structure is the type of sediment on the seabed, relegating depth to a secondary position. This appears to be a fractal problem whereby, on some gradients the grain size could be an important factor in assemblage formation (e.g. Mahon *et al.* 1984).

From the plot of DCA axis 1 versus DCA axis 2 (Figure 14-3), it appears that the first two species assemblages derived from the GTS data, are separated mainly by sediment type - the first on soft bottom and the second on hard bottom. Assemblages 3 and 4 are separated by both sediment type and depth. The last two assemblages are separated from the others mainly by depth.

Consequently, it may be sufficient to regard depth and bottom sediment type as the principal forcing factors determining the structure of fish assemblages on the continental shelf and upper slope off Ghana. As temperature and dissolved oxygen also then become important on the second axis in case of short depth gradient, it appears, therefore, that these physico-chemical parameters are important in the dynamics of the assemblages.

4.2 MFRD3 1981-82

Group 1	Group 2	Group 3
Decapterus rhonchus *Pagellus bellottii* *Pomadasys incisus* *Priacanthus arenatus* *Pseudupeneus prayensis* *Sepia* sp.	*Balistes capriscus* *Boops boops* *Brachydeuterus auritus* *Chromis lineatus* *Dentex angolensis* *Dentex congoensis* *Dentex gibbosus* *Paracubiceps ledanoisi* *Rhizoprionodon acutus* *Trachurus* sp. *Umbrina canariensis*	*Acanthurus monroviae* *Boops boops* *Chaetodon luciae* *Dactylopterus volitans* *Distodon speciosus* *Fistularia villosa* *Lutjanus fulgens* *Lutjanus goreensis* *Rhizoprionodon acutus* *Trigla* sp.
Balistes capriscus **Brachydeuterus auritus** **Chloroscombrus chrysurus** **Dentex canariensis** **Epenephelus aeneus** **Galeoides decadactylus** **Selene dorsalis** **Sparus caeruleostictus**	**Dactylopterus volitans** **Dentex canariensis** **Epenephelus aeneus** **Fistularia villosa** **Pagellus bellottii** **Priacanthus arenatus** **Pseudupeneus prayensis** **Sparus caeruleostictus**	**Balistes capriscus** **Dentex canariensis** **Dentex gibbosus** **Epenephelus aeneus** **Pagellus bellottii** **Priacanthus arenatus** **Pseudupeneus prayensis** **Sepia sp.** **Sparus caeruleostictus**
Elops senegalensis *Engraulis encrasicolus* *Ilisha africana* *Pseudotolithus senegalensis* *Pseudotolithus* sp. *Pteroscion peli* *Scyacium micrurum* *Sphyraena sphyraena*	*Acanthostracion guineensis* *Acanthurus monroviae* *Balistes forcipatus* *Chaetodon* sp. *Chloroscombrus chrysurus* *Decapterus* sp. *Lagocephalus laevigatus* *Lethrinus atlanticus* *Lutjanus fulgens* *Scyacium micrurum* *Sepia* sp.	*Brachydeuterus auritus* *Chloroscombrus chrysurus* *Chromis lineatus* *Decapterus rhonchus* *Lepidotrigla* sp. *Pomadasys incisus* *Raja miraletus* *Sardinella aurita* *Selene dorsalis*

Table 14-4 (contd) Table indicating seasonal membership of species in the various assemblages; names in bold type face are for species found in both seasons (resident) in the indicated assemblages, those at the top were found in the upwelling season only and those below were found in the thermocline season only.

The information on the site-environmental parameters bi-plots (Figure 14-3) correspond with the habitat preferences of various species in the assemblages as described by Longhurst (1969), Williams (1968), Blache *et al.* (1970) and Schneider (1990).

The results obtained in this study, in one way or another, corroborate those of similar studies undertaken elsewhere. For example, it has been shown by several authors (including Fager and Longhurst 1968; Mahon *et al.* 1984; Yáñez-Arancibia *et al.* 1985; McManus 1986; Roel 1987 and Bianchi 1992) that depth is the most important gradient along which faunal changes occur. Working on the entire GTS data collected from Guinea Bissau to Congo, Fager and Longhurst (1968) attributed assemblage boundaries in the Gulf of Guinea to sediment type,

4.3 MFRD5 1989

Group 1	Group 2	Group 3
Lagocephalus laevigatus *Pagellus bellottii* *Pentheroscion mbizi* *Pomadasys incisus* *Priacanthus arenatus* *Pteroscion peli* *Rhizoprionodon acutus* *Trachinocephalus myops* *Trachurus trecae*	*Acanthurus monroviae* *Boops boops* *Decapterus rhonchus* *Pomadasys incisus* *Trachurus* sp. *Trachurus trecae*	*Anthias anthias* *Boops boops* *Chromis* sp. *Decapterus* sp. *Dentex canariensis* *Dentex gibbosus* *Sardinella aurita* *Scomber japonicus*
Brachydeuterus auritus **Galeoides decadactilus** **Penaeus notialis** **Pomadasys jubelini** **Pseudupeneus prayensis** **Sepia officinalis** **Selene dorsalis** **Sparus caeruleostictus** **Trichiurus lepturus**	**Balistes forcipatus** **Brachydeuterus auritus** **Chloroscombrus chrysurus** **Dentex canariensis** **Fistularia villosa** **Lagocephalus laevigatus** **Lutjanus fulgens** **Pagellus bellottii** **Priacanthus arenatus** **Pseudupeneus prayensis** **Sepia officinalis** **Sparus caeruleostictus**	**Dactylopterus volitans** **Dentex congoensis** **Epenephelus aeneus** **Fistularia villosa** **Pagellus bellottii** **Pseudupeneus prayensis** **Rhizoprionodon acutus** **Sparus caeruleostictus**
Chilomycterus spinosus *Chloroscombrus chrysurus* *Dasyatis* sp. *Dentex canariensis* *Drepane africana* *Elops senegalensis* *Epenephelus aeneus* *Eucinostomus melanopterus* *Grammoplites gruveli* *Lutjanus fulgens* *Sardinella maderensis* *Sphyraena sphyraena* *Torpedo* sp.	*Apsilus fuscus* *Balistes capriscus* *Chaetodon* sp. *Chromis lineatus* *Dactylopterus volitans* *Decapterus* sp. *Epinephelus aeneus* *Lethrinus atlanticus* *Sphyraena sphyraena* *Trigla* sp.	*Balistes forcipatus* *Brachydeuterus auritus* *Chloroscombrus chrysurus* *Lagocephalus laevigatus* *Lepidotrigla* sp. *Priacanthus arenatus* *Raja miraletus* *Scyacium micrurum* *Sepia officinalis* *Selene dorsalis* *Serranus accraensis* *Sphyraena sphyraena* *Trachurus* sp. *Umbrina* sp.

Table 14-4 (contd) Table indicating seasonal membership of species in the various assemblages; names in bold type face are for species found in both seasons (resident) in the indicated assemblages, those at the top were found in the upwelling season only and those below were found in the thermocline season only.

the latter of which also changed with depth.

The analyses carried out in this work on temporal and spatial patterns of community structure (using the first three assemblages and the three surveys) does not seem to have resulted in definite or clear assemblage dynamics over the time period under consideration. However, the calculated values of the Jaccard Index and Similarity Ratio (Tables 14-5 and 14-6) indicate that assemblages 2 and 3 showed the closest resemblance to each other during the cold season and 1 and 2 in the warm season. The first situation could be due to fishes in assemblage 3 moving closer inshore during the upwelling and the second perhaps, due to

	GTS			MFRD3			MFRD5			
Group	1	2	3	1	2	3	1	2	3	
1	**1**	0.27	0.35	**1**	0.45	0.37	**1**	0.43	0.33	1
2	**0.45**	**1**	0.35	**0.36**	**1**	0.35	**0.45**	**1**	0.37	2
3	**0.40**	**0.41**	**1**	**0.36**	**0.46**	**1**	**0.39**	**0.46**	**1**	3
	1	2	3	1	2	3	1	2	3	

Table 14-5 Jaccard's index of similarity between pairs of assemblages for the same survey. Figures above shaded diagonal are for the upwelling period, and those below are for the thermocline period.

	GTS			MFRD3			MFRD5			
Group	1	2	3	1	2	3	1	2	3	
1	**1**	0.09	0.01	**1**	0.17	0.10	**1**	0.20	0.06	1
2	**0.18**	**1**	0.23	**0.16**	**1**	0.10	**0.60**	**1**	0.12	2
3	**0.01**	**0.28**	**1**	**0.30**	**0.07**	**1**	**0.01**	**0.04**	**1**	3
	1	2	3	1	2	3	1	2	3	

Table 14-6 Within survey Similarity Ratios among the three assemblages. Figures above shaded diagonal are for the upwelling period, and those below are for the thermocline period.

assemblage 1 fishes moving away from shallow areas during the warm season (Koranteng 1998). It could also be due to seasonal inshore-offshore movement of fishes in assemblage 2.

In general, the properties of the derived assemblages in the MFRD surveys are quite similar to each other and different from the GTS. This is true in both the upwelling and thermocline periods. These differences could be due to (1) fishing (Brown *et al.* 1976; Sherman *et al.* 1981; and Overholtz and Tyler 1985; Mensah and Quaatey this volume), or (2) differential response to changes in environmental forcing factors (Gulland and Garcia 1984; Overholtz and Tyler 1985; Macpherson and Gordoa 1992).

GTS was conducted 17 years before MFRD3 and at a time when commercial trawling on the continental shelf of Ghana and neighbouring countries was much less intense than was the case at the time of the MFRD surveys. Koranteng (1998) and Koranteng and McGlade (this volume) have shown that the period between 1963 and 1992 could be broken down into three time blocks each of which had distinct environmental characteristics in the Gulf of Guinea. In the first and third time blocks (i.e. before 1972 and after 1982) sea temperatures (surface and bottom) were relatively high, salinity was low and the thermocline was below its long-term average depth. Between 1972 and 1982, there was a global decline in sea temperatures and a rise in salinity. The peak of the changing events was between 1975 and 1979. Differences in the assemblages could also be a consequence of the proliferation of triggerfish in this ecosystem between 1972 and 1988 (Koranteng 1998). It appears that these factors conjointly or singly, affected the nature of species assemblages in Ghana's coastal waters.

Conclusions

The analyses of community structure resulted in species assemblages comparable to those described by Fager and Longhurst (1968) and Longhurst (1969) for the whole Gulf of Guinea. The first two, namely the sciaenid and lutjanid assemblages, are predominantly supra-thermocline whilst the two sparid assemblages begin at the thermocline depth of about 40 m and extend offshore. The last two (deep shelf and upper slope) assemblages occur well below the thermocline.

Associated with the soft, muddy substrate which is found in shallow waters (generally less than 40 m deep) is the sciaenid community made up mainly of species of the *Pseudotolithus* and *Galeoides* genera. Lying beyond this belt is a wide area of mixed-to-hard bottom associated with which are the lutjanid and the sparid assemblages (shallow and deep parts). There are clear faunal discontinuities around 30 m, 100 m and 200 m deep with the first ecotone closely related with depth and the thermocline, the second to a steep shelf drop, and the third to significant division between shelf and slope assemblages.

The structure of the assemblages is determined primarily by depth (in the GTS) and sediment type the latter of which is more important when considering a restricted depth gradient as in the MFRD surveys. The dynamics of the assemblages, including seasonal movements of component species, are influenced by physico-chemical properties of the water masses, mainly temperature, salinity and dissolved oxygen. Therefore, upwelling which changes the characteristics of these water masses on the continental shelf, would have effect on the dynamics of the species assemblages.

References

Ball S. 1966 *A comparison of some cluster seeking techniques*. Stanford Research Institute, California.

Bianchi, G. 1992 *Demersal Assemblages of tropical continental shelves. A study based on the data collected through the surveys of the R/V 'Dr. Fridtjof Nansen'*. Thesis for Dr. Scient. degree, University of Bergen, Norway.

Bianchi, G. and T. Hoisaeter 1992 The relative merits of using numbers and biomass in fish community studies. Marine Ecology Progress Series 85: 25 – 33.

Blaber, S.J.M., D.T. Brewer and A.N. Harris 1994 Distribution, biomass and community structure of demersal fishes of the Gulf of Carpentaria, Australia. Australian Journal of Marine and Freshwater Research 45: 375 – 96.

Blache, J., J. Cadenat and A. Stauch 1970 Clés de détermination des poissoins de mer signalés dans l'atlantique oriental entre le 20° parallele nord et le 15° parallele sud. Faune Tropicale XVIII. Paris: ORSTOM Editions. 479p.

Brown, B.E., J.A. Brennan, M.D. Grosslein, E.G. Heyerdahl and R.C. Hennemuth 1976 The effect of fishing on the marine finfish biomass in the Northwest Atlantic from the Gulf of Maine to Cape Haterras. International Commission of Northwest Atlantic Fisheries Research Bulletin 12: 49-68.

Caddy, J.F. and G.D. Sharp 1986 An ecological framework for marine fisheries investigations. FAO Fisheries Technical Paper No. 283.

Digby, P.G.N. and R.A. Kempton 1987 *Multivariate analysis of ecological communities.* Chapman and Hall. 206 p.

Fager, E.W. and A.R. Longhurst 1968 Recurrent Group Analysis of Species Assemblages of Demersal Fish in the Gulf of Guinea. Journal of the Fisheries Research Board of Canada 25 (7): 1405 – 1421.

Fischer, W.G., G. Bianchi G. and W.B. Scott 1981 *FAO Species Identification Sheets for fishery purposes.* Eastern Central Atlantic fishing areas 34, 47 (in part), 7 vols.

FishBase 1996 FishBase 96 CD-ROM. International Centre for Living Aquatic Resources Management (ICLARM), Manila.

Gauch, H.G. Jr. 1994 *Multivariate analysis in community ecology.* Cambridge Studies in Ecology. 298 p.

Gulland, J.A. and S. Garcia 1984 Observed patterns in multispecies fisheries. pp155 – 190. *In*: R.M. May (ed.). *Exploitation of marine communities.* Dahlem Workshop on Exploitation of Marine Communities. Springer-Verlag, Berlin.

Hill, M.O. 1974 Correspondence analysis: A neglected multivariate method. Journal of the Royal Statistical Society Series C 23: 340 – 354.

Hill, M.O. 1979 *TWINSPAN, A FORTRAN Program for Arranging Multivariate Data in an ordered Two-way Table by Classification of the Individuals and Attributes.* Cornell University, Ithaca, N.Y.

Hill, M.O. and H.G. Gauch 1980 Detrended correspondence analysis: and improced ordination technique. Vegetatio 42: 47-58.

Jongman, R.G.H., C.J.F. ter Braak, O.F.R. van Tongeren 1995 *Data analysis in community and landscape ecology.* Cambridge: Cambridge University Press. 299 p.

Koranteng, K.A. 1984 A trawling Survey off Ghana, CECAF/TECH/84/63. Dakar, Senegal: CECAF Project (FAO).72 p.

Koranteng, K.A. 1998 *The impacts of environmental forcing on the dynamics of demersal fishery resources of Ghana.* PhD Thesis, University of Warwick. 377 p.

Koranteng, K.A. and J.M. McGlade (this volume, chapter 8) Physical and chemical changes in continental shelf waters of the Gulf of Guinea and possible impact on resource variability.

Longhurst, A.R. 1969 Species assemblages in tropical demersal fisheries. pp147 – 168. *In:Proceedings of the symposium on the oceanography and fisheries resources of the tropical Atlantic.* Results of ICITA and GTS, Abidjan, Ivory Coast, 20 - 28 October 1966. UNESCO Publications, Paris.

Macpherson, E. and A. Gordoa 1992 Trends in the demersal fish community off Namibia from 1983 to 1990. South African Journal of Marine Science 12: 635 - 649.

Magurran, A.E. 1988 *Ecological diversity and its measurement.* Croom Helm, London 125 p

Mahon, R. 1985 Groundfish Assemblages on the Scotian Shelf. pp153 – 162. *In*: R. Mahon (ed.). *Towards the Inclusion of Fishery Interactions in Management Advice.* Canadian Technical Report of Fisheries Aquatic Science No. 1347.

Mahon, R.R., W. Smith, B.B. Bernstein and J.S. Scott 1984 Spatial and Temporal Patterns of Groudfish Distribution on the Scotian Shelf and in the Bay of Fundy, 1970 - 1981. Canadian Technical Report of Fisheries and Aquatic Science No. 1300.

McManus, J.W. 1986 Depth zonation in a demersal fishery in the Samar Sea, Philippines *In*: L. Maclean, L.B. Dizon and L.V. Hosillos (eds) *The First Asian Fisheries Forum. J.* Manila, Philippines: Asian Fisheries Society.

Mensah, M.A. and S.N.K. Quaatey (this volume, chapter 17) An overview of fishery resources and fishery research in the Gulf of Guinea.

Overholtz, W.J. and A.V. Tyler 1985 Long-term Response of the Demersal Fish Assemblages of Georges Bank. Fishery Bulletin 83(4): 507 – 520.

Oviatt, C.A. and S.W. Nixon 1973 The demersal fish of Narragansett Bay: an analysis of community structure, distribution and abundance. Estuarine and Coastal Marine Science 1: 361 – 378.

Pinkas, L., M.S. Oliphant and I.L.K. Iverson 1971 Food habits of albacore, bluefin tuna and bonito in California waters. Fisheries Bulletin of the Californian Department of Fish and Game 152: 1 – 105.

Ramos, A.R., I.S. Yraola, L.F. Peralta and J.F.G. Jiminez 1990 Informe de la Campana Guinea 90. Instituto Español de Oceanografia, Malaga. Spain.

Rijavec, L. 1980 A survey of the demersal fish resources of Ghana. CECAF/TECH/80/25, Dakar, Senegal: CECAF Project (FAO). 28 p.

Roel, B.A. 1987 Demersal communities off the west coast of South Africa. pp575–584. *In*: A.I.L. Payne, J.A. Gulland and K.H. Brink (eds). *The Benguela and comparable ecosystems.* South African Journal of Marine Science 5.

Schneider, W. 1990 *Field guide to the commercial marine resources of the Gulf of Guinea.* FAO Species Identification Sheets for Fishery Purposes. RAFR/F1/90/2, Rome: FAO. 268 p.

Sherman, K.A., C. Jones, L. Sullivan, W. Smith, P. Berrien and L. Ejsymont 1981 Congruent shifts in sand eel abundance in western and eastern North Atlantic ecosystems. Nature 291: 486 – 489.

Smilauer, P. 1992) *CANODRAW*. Microcomputer Power, Ithaca, N.Y.

Southwood, T.R.E. 1988 The concept and nature of the community. *In: Organisation of Communities: Past and Present.* Blackwell, Oxford.

Stromme, T. 1992 *NAN-SIS, Software for fisheries survey data logging and analysis.* User's manual. FAO Computerized Information Series (Fisheries). No. 4. FAO, Rome. 103p.

ter Braak, C.J.F. 1991 *CANOCO - a FORTRAN program for canonical community ordination by [partial] [detrended] [canonical] correspondence analysis, principal component analysis and redundancy analysis (version 3.10).* ITI-TNO, Wageningen. The Netherlands.9 (Manual and Update Notes)

van Tongeren, O.F.R. 1995 Cluster analysis. pp174-212. *In*: R.H.G. Jongman, C.J.F. ter Braak and O.F.R. van Tongeren (eds). *Data analysis in community and landscape ecology.* Cambridge University Press, Cambridge.

Williams, F. 1968 Report on the Guinean Trawling Survey, Organisation of African Unity Scientific and Technical Research Commission (99).

Yáñez-Arancibia, A., P. Sanchez-Gil, M. Tapia Garcia and M. de la C. Garcia-Abad M. 1985 Ecology, Community Structure and Evaluation of Tropical Demersal Fishes in the Southern Gulf of Mexico. Cahiers de Biologie Marine, Tome XXVI:137-163.

The Gulf of Guinea Large Marine Ecosystem
J.M. McGlade, P. Cury, K.A. Koranteng and N.J. Hardman-Mountford (Editors)

15

Analysis of the Spatial and Temporal Variability of Demersal Communities on the Continental Shelf of Côte-D'Ivoire

T. Joanny and F. Ménard

Abstract

From 1978 to 1995, sixteen scientific bottom trawling surveys were carried out across a depth range of 10-120m on the continental shelf of Côte d'Ivoire, with two sampling methods and then combined using post-stratified sampling. Nine surveys were conducted during the hot season and seven in the cold season. Using the species communities defined by Longhurst (1969), the structure of the demersal resources was studied in terms of global abundance, community proportions, diversity and species composition. The influences of three factors: season, stratum and year was also explored. From the 287 species identified, 4 species: *Balistes capriscus* (shallow sparid), *Brachydeuterus auritus* (eurybatic), *Dentex angolensis* (deep sparid) and *Pagellus bellottii* (shallow sparid) largely determined overall abundance. The two species *B. auritus* and *D. angolensis* did not fluctuate significantly. Opposite combinations were observed between the two communities of sparids during the hot season in the two time blocs 1978-1986 and 1993-1995. The main fluctuation in species diversity occurred above the thermocline mean position in the central part of the shelf, and was related to changes in the abundance of *B. capriscus*.

Introduction

Although the continental shelf of Côte-d'Ivoire is very narrow (Tastet *et al.* 1993), it contains demersal fish of significant commercial importance (Caverivière 1993b). The industrial trawl fishery, which began in 1950 (Ménard *et al.*, this volume), commonly exploits species belonging to the families of Sciaenidae and Sparidae. To support the management of these resources, the fishery research institute Centre de Recherches Oceanographique d'Abidjan (CROA) has conducted sixteen scientific trawl surveys since 1978. This paper presents the analysis of the variability in abundance of the demersal community. The period from 1978 to 1995 and the number of cruises allowed the fluctuation in abundance of the communities to be assessed together with an analysis of the stability and changes in the structure of the communities on a continental shelf scale.

Descriptions of the structure of the demersal communities given in previous studies were restricted to very few species. A more comprehensive and recent work concerning the continental shelf of Côte-d'Ivoire was undertaken by Caverivière (1982, 1993a,b) but the data were very limited in time and so trends in the most important species could not be elucidated. The purpose of this paper is to i) give global results on the structure of the communities with respect to abundance, diversity and species composition details, and ii) to describe the variability of the abundance of the community using year, bathymetric stratum and season and their interactions, as sources of variation.

The data

The data used, derive fom the results of sixteen trawling surveys undertaken between 1978-1995. The first 12 surveys were carried out from 1978 to 1986 (except in 1981 and 1982) according to a stratified random sampling design (Anon. 1978). The remaining 4 surveys were carried out from 1993 to 1995 according to a systematic sampling procedure; these data sets were thus grouped according to the first sampling design in order to extend the analysis and ensure homogeneity amongst samples (Ménard *et al.* this volume).

The continental shelf was post-stratified into two bathymetric bands (10-50 m and 50-120 m) in order to study the variability of abundance and the variance. Three areas were considered for each band: East, Centre, and West according to the bottom substratum (Martin 1973). Six bathymetric strata of unequal sizes were thus obtained (Table 15-1); for each trawl haul, the species were identified, weighed and measured.

Two main seasons influence the continental shelf: the cold season, lasting from July to September and the hot season lasting from October to June. Seven and nine surveys were carried out respectively in the cold and hot seasons (Tables 15-2 and 15-3).

Seven fish communities were defined by Longhurst (1969) in the Gulf of Guinea using the 1964 Guinean Trawling Survey data. Caverivière (1982, 1993a) modified these slightly for the continental shelf of Côte-d'Ivoire.

Bathymetric band	West	REGIONS Center	East	(Total)
10m-50m	st5 (1927)	st3 (1632)	st1 (853)	4412
50m-120m	st6 (3198)	st4 (2143)	st2 (1496)	6837
(Total)	5125	3775	2349	11249

Table 15-1 The bathymetric strata (and their surfaces km^2).

Survey	7801	7901	8301	8302	8401	8501	8601		9402	9503
Date	02-01	03-15	01-11	03-10	01-17	02-20	02-05	09-19	02-26	03-08
	02-11	03-29	01-19	03-18	01-24	02-27	03-11	03-23	03-16	03-25

Table 15-2 Hot season surveys.

Survey	7802	8001	8402	8502	8603	9309	9508
Dates	08-08	08-21	07-03	07-01	07-18	09-19	08-17
	08-11	09-04	07-10	07-10	07-26	10-08	09-02

Table 15-3 Cold season surveys.

In this work the Sciaenid, Eurybathic, shallow Sparid, deep Sparid and deep shelf community were considered. These five communities of the continental shelf (Caverivière, 1993a) made up 85% of the mean weight of the total catch per haul. The Lutjanid (a sixth continental shelf community) was not taken into account because its constituent species were very scarce - the species of this community are found in areas dominated by reef-corals usually avoided by fishing vessels. The seventh, slope, community belongs to a different ecosystem and was therefore not considered here.

Data analysis

The global abundance

The abundance and its variance were estimated by the stratified mean according to Cochran (1977):

$$\overline{y}_{st} = \frac{1}{A}\sum_{h=1}^{L} A_h \overline{y}_h$$

$$\mathrm{Var}\left(\overline{y}_h\right) = \frac{s_h^2}{n_h} \quad \text{with} \quad s_h^2 = \frac{1}{n_h - 1}\sum_{i=1}^{n_h}(y_{hi} - \overline{y}_h)^2 \qquad \textbf{Eq. 15-1}$$

$$y_{hi} = \frac{C_{hi}}{t_i},$$

where L, the number of strata; h, the index of the strata ($h=1, ..., L$); A, the surface area of the continental shelf; A_h, the surface area of stratum h; y_{hi}, the yield in kg.h^{-1} of trawl i in stratum h, and C_{hi} is the weight of trawl h in stratum i and t_i is the duration of the haul. These model-based estimators formally developed for normal distributions are also appropriate for highly skewed distributions (Smith 1990).

The species diversity

The global properties of the biodiversity of the main demersal resources based on species richness (number of species) and the evenness (fishes equidistribution among the species) are measured by the Shannon-Weaver index derived from information theory (Barbault 1983; Frontier 1985; Frontier and Pichod-Viale 1991) and defined for a trawl haul as

$$H' = -\sum_{i=1}^{S} P(i)\mathrm{Log}_2[P(i)] \qquad \textbf{Eq.15-2}$$

where $P(i)=N(i)/N$ is the relative abundance of species i ; N is the abundance of the trawl ; $N(i)$ is the abundance of species i and S is the number of species.

The evenness index is the ratio

$$R = \frac{H'}{H_{max}} \qquad \textbf{Eq.15-3}$$

where $H_{max} = \log_2(S)$.

R is a number between 0 and 1 and the Shannon index is thus a number between 0 and H_{max}. We give details of the species abundance distribution by the shapes of rank-frequency curves. These describe the species distribution in catches and can be interpreted as an index of maturity or regression for the exploited ecosystem.

Effects of season, stratum, year and interactions

Three main factors were examined for the sixteen scientific trawl surveys: season, stratum and year. The surveys were carried out in seasons and strata with different physical and hydrobiological characteristics. These characteristics could vary significantly for some years (for example years of extreme cold) and their effect on species could be reinforced by industrial fishing. The aim was to find out which species were affected by the three main sources of variability on abundance and their interactions.

The multivariate statistical method used for this purpose was a principal component analysis on instrumental variables or PCAIV (Chessel *et al.* 1993; Prodon and Lebreton 1994). This method combines classical tools of regression and principal component analysis (PCA). Pech and Laloe (1997) used this method on artisanal fisheries data from Senegal. An example of the method is described below.

Let us consider four years, two seasons, 6 strata and 7 species (Table 15-4). These three key variables are considered as factors having respectively two, six and four levels coded from 1 to the number of levels. For each trawl, the abundance of the seven species are available. The table of data contains thus 48 lines. This table is split into one (Table F, called species table) containing in its rows, the weights of the species per trawl and another (Table M, called the variable table) containing in its column the numerical level codes. For example trawl number 2 (Table 15-4) is performed in 1984 in hot season and in stratum 2.

Because of the skewness of the distribution of the species abundances, Table F is log-transformed: E = Log (F+1). The resulting matrix E is then regressed on the columns of M according to the usual formula of additive decomposition of analysis of variance:

$$E^i = E^i_{season} + E^i_{stratum} + E^i_{year} + E^i_{season*stratum} + E^i_{season*year} + E^i_{stratum*year} + E^i_{season*year*stratum} \quad \textbf{Eq.15-4}$$

For a species *i*, the left term is the abundance value and the right terms are the main effects of season, stratum, year and interactions season*stratum, season*year, stratum*year, season*year*stratum. Note the absence of error term due to the presence of only one trawl by combination of factors. This decomposition gives 7 matrices that are analysed by PCA.

The factors studied do not have the same influence on the variability of abundance. The total inertia, which is the sum of variances of species, is decomposed with respect to the factors and their interactions in order to identify their contributions. The PCAIV was processed with a SAS macro called *species.sas* developed by Philip Vey of IRD, Montpellier, France.

The results are made up of 10 correlation circles of PCA representing the year, stratum, season and interaction effects, and graphics which are the regression of effect values of a species on principal components for a given factor. They can be considered as the response of abundance to factor levels.

Results

The global abundance per cruise (Figure 15-1a)

In the hot season the stratified mean abundance was between 200 kgh^{-1} and 600 kgh^{-1} except for 8301 where a peak of 836 kgh^{-1} was observed. In the cold season, the maximum abundance was lower (400 kgh^{-1}) and the means less variable. The abundance was weak during 1995 in both seasons.

Sp.	sp1	sp2	...	sp7	
	Year	84	85	86	95
Var.	**Season**	hot	cold		
	Stratum	stratum1	stratum2	...	stratum6

trawl	Year	Season	Stratum	sp1	sp2	...	sp7
t1	84	hot	stratum1	5.7	0.6	...	0.0
t2	84	hot	stratum2	1.6	1.3	.	0.0
.	.	.	.	.	.	...	.
.	.	.	.	.	.	...	.
.	.	.	.	.	.	...	.
t48	95	cold	stratum6	0.0	0.2	...	0.0

F = Species Table   **M = Variable Table**

sp1	sp2	...	sp7		year	season	stratum
5.7	0.6	...	0.0		1	1	1
1.6	1.3	...	0.0		1	1	2
.	.	...	.	<---->	.	.	.
.	.	...	.		.	.	.
.	.	...	.		.	.	.
0.0	0.2	...	0.0		4	2	6

Table 15-4 An example of data description for PCAIV.

The community proportion (Figure 15-1b)

- The shallow sparid

During the hot season between 1978 and 1986 (7801 to 8601) the catches were dominated by sparids (shallow elements, roughly 40%) with the peak in 1983 (70%). These became scarce between 1993 and 1995. In the cold season, the proportion of the group was less important and fluctuated minimally.

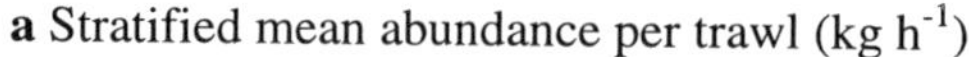

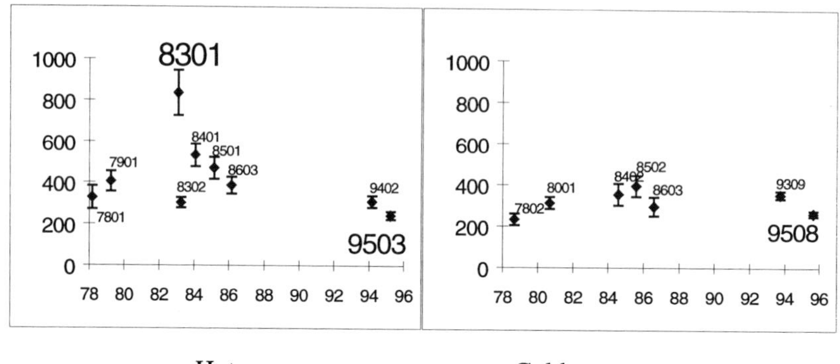

b Mean proportion of community per trawl

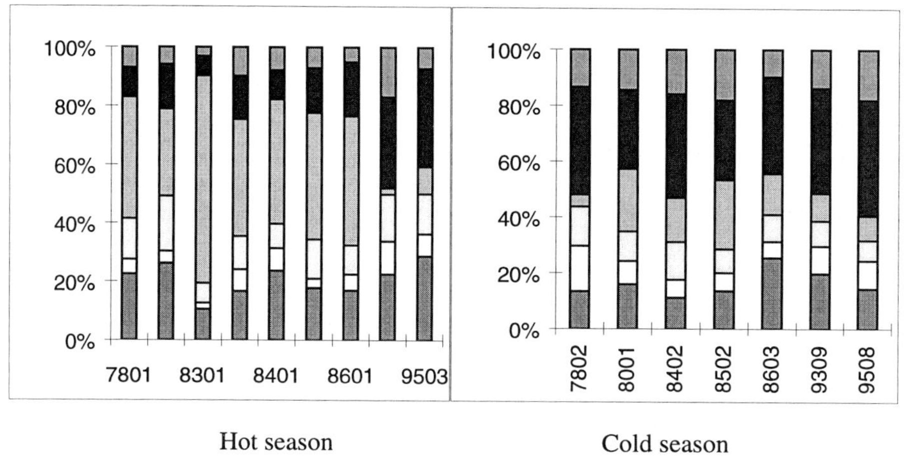

Eurybathic ☐ Deep shelf ☐ Sciaenid ☐ Shalow Sparid ■ Deep Sparid ▨ Remainder sp.

Figure 15-1 Abundance and community composition in trawl by **a** season/ year, and **b** survey/year. Key (left to right) indicates groups in ascending vertical order.

- The deep sparid

In the hot season the proportion was lowest (20%) between 1978 and 1986 when the shallow sparids were most abundant but was almost twice (35%) in the last two years (1994 and 1995) when the proportion of shallow sparids was low. In the cold season, this community dominated the hauls throughout the study period.

- The sciaenids and the deep shelf community

These 2 communities do not exhibit any noticeable changes and their proportions seem stable in both seasons.

- The eurybathic

This community was most abundant in the hot season in 1984, 1985 and 1995. In 1986, the abundance was higher in the cold season. The last five years of data show a slow rising trend in the hot season and a slow decrease in the cold season.

Species diversity

For each of the six strata, the mean Shannon index, the evenness index and their standard errors were calculated. The values were similar to those repeated by previous authors (Frontier and Pichod-Viale 1991) ranging between 1- 4 for the mean Shannon index and higher than 0.5 for the mean evenness index. The major modification in structure related to the 10-50m bathymetric band in the hot season. The fluctuation in the indices was particularly pronounced in stratum 3. The results presented are thus restricted to this central part of the continental shelf.

The Shannon index was very low in 1983 (Figure 15-2a). This was due to a large catch of *Balistes capriscus* (shallow sparid); this was also reflected in the evenness index. For a total number of 18 trawls, 13 of them had a catch rate of more than 1,080 kgh^{-1} but less than the maximum observed weight of 8,800 kgh^{-1}.

Two types of rank-frequency diagrams are shown (Figure 15-2b). In the first (i.e. the first survey of 1983, coded 8301), the proportion of *B. capriscus* was very high (86%) with all other species sparsely represented. In the remaining surveys, the first 3 species regularly present were *Brachydeuterus auritus* (Eurybathic), *Balistes capriscus* and *Pagellus bellotii* (both shallow Sparid) (Table 15-5). Note that without 8301 the difference in species abundance was not significant for all other surveys.

Effects of season, stratum, year and interactions

As the number of trawls per combination of factor levels varied to a high degree, the mean per combination for the species abundance in trawls was used.

Over all the surveys, 287 species were identified. As the series contained many zeros, 50 regular species, making up approximately 80% of the abundance per survey trawl were chosen for the PCAIV analysis.

Because season seems to be a significant factor, two other PCAIV were applied to the years with cold seasons and hot seasons respectively (Table 15-6).

- Main effects

The decomposition of inertia (Table 15-7) shows that the main effects explain 86% of the variability of abundance in all the species. The significance of the seasonal effect can also be seen. Although the effect of interactions is very weak they can be useful for describing the variability of some species.

- Year effect

Two groups are shown (Figure 15-3b) associated with to the two periods 1978-1986 and 1993-1995. In the first group, five species are included. Year effect is large after 1993 and very irregular before 1986. The second class is the opposite; here year effect is very weak in the period 1993-1995, and the species identified belong to all of the communities

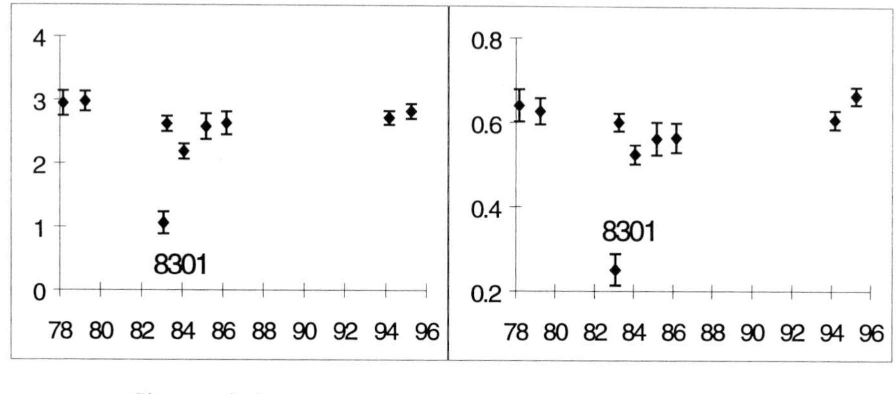

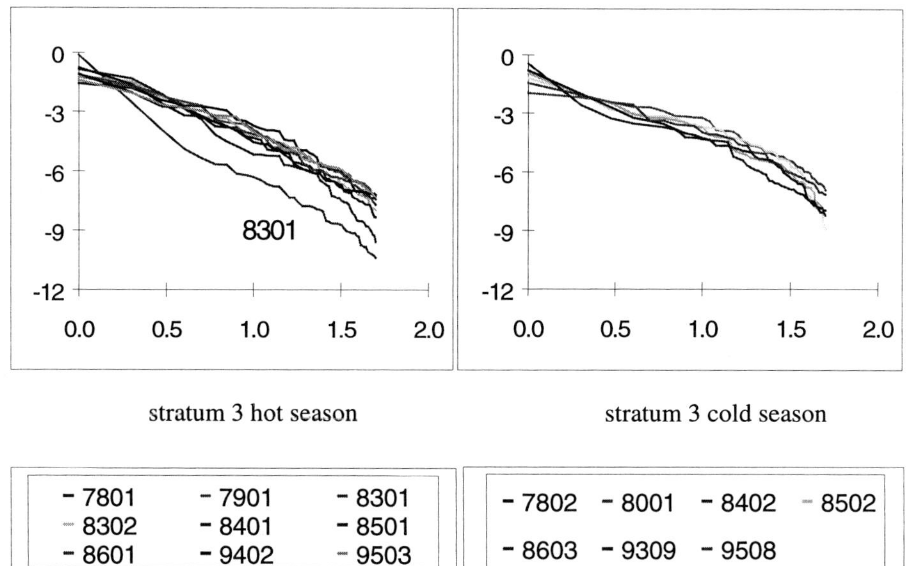

Figure 15-2 Diversity and species abundance distribution. Rank transformed by logarithm for the 50 first species.

- Stratum effect

Two classes can be identified (Figure 15- 4a). The species most abundant in depths above 50m include the Sciaenid, Eurybathic and Sparid shallow communities. The species below 50m include the deep sparid and deep shelf community. However three shallow species of the sparid community *Epinephelus arenatus, Chylomycterus spinosus, Dentex canariensis* are not well classified, and are more abundant below 50m.

A longitudinal effect can also be seen in some species. *Raja miraletus* (Eurybathic) and *Brotula barbata* (deep shelf) are more abundant on the western side of the continental shelf but *Dentex canariensis* and *Syacium micrurum* (deep sparid) are located on the eastern side (Figure 15-4b).

- Season effect

The relative importance of season was calculated using two parameters for each species - the relative variance of the abundance with respect to season and the seasonal effect. The three species most influenced by season are *Sphyraena guachancho* a Sciaenid, *Balistes capriscus* and *Pagellus bellottii* (both shallow sparid elements) (Table 15-8); these 3 species are all associated with the hot season and explain 38% of the seasonal variability.

- Season*stratum interaction

Pteroscion peli (Sciaenid) is very scarce in the 50-120m depth range (Figure 15-5a); in the 10-50 depth range, it is more abundant in the cold season than in the hot season. *Dentex canariensis* is abundant at all depths during the hot season (Figure 15-5b), but in the cold season, it is more abundant in the deep waters.

- Year*stratum interaction

This effect is due to a peak in catches of *Sparus caeruleostictus* in the 10-50m depth range during the two years 1985 and 1986.

- Season*year interaction

Four classes can be seen which correspond to the annual trend of increase or decrease in each season (Table 15-9). Some species have the same trend in both seasons, whilst for others the trend is specific to the season.

Discussion

The global abundance of demersal fish off Côte-d'Ivoire is mainly due to *Balistes capriscus* (shallow sparid), *Brachydeuterus auritus* (Eurybathic), *Dentex angolensis* (deep sparid) and *Pagellus bellottii* (shallow sparid).

B. capriscus, mainly fished by purse seiners that exploit *Sardinella aurita* resources, is usually classified as pelagic, but because of its widespread and regular occurrence in bottom trawls, it is also considered as a demersal species. This species was very abundant in the period from 1983 to 1986 in the hot season; it then disappeared from 1993 (there was no survey between 1987 and 1992). This may explain the low value of abundance in 1995. Cavand (1982) tried to explain the abundance of *B. capriscus* with respect to the usual physical and biological environmental parameters, but he concluded that no obvious link could be found. Conversely, Koranteng (1998) in a study of the seasonal effect on the demersal fisheries on the continental shelf off Ghana, observed that the species only flourished in cold regime and declined when waters warmed up.

a Hot season, 10-50m

Survey	1st Rank	2nd Rank	3rd Rank
7801	*Brachydeuterus auritus*	*Balistes capriscus*	*Pagellus bellottii*
7901	*Brachydeuterus auritus*	*Pagellus bellottii*	*Balistes capriscus*
8301	*Balistes capriscus*	*Brachydeuterus auritus*	*Pagellus bellottii*
8302	*Balistes capriscus*	*Brachydeuterus auritus*	*Pagellus bellottii*
8401	*Balistes capriscus*	*Brachydeuterus auritus*	*Pagellus bellottii*
8501	*Balistes capriscus*	*Brachydeuterus auritus*	*Pagellus bellottii*
8601	*Balistes capriscus*	*Brachydeuterus auritus*	*Pagellus bellottii*
9402	*Brachydeuterus auritus*	*Chloroscombrus chrysurus*	*Pricanthus arenatus*
9503	*Brachydeuterus auritus*	*Pagellus bellottii*	*Galoides decadactylus*

b Hot season, 50-120m

Survey	1st Rank	2nd Rank	3rd Rank
7801	*Balistes capriscus*	*Pagellus bellottii*	*Brachydeuterus auritus*
7901	*Pagellus bellottii*	*Brachydeuterus auritus*	*Dentex congoensis*
8301	*Balistes capriscus*	*Pagellus bellottii*	*Dentex congoensis*
8302	*Balistes capriscus*	*Dentex angolensis*	*Pagellus bellottii*
8401	*Dentex angolensis*	*Pricanthus arenatus*	*Pagellus bellottii*
8501	*Dentex angolensis*	*Balistes capriscus*	*Pagellus bellottii*
8601	*Dentex angolensis*	*Balistes capriscus*	*Pagellus bellottii*
9402	*Dentex angolensis*	*Brachydeuterus auritus*	*Boops boops*
9503	*Brachydeuterus auritus*	*Dentex angolensis*	*Pagellus bellottii*

c Cold season, 10-50m

Survey	1st Rank	2nd Rank	3rd Rank
7802	*Dentex angolensis*	*Dentex congoensis*	*Brachydeuterus auritus*
8001	*Brachydeuterus auritus*	*Pagellus bellottii*	*Balistes capriscus*
8402	*Pagellus bellottii*	*Brachydeuterus auritus*	*Pomadasys jubelini*
8502	*Pagellus bellottii*	*Brachydeuterus auritus*	*Balistes capriscus*
8603	*Brachydeuterus auritus*	*Balistes capriscus*	*Galoides decadactylus*
9309	*Brachydeuterus auritus*	*Pagellus bellottii*	*Galoides decadactylus*
9508	*Brachydeuterus auritus*	*Pagellus bellottii*	*Trachurus trecae*

d Cold season, 50-120m

Survey	1st Rank	2nd Rank	3rd Rank
7802	*Dentex congoensis*	*Dentex angolensis*	*bba*
8001	*Dentex angolensis*	*Epinephelus aeneus*	*Dentex congoensis*
8402	*Dentex angolensis*	*Pricanthus arenatus*	*Pagellus bellottii*
8502	*Dentex angolensis*	*Trachurus trecae*	*Dentex congoensis*
8603	*Dentex angolensis*	*Trachurus trecae*	*Pagellus bellottii*
9309	*Dentex angolensis*	*Dentex congoensis*	*Pricanthus arenatus*
9508	*Brachydeuterus auritus*	*Dentex congoensis*	*Uranoscopus albesca*

Table 15-5 The species of the rank-frequency diagrams in the first 3 ranks.

PCAIV 1: all 3 factors

Factor	Level
year	84 85 86 94
stratum	1 2 3 4 5 6
season	hot cold

PCAIV 2: hot season surveys

Factor	Level
year	78 79 83 84 85 86 94 95
stratum	1 2 3 4 5 6

PCAIV 3: cold season surveys

Factor	Level
year	80 84 85 86 93 95
stratum	1 2 3 4 5 6

Table 15-6 Factors descriptions for the 3 PCA.

PCAIV 1 (3 factors)

Source of variation	Df	Inertia	$\frac{\text{Inertia}}{\text{Df}}$
strata	5	15.43	3.08
season	1	2.06	2.06
year	4	4.89	1.63
year*season	3	2.75	0.91
strata*season	5	2.46	0.49
year*strata	15	6.53	0.43
strata*year*season	15	5.88	0.39

PCAIV 2 (hot season survey)

Source of variation	Df	Inertia	$\frac{\text{Inertia}}{\text{Df}}$
strata	5	23.6	4.72
year	7	11.1	1.57
year*strata	35	19.2	0.54

PCAIV 3 (cold season survey)

Source of variation	Df	Inertia	$\frac{\text{Inertia}}{\text{Df}}$
strata	5	22.9	4.58
year	5	9.5	1.9
year*strata	25	16.6	0.66

Table 15-7 Inertia decomposition for each PCAIV.

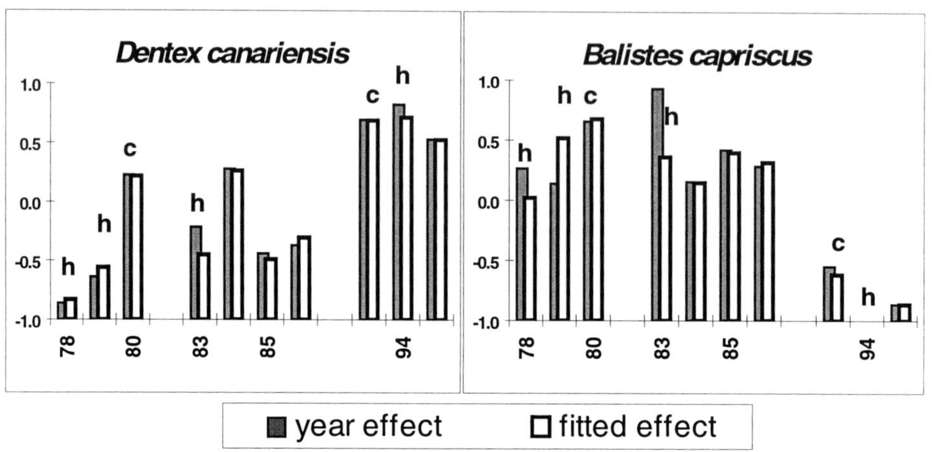

b Species corresponding to the 2 trends

Group A1	**Group A2**
Sciaenid	**Eurybathic**
Sphyraena guachencho	*Torpedo torpedo*
Shallow Sparid	*Cynoglossus canariensis*
Dentex canariensis	**Shallow Sparid**
Deep Sparid	*Chilomycterus spinosus*
Saurida brasiliensis	*Balistes capriscus*
Trachurus trecae	**Deep Sparid**
Deep shelf	*Chelidonichthys gabonensis*
Lepidotrigla cadmani	*Syacium microrum*
	Fistularia petimba
	Deep shelf
	Priacanthus arenatus

Figure 15-3 a The year effect and **b** groups of species of the same trend. Abscisse: year, c = cold season, h = hot season.

a Bathymetric distribution

    Sciaenid community   Sparid (deep)

    Eurybathic   Deep shelf community

    Sparid (shallow)

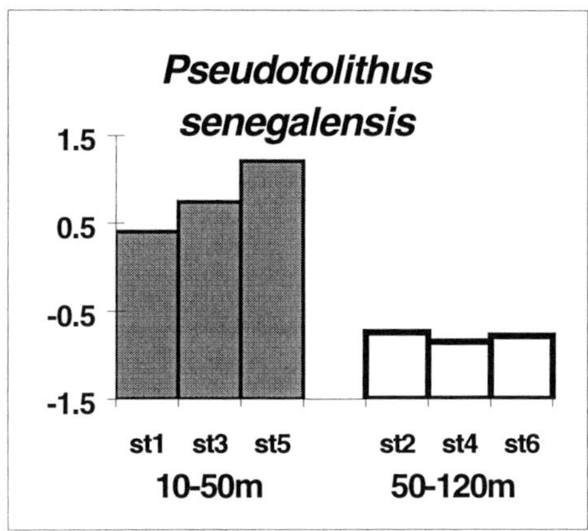

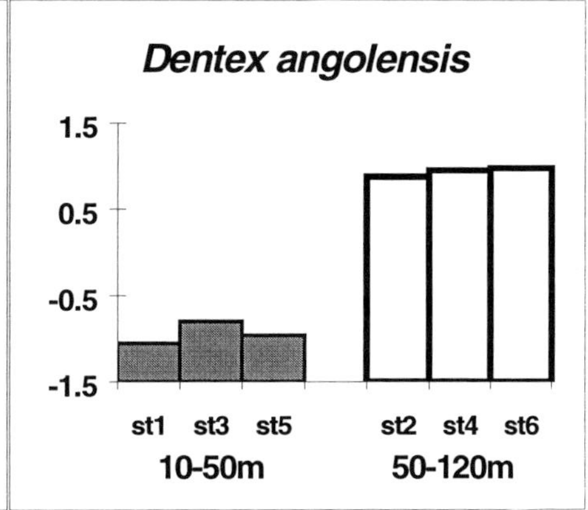

b Not well classed

    Epinephelus arenatus   (Sparid shalow element)

    Chylomycterus spinosus   (//   )

    Dentex canariensis   (//   )

c Longitude effect

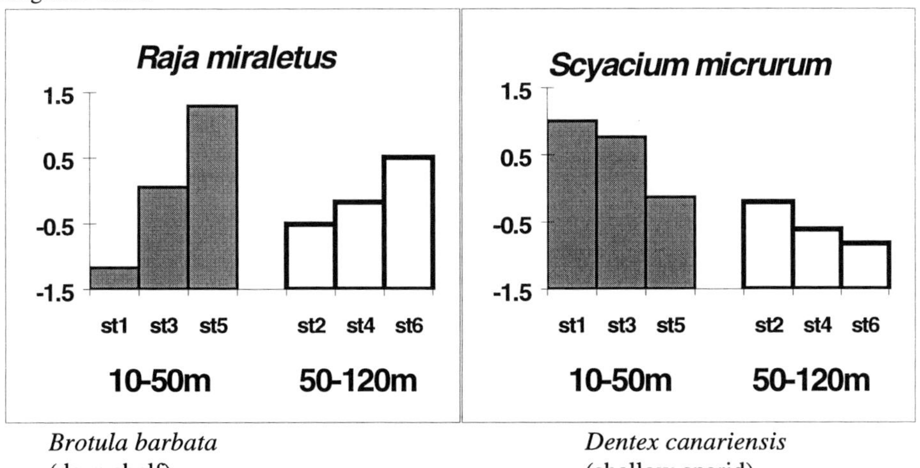

    Brotula barbata   *Dentex canariensis*

    (deep shelf)   (shallow sparid)

Figure 15-4 The spatial distribution of species. Distribution according to **a** the 2 bathymetric band 10-50m and 50-120m, **b** compared to Longhurst classes, and **c** longitude.

Regarding the increase in abundance observed in 1983, Caverivière (1982, 1993a) assumed that the survey was carried out at the end of the reproduction period. This hypothesis is very plausible, because the species compositions are very similar before and after 1983 in the same bathymetric band.

The catches of the second and third species are very stable. Note that *B. auritus* is semi-pelagic and tolerates a wide variability in temperature and depth. *D. angolensis* lives in the 50-120m bathymetric band which is not fished very much by the industrial trawlers.

In the two different time periods (1978-1986 and 1993-1995), community proportions in trawl hauls showed an opposite trend between the two sparid communities, in the hot season. Species did not have the same increasing or decreasing trends within the same community, and the catches of most of the species which explain the inter-annual variability were very low in the period 1993-1995. It was also noticeable that sea surface temperature (SST) anomalies were high all along the Ivorian coast in 1995 (CRO unpublished data).

With the PCAIV method, two large sets of species separated by the mean position of the thermocline at the 50m depth can be identified. The slight modification of this structure concerns 3 shallow species which occur in large quantities below the thermocline. This is probably due to the size distribution of these species with respect to depth. According to Caverivière (1982) the occurrence in number is high below 50m but larger individuals migrate to deeper waters. Nurseries may be located in shallower areas.

From Martin (1973) and Tastet *et al.* (1993) the nature of the deposits that cover the continent shelf is well known. The main difference between the 3 regions East, Centre and West is the muddy deposits associated with main rivers. These deposits are more important in Centre and much more in the West. We can suggest that *Raja miraletus* (Eurybathic) and *Brotula barbata* (deep shelf) avoid the sandy deposits located in East but *Dentex canariensis* and *Syacium micrurum* (deep sparid) live on them.

Species	Season	Effect	variance (%)
Scieanid community			
Selene dorsalis	Hot	0.21	6.3
Sphyraena guachencho	Hot	0.13	15.4
Eurybathic			
Brachydeuterus auritus	Hot	0.32	2.8
Shallow Sparid community			
Pagellus bellottii	Hot	0.49	12.03
Balistes capriscus	Hot	0.77	11.05
Deep Sparid community			
Trachurus trecae	Hot	-0.2	2.4
Deep shelf community			
Priacanthus arenatus	Hot	0.22	4.0
Brotula barbata	Cold	-0.29	7.6

Table 15-8 Season effect and variance contribution (>2%).

a

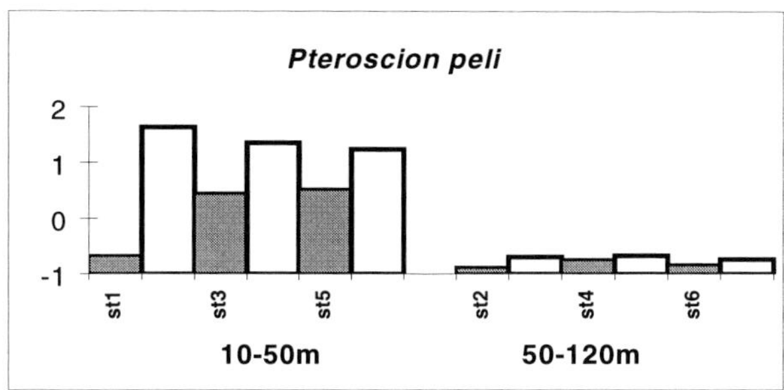

b

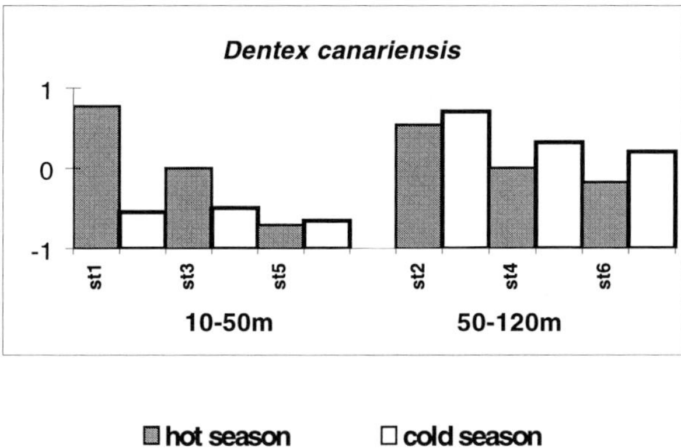

Figure 15-5 Interaction season*stratum. Fluctuation of strata mean effect by season for **a** *Pteroscion peli* and **b** *Dentex canariensis*.

Conclusions

The spatial distribution of abundance of the species examined was observed to be very stable, although this structure should be studied in more detail by considering environmental parameters such as temperature and salinity. Three trends in the annual fluctuation of catches of assemblages were identified. The deep sparid community was seen to be increasing, the shallow sparid community decreasing and the other three communities roughly stable. The main fluctuations of the species diversity were observed in the 10-50m depth range. The decrease of species abundance did not only involve *Balistes capriscus* but also involved a further seven species.

a Cold season

Increasing trend	**Decreasing trend**
Sciaenid	Sciaenid
Chloroscombrus chrysurus *	*Dasyatis margarita*
lagocephalus laevigatus	*Pomadasys jubelini*
Sphyraena guachencho *	*Pseudotolithus senegalensis*
Eurybathic	*Epiphyon gutifer*
Mustelus mustelus	Eurybathic
Deep Sparid	*Cynoglossus canariensis* *
Saurida brasiliensis	*Torpedo torpedo*
Shallow Sparid	Deep Sparid
Dentex canariensis *	*Syacium micrurum* *
Deep shelf	*Fistularia petimba* *
Zeus faber	Shallow Sparid
	Sparus caeruleosticus
	Balistes capriscus *
	Chylomicterus spinosus

The species tagged with '*' appear in both season

b Hot season

Increasing trend	**Decreasing trend**
Sciaenid	Sciaenid
Chloroscombrus chrysurus	*Lagocephalus laevigatus*
Sphyraena guachencho	Eurybathic
Deep Sparid	*Mustelus mustelus*
Boops boops	*Cynoglossus canariensis*
Scomber japonicus	Deep Sparid
Shallow Sparid	*Fistularia petimba*
Dentex canariensis	*Syacium micrurum*
Trachurus trecae	*Chelidonichthys gabonensis*
Deep shelf	*Pontinus kuhlii*
Priacanthus arenatus	*Saurida brasiliensisc*
	Shallow Sparid
	Epinephelus aeneus
	Balistes capriscus
	Pagelus bellottii
	Saurida brasiliensis

Table 15-9 Trends of species for each season.

References

Anon., 1978 Campagne CHALCI 78.01. 30/01/1978 - 12/2/1978. Résultats des chalutages. Arch. Sc. Cent. Rech. Océanogr. Abidjan 4 (1).

Barbault, R. 1983 *Ecologie générale*. Edition Masson. 224p.

Caverivière, A. 1982 *Les espèces démersales du plateau continental ivoirien, Biologie et exploitation*. Thèse Université d'Aix-Marseille II, Faculté des Sciences de Lumny, Volume I. 415p.

Caverivière, A. 1993a Les peuplements ichtyologiques démersaux. Ecologie et biologie. pp427-448. *In*: P. Le Loeuff, E. Marchal and J.B. Amon Kothias (eds). *Environnement et ressources aquatiques de Côte d'Ivoire* Tome 1 *Le milieu marin*. ORSTOM editions, Paris.

Caverivière, A. 1993b Les peuplements ichtyologiques démersaux. Ecologie et biologie. pp271-220. *In*: P. Le Loeuff, E. Marchal, J.B. Amon Kothias (eds). *Environnement et ressources aquatiques de Côte d'Ivoire* Tome 1 *Le milieu marin*. ORSTOM editions, Paris.

Chessel, D. and P. Mercier 1993 Couplages de triplets statistiques et liaisons espèces-environnement. *In* : J.D.Lebreton and B. Asselain (eds). Biometrie et Environnement Masson.

Cochran, W. G. 1977 *Sampling techniques* (3rd Edition).Wiley, New York. 422p.

Frontier, S. 1985 Diversity and structure in aquatic ecosystems. Oceanography and Marine Biology Annual Review 23: 253-312

Frontier, S. and D. Pichod-Viale 1991 *Ecosystèmes, structures-fonctionnement-évolution*. Editions Masson. 392 p.

Koranteng, K. 1998 *The impact of environmental forcing on the dynamics of demersal fishery resources of Ghana*. Ph.D. thesis, University of Warwick, United Kingdom. 377 p.

Longhurst, A. 1969 Species assemblage in the tropical demersal fisheries. Proc. Symp. UNESCO: Oceanography and Fisheries Resources of tropical Atlantic: 147-170.

Martin, L. 1973. Carte sédimentologique du plateau continental de Côte-d'Ivoire. Orstom, Notice explicative: 48. 22p.

Ménard, F., V. Nordström, J. Hoepffner and J. Konan (this volume, chapter 20). A database for the trawling fisheries of Côte-d'Ivoire: structure and use.

Prodon, R. and J.D. Lebreton 1994 Analyses multivariées des relations espèces-milieu: structure et interpretation écologique. Vie et milieu 44 (1): 69-91.

Pech, N. and F. Laloe 1997 Use of principal component analysis with instrumental Variables (PCAIV) to analyse fisheries catch data. ICES Journal of Marine Science 34:32-47.

Smith, S.J. 1990 Use of statistical Models for the Estimation of abundance from Groundfish Trawl Survey Data. Canadian Journal of Fisheries and Aquatic Science 47: 894-903.

Tastet, J.P., L. Martin, and K. Aka 1993 Géologie et environnements sédimentaires de la marge continentale de Côte-d'Ivoire pp23-61. *In*: P. Le Loeuff, E. Marchal, J.B. Amon Kothias (eds). *Environnement et ressources aquatiques de Côte d'Ivoire* Tome 1 *Le milieu marin*. ORSTOM editions, Paris.

16

Population Structure of Two Commercially Important Marine Species In and Around the Gulf of Guinea, West Africa

A. Lovell and J. M. McGlade

Abstract

This paper presents the results of a study in which molecular markers were used to investigate the population structure of two commercially important marine species in the Gulf of Guinea, namely *Trachurus trecae* (Carangidae) and *Sepia officinalis* (Sepiidae). Sequence analysis of the cytochrome *b* gene was used for the analysis of *Trachurus trecae*, while seven microsatellite loci were screened for the analysis of *Sepia officinalis*. Preliminary results across LMEs (Large Marine Ecosystems) indicate deep evolutionary divergence between the Gulf of Guinea and other regions, possibly indicating taxonomic structures such as independent species / subspecies. However within the Gulf of Guinea there is evidence for high gene flow across sampled regions, leading to no clearly defined spatial structure. Such results highlight the need for regional management structures. In the *Trachurus* data set there is some evidence for genetic differentiation within the Gulf of Guinea between age classes.

Introduction

There are many commercially important marine species in the Gulf of Guinea. In order to implement a successful conservation and management regime, it is important to understand not only their population structure but also their reproductive behaviour. Although some behavioural details are known, for example the tendency to move inshore to spawn or to have spawning peaks around the time of upwelling, little is known of whether different species remain loyal in time and space to particular spawning sites, and whether this in turn leads to population differentiation.

The two species studied in this paper are *Trachurus trecae*: Cadenat (Carangidae) and *Sepia officinalis*: Rong (Sepiidae). *Trachurus trecae* (the cunene horse mackerel or chinchard cunène) is a semi-pelagic schooling species. In the tropical Gulf of Guinea it is found in deeper waters, is less abundant and of smaller size than *Trachurus* spp. in the Canary and Benguela Current systems. It is predominately a cold water species, so the sharp thermal profile of the Gulf of Guinea forces it to live below the thermocline, moving to shallower waters during periods of upwelling. Off South Africa and Namibia there is evidence that *Trachurus trachurus* has two spawning peaks during the reproductive season (Hecht 1990) and undergoes a spawning migration. All species of *Trachurus* are known to have a pelagic egg phase (Whitehead *et al.* 1986). The family Carangidae, of which *Trachurus trecae* is a member, is an important part of the fisheries of the East Central Atlantic Ocean Area (Anon. 1981). *Trachurus* spp. catches peaked in the 1970's, and have been falling ever since.

The second species, *Sepia officinalis* (the common cuttlefish or seiche commune) has a wide distribution along the coast of Western Africa and into the North Sea. They are most abundant in shallow waters of less than 100m depth, though are observed down to a depth of about 200m, and are most often found over sandy bottomed areas of the shelf. The larger specimens are generally caught at greater depth. Seasonal migration (mainly vertical) has frequently been observed. As with most other species of cephalopods they have a short life span, with one spawning event followed by death. Population spawning occurs all year round with a peak around May to August at depths of generally less than 40m and at temperatures of between 13 to 15°C (Anon. 1981). Females lay between 150 to 4,000 eggs in grape-like clusters on rocks and shells on the sandy bottom which, after 30 to 70 days of incubation, give birth to hatchlings of 7 to 8 mm in length. The reason for choosing these two species is not only because of their commercial importance but also because they are relatively common and easy to find, thus aiding sample collection. They also make an interesting comparative study, given their alternative life histories.

Population differentiation was measured with two molecular markers: mitochondrial DNA and microsatellite DNA. Mitochondrial DNA (mtDNA) has highly conserved regions across many species which aids relatively rapid analysis of newly sampled species. The circular mtDNA genome has been mapped in a number of species, including humans, and is probably the currently most favoured DNA for molecular ecology work. The discovery and development of microsatellite DNA as a molecular marker is a more recent development. More preliminary work is required for each newly sampled species than with mtDNA. Microsatellite DNA is part of the nuclear genome and takes the form of many tens or hundreds of repeats of a short two, three or four base sequence. They do not code for proteins and are therefore likely to be selectively neutral: indeed it is often postulated that they are examples of 'junk DNA' that serve no purpose. They have a high mutation rate, because of their sequence repeat nature, and show great potential as molecular markers in population work (for good reviews on mtDNA and microsatellites in fisheries see Carvalho and Hauser 1994; O'Connell and Wright 1997).

Methods

The molecular analyses were based on sequence data of the cytochrome *b* region of the mitochondrial genome for *Trachurus*, and seven microsatellite loci for *Sepia*. The first section on Sampling and DNA extraction applies to both *Trachurus* and *Sepia* samples. The next three sections then apply to *Trachurus* (i.e. amplification, sequencing and sequence analysis), and the final three to *Sepia* (i.e. amplification, microsatellites and microsatellite analysis).

Sampling and DNA extraction

Samples were collected by A. Lovell while on the Dr Fridtjof Nansen survey of the pelagic and demersal resources of the western Gulf of Guinea, in the waters of Benin, Togo, Ghana and Côte d'Ivoire from 19th April - 6th May, 1999) (NORAD - FAO/UNDP Project GLO 92/013). Additional *Sepia* samples were provided by E. Morize from the coastal waters of Guinea. On collection muscle tissue was placed in 95% ethanol to preserve the DNA.

Station	Sample size	Co-ordinates & Depth	Description	Date
tt44	41	5°31'N 0°12'E 105m	unimodal length frequency	May 1999
tt51	80	5°20'N 0°11'W 60m	unimodal length frequency	May 1999
tt70	63	4°29'N 2°10'W 111m	unimodal length frequency	May 1999
tt99	60	4°59'N 4°58'E 85m	bimodal length frequency	May 1999
tt104	50	5°04'N 4°34'E 85m	bimodal length frequency	May 1999

Table 16-1 Shows sample size, co-ordinates, depth and description of each station used for collection of *T. trecae* from the Gulf of Guinea. The sample names refer to the sampling station from which they were collected from (e.g. *tt44* are those *T. trecae* individuals collected from sampling station 44). *tt99* and *tt104* showed two clear length frequency peaks and samples were taken from both large and small individuals to allow investigation of variation between length classes.

Table 16-1 shows the co-ordinates and depth of each station used for collection of *Trachurus* from the Gulf of Guinea. Also given are the sample sizes collected and a brief description of the associated length frequency data. All *Trachurus* samples were taken with a demersal trawl. *Trachurus* samples from Namibiam, South African and United Kingdom waters were used as outgroups in this study (data for the outgroup samples are available from A. Lovell).

Sepia individuals were grouped into three main samples designated as GTB (for Ghana, Togo and Benin, i.e. samples from the east of Cape Three Points), CdI (for Cote d'Ivoire, i.e. samples collected between Cape Three Points to the east and Cape Palmas to the west) and GUI (for Guinea, i.e. samples collected by E. Morize off the coast of Guinea). A *Sepia officinalis* data set from two samples collected from the Atlantic and Mediterranean coasts of Spain, analysed using the same microsatellite loci, were used as outgroups (as with *Trachurus*, data for the outgroup samples are available from the first author).

Once returned to the laboratory the animal tissue was transferred to fresh 99% high grade ethanol and kept in a 4°C refrigerator. The DNA was then extracted by a standard Proteinase *K* digestion followed by purification through phenol-chloroform extraction and ethanol precipitation overnight at -20°C. After washing in 70% ethanol, resuspension was in $50\mu l$ of TE.

Amplification (Trachurus)

Each individual PCR (polymerase chain reaction) contained $2.5\mu l$ of MgCl (25mM), $5\mu l$ of 10x Taq buffer (supplied by GIBCOBRL, the manufacturers of the Taq polymerase), $1\mu l$ of each primer (Figure 16-1), 1μl of resuspended DNA as template, bought up to $50\mu l$ with ultra pure H_2O. The reactions were overlaid with $45-50\mu l$ of mineral oil to prevent evaporation during thermal cycling. PCR conditions consisted of an initial step of 94°C for 5 minutes, followed by a hold at 82°C. At this point the $2.5\mu l$ of Taq polymerase was added. Once added, the thermocycler continued for 35 cycles of denaturation (94°C, 1 min), annealing (50°C, 1min) and extension (72°C, 1min). The final stage consisted of 10 minutes at 72°C, ensuring that all annealed template was fully polymerised.

```
cytb h (CB2-3'):   CCC CTC AGA ATG ATA TTT GTC CTC A

cytb l (CB1-5'):   CCA TCC AAC ATC TCA GCA TGA TGA AA
```

Figure 16-1 Primers used for sequencing analysis of a section of the cytochrome *b* gene in *Trachurus* spp. The 3' and 5' notation indicate the direction of the primer along the mtDNA strand.

Sof 1, 2, 7   Annealing Temperature 61-51°C   x30 cycles

Sof 3, 4, 5, 6   Annealing Temperature 65-55°C   x30 cycles

Table 16-2 PCR conditions for amplification of the seven variable loci used for screening of *Sepia officinalis* samples. Annealing temperature starts at the lower temperature and steps up through the sequence during the reaction.

Sequencing

$4\mu l$ of the amplified PCR product was run on a 0.8% agarose gel. Following a successful result the remaining PCR product, approximately $46\mu l$, was cleaned using Microcon centrifugal cleaners, following the manufacturers instructions. Approximately 2ng of the purified DNA was used in a sequencing reaction with 3ppmol of primer. The sequencing product was loaded onto a Perkin Elmer ABI Prism 377 DNA sequencer (or an Applied Biosystems 373A DNA sequencer) and run overnight.

Sequence analysis

Chromatogram files for every nucleotide sequence were viewed using the Chromas software, and sequences were aligned using ClustalW (Thompson *et al.* 1994). DnaSP (DNA Sequence Polymorphism; Rozas and Rozas 1999) and Arlequin software (Schneider *et al.* 1997) were used to produce statistics on the level of polymorphism in and among the sequences (e.g. nucleotide diversity, transition to transversion ratio and estimates of theta, θ, the product of the effective population size and mutation rate). Genetic distances between the sampled stations wcre cstimatcd by thc Analysis of MOlecular VAriance (AMOVA) approach as described by Excoffier *et al.* (1992). Such an approach is essentially similar to other approaches based on analysis of variance of gene frequencies, but it takes into account the number of mutations between molecular haplotypes. Pairwise fixation indices (F_{ST}) were also calculated among samples, with a slight transformation to linearize the distance with population divergence time (Reynolds *et al.* 1983). Both analyses were performed with the Arlequin software package.

Amplification (Sepia)

Individuals were screened for variation at a total of seven variable loci (Sof1 to Sof7) previously isolated and characterised for *Sepia officinalis* by Shaw & Perez-Losada (2000). PCR conditions, i.e. annealing temperatures and number of cycles, are given in Table 16-2. The reaction mixes contained 20ng of template DNA, 1.5-2.5mM MgCl, 0.2 mM of each nucleotide, 0.2μM of each primer (forward primer 5' end-labelled with a Cy5 fluorescent dye group), 0.2 U Taq polymerase (Bioline UK) with the manufacturers supplied 1x NH_4 buffer, in a final reaction volume of 10 μl.

Microsatellite screening

Amplified products were resolved on a 6% denaturing polyacrylamide gel run on an ALFexpress TM (Pharmacia Biotech) automated sequencer. Product sizes were determined by comparison to an internal series of standard size markers using FRAGMENT MANAGER v1.2 software (Pharmacia Biotech).

Microsatellite analysis

Genotypes at all pairs of loci were tested for genotypic disequilibrium and deviations from Hardy Weinberg using exact tests with significance determined by a Markov chain method. Levels of non-random association of alleles within samples were also estimated for each locus by calculating F_{IS}, (Weir & Cockerham 1984) with the Genepop package (GENEPOP v3.1d, Raymond and Rousset 1995). F_{IS}, the fixation index, is a general measure of inbreeding and quantifies the level of reduction in heterozygosity expected with population subdivision. Differentiation between samples can be estimated using one of two alternative models of microsatellite evolution, the infinite alleles model (IAM, Kimura and Crow 1964) or a stepwise mutation model (SMM, Kimura and Ohta 1978). Given that *Sof 3* displays two parallel sets of allelic arrays differing by one base pair it is likely that a non-SMM would better reflect the data. Therefore non-SMM based analyses (F_{ST}) were used in addition to analysis of molecular variance (AMOVA), though future analyses will compare non-SMM and SMM methods of analysis.

Results

Trachurus

Molecular nature of variation

The 156 copies of the 211 base pair sequence in the Gulf of Guinea individuals sampled contained 40 polymorphic sites and a total of 62 haplotypes. Transitions outweighed transversions by about 2.5 to 1. Table 16-3 lists the sample sizes analysed from the stations used during sampling, and for each population gives the number of polymorphic sites, the nucleotide diversity and an estimation of theta. Samples *tt44-tt104* are Gulf of Guinea stations, *ttnam2-ttnam5* are from the Benguela LME (mostly from Namibian waters), and *ttplym* is from the English Channel off Plymouth on the southwest coast of the UK. Samples *gog all* and *nam all* give estimations of nucleotide diversity and theta for all populations of the Gulf of Guinea and Benguela samples respectively, while *tt all* gives estimates using all sampled *Trachurus* individuals. Note that samples *tt99* and *tt104* have each been split into two samples based on the bimodal length frequency distribution. Samples *tt99s* and *tt104s*

sample location	sample size	number of polymorphic sites	nucleotide diversity	theta (θ)
tt44	21	9	0.005	1.086
tt51	40	18	0.009	1.855
tt70	25	11	0.008	1.610
tt99s	24	14	0.010	2.083
tt99b	17	11	0.011	2.250
tt104s	14	5	0.003	0.714
tt104b	15	7	0.007	1.505
gog all	156	40	0.008	1.741
ttnam2	20	7	0.003	0.700
ttnam3	18	10	0.009	1.869
ttnam4	13	3	0.003	0.718
ttnam5	20	3	0.001	0.300
nam all	71	16	0.004	0.907
ttplym	17	10	0.011	2.333
tt all	244	47	0.013	2.832

Table 16-3 Sample location, sample size, the number of polymorphic sites, nucleotide diversity and theta calculated for all sampled locations of *Trachurus* spp. "gog" stands for Gulf of Guinea while "nam" for Benguela samples (mostly collected from Namibia). Gulf of Guinea samples generally show greater diversity, and higher values of theta, than Benguela samples.

consist of smaller individuals (*s* denotes small) from the sample stations while *tt99b* and *tt104b* consist of the bigger individuals (*b* denotes big).

Nucleotide diversity varied considerably between localities, though a trend of higher diversity in the Gulf of Guinea LME over the Benguela LME is apparent. The highest diversity was found in the outgroup sample from off the UK coast and the sample of larger fish from station *tt99* (nucleotide diversity 0.11). The estimate of θ (the product of the mutation rate and the effective population size) also showed a trend of greater values in the Gulf of Guinea, though there was considerable variation between localities within LMEs.

Patterns of variation across and within LMEs using AMOVA

Levels of genetic divergence between samples tested with AMOVA are given in Table 16-4. Between The Gulf of Guinea LME, the Benguela LME, and the samples from the English Channel, part of the North Sea LME (McGlade 2002) over 62% of the variation is explained due to differences among the LMEs, and only 1.2% due to differences among populations within the LMEs. The remaining 36% of the variation is due to genetic variation within the individual populations. The *P* value gives the probability of such a pattern of variance happening by chance alone.

Source of variation	d.f.	Sum of squares	Variance components	Percentage of variation
Among LMEs	2	158.52	1.29	62.48
Among populations within LMEs	7	9.62	0.02	1.21
Within populations	233	174.52	0.75	36.31
Total	242	342.65	2.06	
Fixation index	F_{ST}:		**0.637**	**P=0.00**

Table 16-4 Patterns of variation of a 211-bp fragment of the mitochondrial cytcohrome *b* gene of *Trachurus* spp. from the Gulf of Guinea and Benguela LMEs, and an outgroup from the UK (designated for this test as a third LME with a single population; McGlade 2002). Gulf of Guinea populations are only split spatially (see text). Over 62% of the total variation can be explained by variance between LMEs, indicating a deep evolutionary divergence.

Source of variation	d.f.	Sum of squares	Variance components	Percentage of variation
Among populations	6	11.411	0.049	5.61
Within populations	149	123.487	0.829	94.39
Total	155	134.897	0.878	
Fixation index	F_{ST}:		**0.056**	**P=0.00**

Table 16-5 Patterns of variation of a 211-bp fragment of the mitochondrial cytcohrome *b* gene *of T. trecae* within the Gulf of Guinea LME. Populations are split both spatially and temporally. 5.6% of the variation is explained due to variance between population's, indicating moderate but significant (*P*=0.00) population subdivision.

With the sampled localities from the Gulf of Guinea then split into age classes according to their bimodal length frequency distribution (i.e. sample *tt99* into *tt99s* and *tt99b*, and sample *tt104* into *tt104s* and *tt104b*), all samples from the Gulf of Guinea were then tested again with AMOVA. The results, as displayed in Table 16-5, show that just over 94.4% of the variation was found within the sampled populations, leaving 5.6% of the variation between the spatially and temporally split populations. The F_{ST} value of 0.056 reflects a degree of moderate genetic differentiation (Wright 1978), and as such indicates a high probability of non-random mating (*P*=0.00). Thus populations split according to age class accounts for more variation than spatial patterning alone.

Pairwise comparisons

Pairwise comparisons across all samples (*tt99* and *tt104* not split into age classes) as shown in Table 16-6 indicate very deep divisions between the Gulf of Guinea and the outgroups from the Benguela LME and the English Channel. The figures above the diagonal give the actual F_{ST} pairwise comparisons, while the asterisks below the diagonal give the level of significant difference between pairs of samples. It is notable that the pairwise differences between the

	tt44	tt51	tt70	tt99	tt104	ttnam2	ttnam3	ttnam4	ttnam5	ttplym
tt44		0.010	0.104	0.014	0.021	0.759	0.629	0.736	0.807	0.649
tt51	-		0.058	0.021	0.006	0.663	0.576	0.632	0.692	0.598
tt70	***	**		0.038	0.125	0.713	0.603	0.683	0.753	0.587
tt99	-	*	*		0.048	0.625	0.543	0.587	0.653	0.558
tt104	-	-	***	***		0.737	0.625	0.713	0.777	0.649
ttnam2	***	***	***	***	***		0.027	0.015	0.000	0.218
ttnam3	***	***	***	***	***	-		0.019	0.051	0.137
ttnam4	***	***	***	***	***	-	-		0.064	0.226
ttnam5	***	***	***	***	***	-	**	-		0.274
ttplym	***	***	***	***	***	***	***	***	***	

Table 16-6 Matrix of pairwise F_{ST} values (above diagonal) and associated P values (below diagonal) for *Trachurus* spp. samples from The Gulf of Guinea, Benguela LME and English Channel. Sample codes are indicated in Table 16-3. *=$P<0.05$; **=$P<0.01$; ***=$P<0.001$.

	tt44	tt51	tt70	tt99s	tt99b	tt104s	tt104b
tt44		0.017	0.104	0.011	0.089	0.018	0.067
tt51	-		0.058	0.009	0.080	0.011	0.019
tt70	***	**		0.091	0.015	0.122	0.138
tt99s		-	***		0.061	0.038	0.035
tt99b	**	***	-	**		0.149	0.121
tt104s		-	**	*	***		0.091
tt104b	*	-	***	-	***	**	

Table 16-7 Matrix of Pairwise F_{ST} values (above diagonal) and associated P values (below diagonal) for *T.trecae* samples from The Gulf of Guinea. Samples from station 99 and 104 have been split into 'big' and 'small', following their binomial distribution. Sample codes are indicated in Table 16-3. *=$P<0.05$; **=$P<0.01$; ***=$P<0.001$.

two outgroup regions, i.e. the Benguela LME samples and those from the English Channel, are much less than between either outgroup and the Gulf of Guinea, even though the two outgroups are twice as far apart physically from one another than from the Gulf of Guinea.

Pairwise comparisons between the Gulf of Guinea samples based on both spatial and temporal groups are given in Table 16-7. As in Table 16-6, the figures above the diagonal give the actual F_{ST} pairwise comparisons, while the asterisks below the diagonal give the level of significant difference between pairs of samples. Perhaps surprisingly, highly significant differences are found between different length classes from the same trawl (i.e. for stations 99 and 104). The most highly significant differences are found to involve samples *tt99b* and *tt70*, which themselves show no significant pairwise difference from one another.

Figure 16-2 shows a UPGMA tree using the pairwise F_{ST} samples from Table 16-7, with a slight transformation to linearize the distance to divergence time (Reynolds *et al.* 1983). The younger fish form one cluster of samples (*tt44, tt51, tt99s* and *tt104s*), while those from an older age class form a second cluster (particularly *tt99b* and *tt70*), though less distinct. Therefore samples from a similar age class show greater similarity, regardless of how far away from one another they were caught Thus there is no evidence of clustering due to geographic closeness, only evidence of temporal structure.

Sepia

Molecular nature of variation

The Gulf of Guinea sample locations are as introduced in the methods section: GTB - samples from Ghana, Togo and Benin, i.e. those east of Cape Three Points, CdI - samples

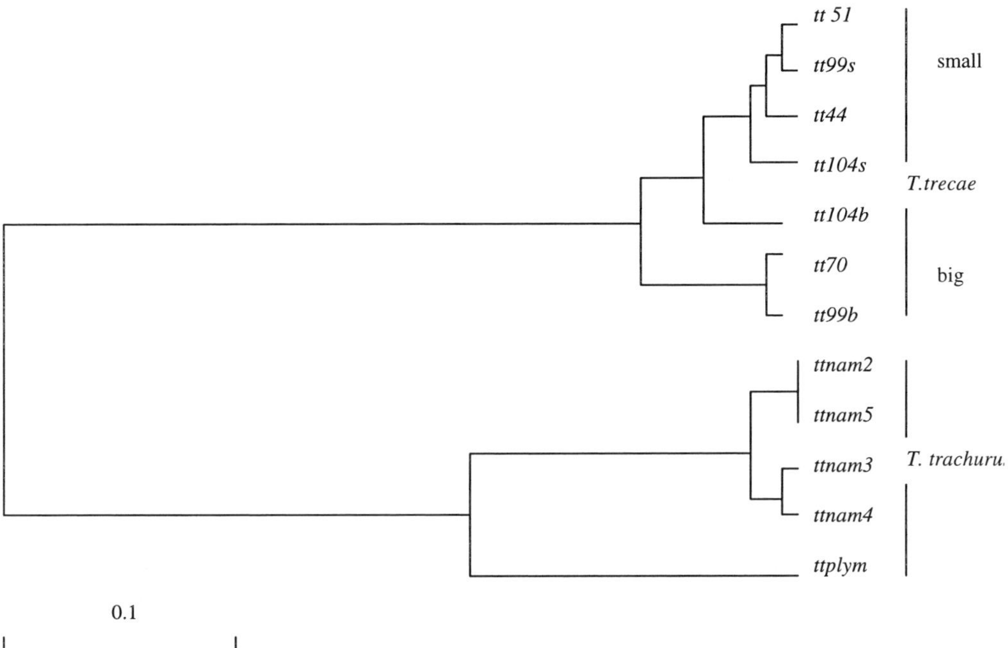

Figure 16-2 UPGMA plot of the F_{ST} pairwise comparisons displayed in Tables 16-6 and 16-7. The two distinct clusters of *T.trachurus* and *T. trecae* are marked. Within the *T.trecae* cluster of the Gulf of Guinea there is a further noticeable clustering of samples of small indivudals (*tt44, 51, 99s* and *104s*) and a second, though less distinct, cluster of samples of large indivudals (*tt70, 99b* and *104b*).

from Côte d'Ivoire, i.e. those between Cape Palmas and Cape Three points, and GUI - samples from Guinea. Samples RIV and AL are samples from Perez-Losada *et al.* (2000) collected from around the Iberian Peninsula, and act as outgroups to the Gulf of Guinea *Sepia* studied in this project. RIV is from the Atlantic coast of Spain, while AL is from the Mediterranean coast. The remaining sample, categorised as "odd", contains four Gulf of Guinea individuals that had allele sizes and frequencies out of the range of the rest of the Gulf of Guinea samples and so were placed in their own group.

Table 16-8 displays information for all loci across all samples: allele sizes, number of alleles (*Na*), estimation of F_{IS} and the probability of the locus meeting Hardy Weinberg expectations (*P*).

Differences in product sizes were consistent with all alleles at all loci representing a simple 3bp repeat motif, except for locus *Sof3*. As described by Shaw and Perez-Losada (2000) and Perez-Losada *et al.* (2000) *Sof3* displays a two parallel sets of allelic arrays differing by 1 bp: a "long" series and a "short" series of alleles. For ease these "long" and "short" series were combined. Exact tests for genotypic disequilibrium between pairs of loci within samples resulted in 4 significant values (*P*<0.05), after the Bonferroni sequential test, out of a table of 45 comparisons. The significant values were found for loci *Sof1-Sof3*, *Sof1-Sof4*, *Sof3-Sof6* and *Sof1-Sof6*, all in the CdI sample set. There was little variation in average gene diversity between samples.

Patterns of variation across and within LMEs using AMOVA

The among sample component of genetic variation, estimated by F_{ST} indicated no significant genetic variation within the Gulf of Guinea data set, but showed very significant variation when the Iberian samples were included (F_{ST} = 0.3). Table 16-9 gives the results from an AMOVA tested on *Sepia* populations from the Iberian peninsula and the Gulf of Guinea. As with *Trachurus* spp. there is very little population differentiation evident within the Gulf of Guinea samples, but highly significant divisions between West African and European populations.

Pairwise comparisons

Pairwise estimates for all populations are given in Table 16-10, and the corresponding UPGMA tree in Figure 16-3. Deep evolutionary divergence can be clearly seen between the west African and Iberian populations. Within the two clusters, further divergence can be seen between the Iberian populations, as described in detail in Perez-Losada *et al.* (2000), but no divergence exists between the Gulf of Guinea populations. As with *Trachurus trecae*, *Sepia officinalis* displays no spatial population structure within the Gulf of Guinea. Interestingly, however, individuals classified as "odd" from west Africa showed greater similarity to Iberian samples than to west African samples, and clearly cluster with the Iberian populations as shown in Figure 16-3. The allele size ranges that were found further illustrated the similarity between west African "odd" individuals and the Iberian samples, most particularly at loci *Sof* 2 and *Sof* 5, where the dominant number of west African samples have far smaller allele sizes (Figure 16-4).

Locus	GTB	CdI	Samples GUI	Odd	Mean (SD)
Sof1 (ATT)					
Allele size	211-271	217-271	217-253	253-253	211-274
Na	20	17	9	1	8.54
F_{IS}	0.018	0.030	0.200	-	0.100
P	0.290	0.060	0.227	-	-
Sof2 (AAT)					
Allele size	130-139	130-139	130-136	157-187	130-187
Na	4	3	2	6	1.71
F_{IS}	0.123	0.085	-0.110	-0.143	0.130
P	0.513	0.810	1.000	1.000	-
Sof3 (AAT)					
Allele size	169-220	175-214	190-206		169-220
Na	17	14	5	0	7.87
F_{IS}	0.351	0.506	0.434		0.080
P	0.000***	0.000***	0.153		-
Sof4 (ATT)					
Allele size	99-153	102-156	102-159		99-159
Na	18	14	8	0	7.83
F_{IS}	0.482	0.347	0.492		0.080
P	0.000***	0.000***	0.007**		-
Sof5 (ATT)					
Allele size	111-114	111-120	111-114	114-198	111-120
Na	2	3	2	7	2.65
F_{IS}	0.217	0.149	0.407	-0.043	0.190
P	0.060	0.569	0.341	1.000	-
Sof6 (AAT)					
Allele size	227-278	236-278	239-266		227-278
Na	16	13	10	0	6.81
F_{IS}	0.182	0.175	0.023		0.090
P	0.000***	0.004**	0.817		-
Sof7 (ATT)					
Allele size	165-177	171-174	171-174	171-174	165-177
Na	4	2	2	2	0
F_{IS}	-0.09	-	-	-0.5	0.29
P	1	-	-	1	-

Table 16-8 Levels of genetic variation observed at seven microsatellite loci from *S. officinalis* samples taken from the Gulf of Guinea: allele size (in base pairs); number of alleles (*Na*); Weir and Cockerham's (1984) estimate of F_{IS} and the P value of the test using the Guo and Thompson (1992) Markov chain algorithm. The *P* value is associated with the H_o (i.e. Hardy Weinberg equilibrium). Deviations from HW expectations within loci: *=*P*<0.05; **=*P*<0.01; ***=*P*<0.001.

Source of variation	d.f.	Sum of squares	Variance components	Percentage of variation
Among groups	3	210.496	0.623	30.860
Among populations within the Gulf of Guinea	2	0.880	-0.015	-0.720
Within populations	524	738.810	1.410	69.870
Total	529	950.187	2.018	
Fixation index	F_{ST}:		**0.301**	**P=0.00**

Table16-9 Patterns of variation over all microsatellite loci of *Sepia officinalis* from the Gulf of Guinea, GTB, CdI and GUI (one group), and the Iberian peninsula samples (both Iberian populations assigned as a separate group). The F_{ST} of 0.3 indicates very great genetic differentiation amongst all groups. The negative value (effectively zero) for percentage of variation among populations within groups displays the lack of differentiation within the Gulf of Guinea.

	GTB	CdI	GUI	odd	RIV	AL
GTB		-0.002	0.002	0.444	0.351	0.413
CdI	-0.002		-0.007	0.451	0.343	0.413
GUI	-0.008	-0.015		0.335	0.296	0.405
odd	0.444	0.451	0.335		0.110	0.286
RIV	0.248	0.249	0.223	0.110		0.279
AL	0.255	0.250	0.249	0.286	0.179	

Table 16-10 Matrix of Pairwise F_{ST} values for all populations of *Sepia officinalis* (Gulf of Guinea samples and Iberian peninsula samples). Below diagonal are pairwise estimates using all loci, above diagonal shows pairwise estimates using only loci *Sof* 1, 2, 5 and 7, leaving out *Sof* 3, 4 and 6 which showed deviations from H-W expectations.

Given that microsatellite mutations are thought to accumulate in length over time, the smaller allele sizes of the Gulf of Guinea samples, particularly at loci *Sof* 2, *Sof* 4 and *Sof* 5, possibly indicate that the west African populations are ancestral to the Iberian samples.

Discussion and Conclusions

Fishery scientists have known for some time that many marine species undergo spawning migrations and that the young develop in nursery grounds before they join the main adult stock (Laegdsgaard and Johnson 1995; Harden Jones 1966). Thus the next stage of understanding population structure relates to the question of migration at the level of the individual. Comprehension of such dynamics would significantly aid fishery management and help researchers understand the response of populations to environmental degradation. Questions of population structuring therefore are of key interest if management of exploited

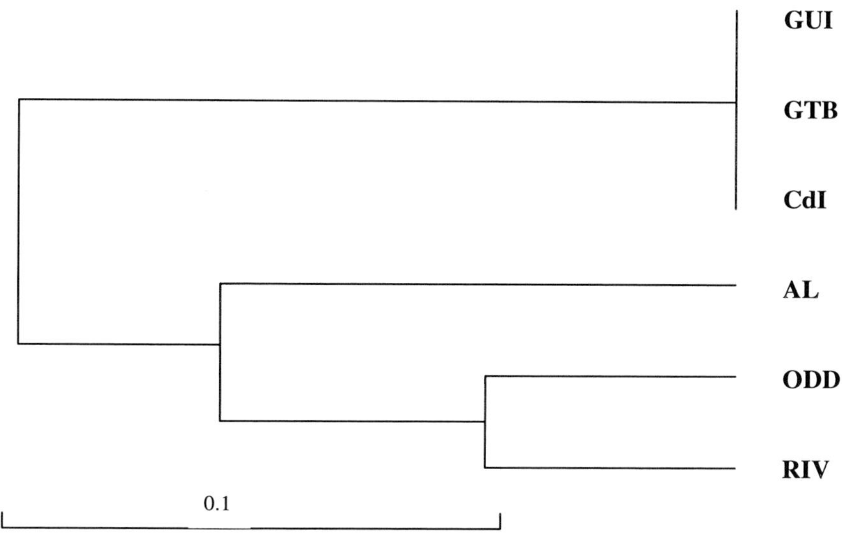

Figure 16-3 UPGMA tree of *Sepia* spp. F_{ST} pairwise comparisons from Table 16-10. Divergence is clearly shown between west African and Iberian populations. Within the two clusters, further divergence is seen between the Iberian populations, but none is found between the Gulf of Guinea populations. Individuals classified as "odd" from West Africa show greater similarity to the Iberian samples than to other west African samples.

marine species is to be a success. The initial results presented here are a first step in understanding the population dynamics of two commercially important species in the Gulf of Guinea.

The two species discussed in this paper, *Trachurus trecae* and *Sepia officinalis*, display no spatial structuring of populations within the Gulf of Guinea at the level sampled for this study. However the subsystems within the Gulf of Guinea were not all covered by the sampling process. Most of the samples were from the Central upwelling system, with only a few *Sepia* samples from the Western subsystem and none from the Eastern subsystem, though the few *Sepia* samples from the Western subsystem that were analysed showed no sign of significant differentiation. It would be interesting to take samples from the Western and Eastern extremes of the Gulf of Guinea LME and examine their relationships with neighbouring LME populations, to determine whether there is any evidence of introgression or a sharp genetic discontinuity.

Between LMEs there are highly significant population divisions, representing deep divergence. The *Trachurus* samples from the outgroups used in the project (samples from Namibian, South African and UK waters) are far more closely genetically related to one another than to any of the Gulf of Guinea samples, even though they are over twice as far

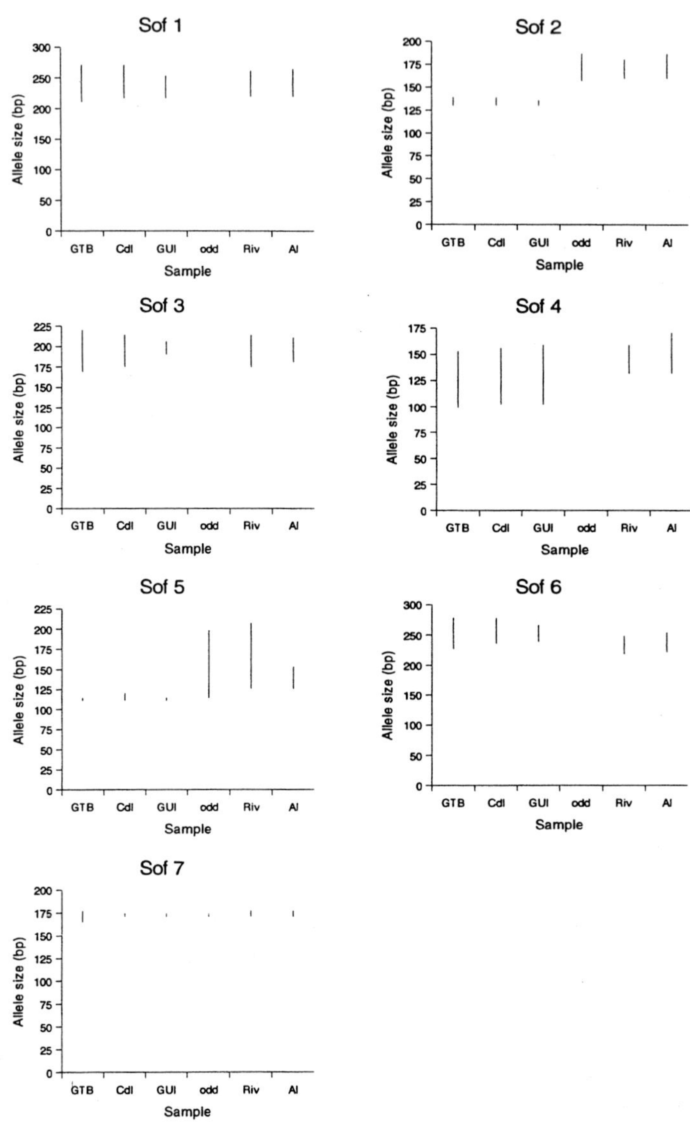

Figure 16-4 Allele sizes found in the analysis of *Sepia* spp.. Microsatellite loci *Sof* 1-7, from the West African and Iberian samples. Alleles sizes are given in base pair length of the PCR product.

apart than from the Gulf of Guinea samples. Similarly for the *Sepia* samples, while showing a noticeable lack of differentiation within the Gulf of Guinea, they were very significantly different from Atlantic and Mediterranean *Sepia* samples. The significant genetic differences found may be indicative of divergence to a species/subspecies level, giving support to the naming of *T. trecae* and possibly *S. hierredda*.

While there is no evidence for spatial structuring within the Gulf of Guinea there is some indication of temporal structuring. The *Trachurus* samples display far stronger genetic relatedness among individuals of a certain age, regardless of the geographic distance between the samples. Samples with a bimodal length frequency distribution showed significant genetic differences between age classes, even though they were caught in the same trawl. An initial analysis indicates similar temporal structuring in *Sepia*, with more variation being explained by mantle length than by geographic location. The causes of such temporal structure are currently unknown, though there are many possible explanations such as natal fidelity (Utter *et al.* 1980; Cury 1994; Iles and Sinclair, 1982), seasonal reproduction, sweepstakes effect (Hedgecock 1994), variable selection differentials etc. Further analyses will attempt to resolve such issues and provide suggestions for further work on the theme of temporal structuring of populations.

The preliminary results therefore show high gene flow throughout the Gulf of Guinea, with deep evolutionary divergence between west African and southern European waters and, in the case of *Trachurus*, Southern African waters. These results are in agreement with the findings of Lounes Chikhi and others (Chikhi 1995; Chikhi *et al.* 1998) from their studies of *Sardinella* spp. between west Africa, the Mediterranean and South America where they too found evidence of some structuring amongst age classes. Given that such results have been found in three species with very different life histories, it is likely that population partitioning is due to external (e.g. environmental) characteristics, rather than individual behaviour, and highlights the need for management at a regional level for not only pelagic, but also demersal resources. Further analyses are underway in order to resolving issues on a finer temporal and spatial scale, as well as introducing the results of studies on a third species, *Pagellus bellottii.*

Acknowledgements

The authors would like to thank the following people and institutions for their invaluable help in collecting samples: Dr Kwame Koranteng and the staff of the Fishery Division, Tema, Ghana and Professor Chidi Ibe, Dr Jaques Abe and staff at the LME-GOG Project based at the CRO, Abidjan. Secondly, the staff and crew of the RV Dr. Fridtjof Nansen on their Gulf of Guinea survey, 1999, in particular the invaluable help of Yaovi Acakpo-Addraof and Joseph Teye, and of course the indefatigable Oddgeir Alvheim. Ekkehard Klingelhoeffer collected the Namibian samples and Eric Morize the Guinean samples. The primers for the microsatellite loci in *Sepia* were developed by Paul Shaw of the University of Hull. This paper was improved thanks to the comments of an anonymous reviewer. AL would also like to thank Jack Cohen for valuable discussion and comment. J.M.McGlade would like to acknowledge the support provided by NERC (GT5/00/MS/1).

References

Anon., 1981 *FAO species identification sheets: fishing areas 34, 47 (in part) (E.C. Atlantic).* FAO, Rome.

Carvalho, G. R. and L. Hauser 1994 Molecular genetics and the stock concept in fisheries. Reviews in Fish Biology and Fisheries 4: 326-350.

Chikhi, L. 1995 *Differenciation genetique chez Sardinella aurita et S. maderensis, allozymes et ADN mitochondrial.* PhD thesis, L'Universite Pierre et Marie Curie, Paris VI.

Chikhi, L., F. Bonhomme and J.F. Agnese 1998 Low genetic variability in a widely distributed and abundant clupeid species, *Sardinella aurita.* New empirical results and interpretations. Journal of Fish Biology.

Cury, P. 1994 Obstinate nature: an ecology of individuals. Thoughts on reproductive behaviour and biodiversity. Canadian Journal of Fisheries and Aquatic Sciences 51: 1664-1673.

Excoffier, L., Smouse, P. E. and Quattro, J. 1992. Analysis of molecular variance inferred from metric distances among DNA haplotypes: applications to human mitochondrial DNA restriction data. Genetics 131: 479-491.

Guo, S. W. and E.A. Thompson 1992 Performing the exact test of Hardy-Weinberg proportions for multiple alleles. Biometrics 48: 361-372.

Harden Jones, F.R. 1966 *Fish migration.* Edward Arnold, London.

Hecht, T. 1990 On the life history of Cape horse mackerel *Trachurus trachurus capensis* off the south-east coast of South Africa. South African Journal of Marine Science 9: 317-326.

Hedgecock, D. 1994 Does variance in reproductive success limit effective population size of marine organisms? pp122-134 *In* A.R.Beaumont (ed.) *Genetics and Evolution of Aquatic Organisms.* Chapman & Hall, London.

Iles, T. D. and M. Sinclair 1982 Atlantic Herring: stock discreteness and abundance. Science 215: 627-633.

Kimura, M and J.F. Crow 1964 The number of alleles that can be maintained in a finite population. Genetics 49: 725-738.

Kimura, M and T. Ohta 1978 Stepwise mutation model and distribution of allelic frequencies in a finite population. Proceedings of the National Academy of Sciences, USA 75: 2868-2872.

Laegdsgaard, P. and C.R. Johnson 1995 Mangrove habitats as nurseries: unique assemblages of juvenile fish in subtropical mangroves in eastern Australia. Marine Ecology Progress Series 126: 67-81.

McGlade, J.M. 2002 The North Sea Large Marine Ecosystem. *In*: K.Sherman *Large Marine Ecosystems of the North Atlantic.* Elsevier Press, Amsterdam

O'Connell, M. and J.M. Wright 1997 Microsatellite DNA in fishes. Reviews in Fish Biology and Fisheries 7: 331-363.

Perez-Losada, M., P.W. Shaw, A. Guerra, G.R. Carvalho and A. Sanjuan 2002 Fine population structuring in the cuttlefish *Sepia officinalis* (Mollusca: Cephalopoda) around the Iberian Peninsula revealed by microsatellie DNA markers. (In prep.).

Raymond, M. and F. Rousset 1995 GENEPOP (version 1.2): population genetics software for exact tests and ecumenicism. Journal of Heredity 86: 248-249.

Reynolds, J., B.S. Weir and C.C. Cockerham 1983 Estimation for the coancestry coefficient: basis for a short term genetic distance. Genetics 105: 767-779.

Rozas, J. and R. Rozas 1999 DnaSP version 3: an integrated program for molecular population genetics and molecular evolution analysis. Bioinformatics 15: 174-175.

Schneider, S., J-M. Kueffer, D. Roessli and L. Excoffier 1997 *Arlequin ver 1.1: A software for population genetic analysis.* Genetics and Biometry Laboratory, University of Geneva, Switzerland.

Shaw, P. W. and M. Perez-Losada 2000 Polymorphic microsatellites in the common cuttlefish, Sepia officinalis (Cephalopoda). Molecular Ecology 9: 237-238

Thompson, J.D., D.G. Higgins and T.J. Gibson 1994 CLUSTAL W: improving the sensitivity of progressive multiple sequence alignment through sequence weighting, positions-specific gap penalties and weight matrix choice. Nucleic Acids Research 22: 4673-4680.

Utter, F.M., D. Campton, S. Grant, G.B. Milner, J. Seeb and L.N. Wishard 1980 Population structures of indigenous salmonid species of the Pacific Northwest. pp285-304. *In*: W.J. McNeil and D.C. Hinsworth (eds). *Salmonid ecosystems of the Pacific Northwest.* Oregon State University Press, Corvallis.

Weir, B. S. and C.C. Cockerham 1984 Estimating F-statistics for the analysis of population structure. Evolution 38: 1358-1370.

Whitehead, P. J. P., M.L. Bauchot, J.C. Hureau, J. Nielsen and E. Tortonese (eds) 1986 *Fishes of the North-eastern Atlantic and the Mediterranean.* UNESCO, Paris.

Wright, S. 1978. *Evolution and the Genetics of Populations. Volume 4. Variability within and among natural populations.* University of Chicago Press, Chicago. 578p.

Fisheries

17

An Overview of the Fishery Resources and Fishery Research in the Gulf of Guinea

M. A. Mensah and S.N.K. Quaatey

Abstract

The countries of the region have been collaborating in fishery research activities mainly through the activities and support of international organizations such as the Fishery Committee for the Eastern Central Atlantic (CECAF), the International Commission for the Conservation of Atlantic Tunas (ICCAT), and Institut de Recherche pour le Développement (IRD, ex-ORSTOM) and regional projects such as the Gulf of Guinea Large Marine Ecosystem Project.

As a result of these activities, a number of resource surveys have been conducted in the Gulf of Guinea over the past four decades, the results of which are examined in this paper. The fishery resources are classified broadly as small pelagics, large pelagics, demersal, crustaceans and molluscs and are currently under various degrees of exploitation. While the small pelagic and coastal demersal resources are over-exploited, deep water demersal and large pelagic resources are either under-exploited or are being exploited close to their maximum sustainable yield.

Changes have occurred in the fishery resources over the last two and half decades. The *Sardinella aurita* stock of the western Gulf of Guinea nearly collapsed in early 1970s and an increase in the abundance of *Balistes capriscus* was observed during the same period. *B. carolinensis* which became the most abundant demersal/semi-pelagic species in the 1980s has virtually disappeared from the entire region. Increases in the abundances of small pelagics and cephalopods was observed in the 1990s.

Introduction

The Gulf of Guinea is a Large Marine Ecosystem extending from Cape Roxo (Lat. 12° 30'N) in Guinea Bissau to Cape Lopez (Lat. 0° 41'S) in Gabon (Map 2, colour plates). It is rich in living marine resources and well endowed with commercially valuable fishes. It does not only ensure the livelihood of fishermen but is also a source of foreign exchange for the countries in the region. The fishery resources of the Gulf of Guinea which include the small pelagics (e.g. sardinellas and shad), large pelagics (e.g. tunas and billfishes), crustaceans and molluscs (e.g. shrimps, lobsters and cuttle-fish) and demersal species (e.g. sparids and croakers) are exploited by both artisanal and industrial fleets. It is estimated that on the average over one million metric tons of fish are caught annually from the Gulf of Guinea of which about a third is exported. It must, however, be noted that these same countries also import fish to supplement local production

Several resource surveys have been conducted in the Gulf of Guinea over the last four decades. Van der Knaap (1985) has listed the surveys that were carried out prior to 1984. Surveys carried out in the region after 1984 include R/V Cornide de Saavedra surveys of 1986 and 1987 (Oliver *et al.* 1987a, b), R/V Dr. Fridtjof Nansen survey of 1989 (IMR 1989) and "Guinea 90" survey of 1990 (Martos *et al.* 1991). The main objective of these surveys was to estimate the biomass and potential yield of the fishery resources for the purpose of managing them sustainably. Other objectives included the study of the abundance and distribution of species in relation to environmental conditions and nature of the bottom.

The hydrographic regimes in the Gulf of Guinea have been described by Longhurst (1962). Except for the occurrence of the seasonal coastal upwellings in the western part (ie Cote d'Ivoire to Benin), there is no great variation in the hydrographical conditions in the entire Gulf of Guinea (Longhurst 1962; Williams 1968). The hydrographic regimes in the Gulf of Guinea are the major factors which determine fish stock abundance and distribution in the region (Williams 1968; Fager and Longhurst 1968; Longhurst 1969; Binet 1982; Martos *et al.* 1991; Koranteng *et al.* 1996, Koranteng and McGlade this volume). The abundance and distribution of the small pelagics in the western Gulf of Guinea are controlled mainly by the intensity of the seasonal coastal upwellings that occur in the sub-region (Mensah 1973; Anon. 1976; Houghton and Mensah 1978; Binet 1982; Koranteng *et al.*1996).

Changes in oceanographic conditions and fishing effort would invariably affect the stock size and species composition of fishery resources. Over the last two and half decades, changes have occurred in the fishery resources in the Gulf of Guinea due to a combination of changes in environmental conditions and fishing effort. Near collapse of the *Sardinella aurita* stock in the western Gulf of Guinea was observed in the early 1970's. During the same period an increase in the abundance of *Balistes capriscus* was also observed. *B. capriscus* which was the most abundant demersal species in the 1980's (Stromme 1984; Oliver *et al.* 1987a) has disappeared in the entire region. There is evidence of overexploitation of demersal resources in coastal waters (ie. 0 - 50 m) (Poinsard and Garcia 1984; CECAF 1990, 1994; Koranteng this volume). In recent years, increases in the abundance of *Sepia officinalis* (Martos *et al.* 1991) and small pelagics (IMR 1989) have been observed.

Some fishery resources in the Gulf of Guinea are shared stocks (e.g. small pelegics in the western Gulf of Guinea, tunas in the eastern Atlantic) while the demersal fauna is homogeneous over the entire Gulf of Guinea (Williams 1968; Longhurt 1969; Martos *et al.* 1991). Thus, joint research programmes for formulation of common management strategies for exploitation of the resources have been adopted by the countries in the region. The countries have been collaborating in fishery research activities through bilateral and multilateral agreements and international organizations such as the Fishery Committee for the Eastern Central Atlantic (CECAF), the International Commission for the Conservation of Atlantic Tunas (ICCAT) and Office de la Recherche Scientifique et Technique Outre-Mer (ORSTOM) and regional projects such as the Gulf of Guinea Large Marine Ecosystem Project.

In this paper, we present an overview of the fishery resources and fishery research in the Gulf of Guinea.

Fishery Resources of the Gulf of Guinea

The fishery resources of the Gulf of Guinea have been documented by a number authors (Williams 1968; Longhurst 1969; Troadec and Garcia 1980; Fischer *et al.* 1981; Villegas and Garcia 1983; Stromme 1984; Oliver *et al.* 1987a,b; Martos *et al.* 1991). The resources may be classified as follows: Small pelagics, Large pelagics, Demersal, Crustaceans and Molluscs

Small Pelagics

There is a great wealth and diversity of small pelagics in the Gulf of Guinea which supports both the artisanal and industrial fisheries in the region. The small pelagics are the most abundant group of species in the Gulf of Guinea; they belong to the families of Clupeidae (ie *Sardinella aurita, S. maderensis, Ethmalosa fimbriata, Ilisha africana*), Engraulidae (ie. *Engraulis encrasicolus*) Carangidae (i.e. *Caranx spp., Decapterus spp., Trachurus trecae, T. trachurus*) and Scombridae (i.e. *Scomber japonicus*).

The composition and dominance of the small pelagics vary from one sub-region to the other. The western Gulf of Guinea is dominated by *S. aurita, S. maderensis, E. encrasicolus* and *S. japonicus* with *S. aurita* being the most abundant. These species normally occur in coastal waters and their abundance and distribution are controlled mainly by the seasonal coastal upwellings that occur in the area. In the central Gulf of Guinea (i.e. Nigeria to Gabon), *E. fimbriata, S. maderensis* and *I. africana* dominate this category of fishery resources with *E. fimbriata* the most abundant. *E. fimbriata* normally occurs in "estuarine" waters. In the northern Gulf of Guinea (i.e. Guinea Bissau to Liberia), the resource is dominated by *Caranx spp., Scomberomorous tritor* (also known as *Apolectus immunis*), *E. fimbriata* and *S. maderensis* in coastal and inshore waters while *E. encrasicolus, S. aurita* and *Decapterus* spp. dominate the offshore waters.

The small pelagic resources in the western and central parts of the Gulf of Guinea are exploited by the artisanal and industrial (including semi-industrial) fleets while the artisanal fleet is the main exploiter of the resource in the northern sector.

The abundance of the clupeoids is known to fluctuate greatly on a decade time scale in the region (Longhurst and Pauly 1987). For example the *S. aurita* stock in the western Gulf of Guinea nearly collapsed in 1972 but has, since the late 1980's, increased in abundance. Yearly variation in abundance of these clupeoids appears to be driven at least by environmental variability (*ibid.*).

Large Pelagics

The most abundant and economically important tuna species in the Gulf of Guinea are the skipjack (*Katsuwonus pelamis*), yellow-fin (*Thunnus albacares*) and big-eye (*Thunnus obesus*). These species are part of a larger tuna community that occurs in the whole of the eastern Atlantic. They occur throughout the whole year and are exploited by the tuna bait boats and purse seiners. The Gulf of Guinea is the spawning ground for these tunas and so most of them caught in the region are young fishes, especially yellow-fin and big-eye. They are known to undertake large scale migration over long distances often staying outside the Exclusive Economic Zone (EEZ) of the coastal states (Blackburn and Williams 1975).

Tuna-like species also occur in the Gulf of Guinea these include the little atlantic tuna (*Euthynnus alletteratus*), atlantic bonito (*Sarda sarda*) and the billfishes. The billfishes include the atlantic sailfish (*Istiophorus albicans*), swordfish (*Xiphias gladius*), blue marlin (*Makaira nigricans*) and white marlin (*Tetrapturus albidus*). The atlantic sailfish is the most abundant of the billfishes and has the tendency to occur in high abundance in coastal waters (Longhurst and Pauly 1987).

Demersal Resources

The demersal resources of the Gulf of Guinea have been studied by many authors (Williams 1968; Fager and Longhurst 1968; Longhurst 1969; Troadec and Garcia 1980; Villegas and Garcia, 1983; Longhurst and Pauly 1987; Martos *et al.* 1991; Koranteng 1998, this volume). The resources may be classified into six different communities. Results of resource surveys conducted in the region over the years indicate that same communities are represented over similar bottom types and water depths throughout the entire Gulf of Guinea. Environmental factors are known to determine which fish occur in a particular area in the Gulf of Guinea (Longhurst and Pauly 1987) and these include the amount of organic mud in the bottom deposits, the occurrence of isolated patches of rocky bottom, the occurrence of estuarine conditions associated with lagoons and rivers and the nature of oceanic water masses lying over the continental shelf.

The species composition and distribution of the six communities are as follows:
- Sciaenid Community

This fish community is represented by species of the families Sciaenidae (*Pseudotolithus senegalensis, P. typus*), Ariidae (*Arius heudeloti, A. mercatoris, A. latiscutatus*), Drepanidae (*Drepane africana*), Polynemidae (*Galeoides decadactylus*), Cynoglossidae (*Cynoglossus senegalensis, C. monodi, C. browni*) and Pomadasyidae (*Pomadasys jubelini, P. incisus, Brachydeuterus auritus*).This community inhabits the soft, sandy and muddy bottoms at depths of between 15 and 50 m. *B. auritus* and the Pseudotolithus species dominate this community.

- Lutjanid Community

This community consists mainly of species of the families Lutjanidae (*Lutjanus fulgens, L. goreensis, L. agennes, Apsilus fuscus*), Acanthuridae (*Acanthurus monroviae*) and Chaetodontidae (butterfishes). These species occupy rocky and fossil bottoms at depths ranging between 15 and 40 m. The lutjanids are the most commercially valuable species of the community.

- Coastal Sparid Community

This community comprises, among others, *Pagellus bellottii, Pagrus caeruleostictus* and *Dactylopterus volitans*. They inhabit hard or sandy substrates between 15 and 70 m depths of water.

- Deep Water Sparid Community

This is the dominant community below the thermocline on soft bottoms and in waters between 40 and 200 m deep. The species composition includes the Sparidae (*Dentex angolensis, D. congolensis* and *P. bellottii*), Triglidae (*Lepidotrigla cadmani*),

Uranoscopidae, Brotulidae (*Brotula barbata*), the deep water sciaenid, *Pentheroscion mbizi*, *Epinephelus aenus*, *Raja miraletus* and *Pseudupeneus prayensis*.

- Deep Shelf Community

These are species that occur on the shelf and at depths of between 200 and 300 m. They include *Peristedion cataphractum, Antigonia capros, Zenopsis spp. Chloropthalmus spp.* and *Bembrops heterurus*.

- Continental Slope Community

These are species that inhabit waters deeper than 400 m and include *Trigla lyra, Merluccius cadeneti, Halosaurus spp.*, and others.

- Crustacean and Mollusc Resources

The shrimps and lobsters are the most important and valuable crustacean species in the Gulf of Guinea. The shrimp resource is the most abundant and is dominated by four species namely, the pink shrimp (*Penaeus notialis*), the tiger shrimp (*Penaeus kerathurus*), the rose shrimp (*Parapenaeus longistrostris*) and the Guinea or white shrimp (*Parapeneopsis atlantica*) (Garcia and Lhomme 1980). The resources are localized (Garcia and Lhomme 1980). *P. atlantica* is the most abundant species in the estuaries and coastal waters and is exploited mainly by the artisanal fleet. *P. notialis* dominates the inshore waters of between 20 and 60 m and is targeted by the commercial shrimpers for export. The species is caught mixed with *P. kerathurus* in the same depth range. *P. longistrostris* occurs in deep waters of between 70 and 300 m (Martos *et al.* 1991).

Two species of lobsters occur in the Gulf of Guinea namely, the royal spiny lobster (*Panulirus regius*) and the red slipper lobster (*Scyllarides herklotsii*) (Schneider 1990). *P. regius* which is the more abundant and valuable of the two occurs in shallow waters ranging between 5 and 40 m and on rocky and sandy bottoms. *S. herklotsii* also occurs on rocky and sandy bottoms and in waters between 5 and 7 m deep.

The most abundant species of molluscs in the Gulf of Guinea are the cephalopods (Martos *et al.* 1991). The cephalopods are dominated by the Sepiidae family with Sepia officinalis hierredda and Sepia bertheloti as the most abundant species. These species occur in waters of between 10 and 100 m deep (Rijavec 1980; Koranteng 1984; Martos *et al.* 1991). While *S. officinalis hierredda* is most abundant in the western Gulf of Guinea, *S. bertheloti* is evenly distributed in the entire area.

Status of the Fishery Resources

The status of the fishery resources of the Gulf of Guinea has been reviewed by many authors (Gulland 1971; Troadec and Garcia 1980; FAO 1992; Ajayi 1994). The subject has also been one of targets of the CECAF Working Party on Resource Evaluation, the latest published assessment is documented in CECAF (1994). Since the biomass of the various resources and their respective levels of exploitation vary from one sub-region to the other, it is necessary to assess the status of the resources on sub-regional basis.

Small Pelagics

In the northern Gulf of Guinea, biomass estimates for the small pelagic resource ranged between 72,000 and 220,000 mt (CECAF 1994) while landings of the resource were about

30,000 mt. Thus, the resource in the northern Gulf of Guinea is grossly under-exploited. CECAF (1994) assessed the biomass of the small pelagics in the western and central Gulf of Guinea as 329,000 mt. The current level of exploitation in the area is about 257,000 mt annually. Clearly, the small pelagics are being over-exploited in the western and central Gulf of Guinea. The observed recent high catches of the resource (which exceed the estimated potential yield) are due mainly to the intensity of upwellings in the area.

Large Pelagics

ICCAT is the body responsible for managing the tuna resources in the whole of the Atlantic Ocean. The maximum sustainable yield (MSY) of the skipjack tuna, yellow-fin tuna and big-eye tuna in the eastern Atlantic were estimated as 200,000 mt, 149,000 mt and 79,000 mt respectively (ICCAT 1997a,b). Current landings of the three species are 175,000 mt, 149,000 mt and 89,000 mt respectively. It is clear that while the skipjack tuna is under-exploited, that of yellow-fin and big-eye tunas are being heavily exploited in the region.

Demersal Resources

Due to the untrawlable nature of the bottom in water depths of 75 m and deeper (Williams 1968), exploitation of demersal resources is not done in water depths greater than 75 m in the Gulf of Guinea. Thus, demersal stocks in offshore waters (i.e. 75 to 200 m) remain grossly under-exploited (Martos *et al.* 1991) in the whole of the Gulf of Guinea.

The MSY of coastal demersal resources in the western and central Gulf of Guinea ranged between 64,000 and 104,000 mt while the annual landing was about 105,000 mt. (CECAF 1994). In the northern Gulf of Guinea, MSY for the coastal demersal resources ranged between 18,000 and 95,000 mt with annual landings around the MSY (CECAF 1994). Thus, coastal demersal resources in the whole of the Gulf of Guinea are either fully exploited or over-exploited. Poinsard and Garcia reported in 1984 the first sign of over-exploitation of coastal demersal resources in the Gulf of Guinea and this has continued up to date (CECAF 1990, 1994).

Balistes capriscus

The triggerfish, *Balistes capriscus* which in the 1960s formed only 5 % of landings of demersal species in the Gulf of Guinea (Longhurst and Pauly 1987) assumed greater importance in the mid 1970s (Caverivière 1982). By 1978, the species was the most abundant demersal resource in the whole of the Gulf of Guinea, replacing the sciaenid community (Longhurst and Pauly 1987; Koranteng this volume).

Two separate stocks of triggerfish were identified in the Gulf of Guinea (Stromme 1984) with no apparent inter-change between them. These were the western stock with its center off Ghana and the northern stock with its center off Guinea Bissau and Guinea. The biomass of the western and northern stocks were estimated as 500,000 and 1,050,000 mt respectively in 1981 (*ibid.*). The biomass for the western stock was estimated to be about 140,000 mt in 1986 (Oliver *et al.* 1987a), indicating a drastic decline in the resource between 1981 and 1986. Recent surveys by R/V Dr. Fridtjof Nansen (IMR 1989) and M/V Lagoa Pesca (Martos *et al.* 1991) only located traces of the species in the region, indicating its disappearance in the area. Landings of the species in the region during the 1980s were far below its estimated potential yield. Thus, the resource was not over-exploited in the area. Changes in oceanographic parameters, especially physical conditions have been attributed for the disappearance of the species (Koranteng *et al.* 1996).

Crustacean and Mollusc Resources

The MSY of the shrimp resource in the northern part was estimated as between 2,600 mt and 3,500 mt (CECAF 1994). Current landing of the resource in the sub-region is about 2,500 mt annually, indicating that the level of exploitation is close to the MSY. In the western and central parts, the MSY of the resource (mainly *P. notialis*) was estimated to be 4,700 mt (CECAF 1994). Annual landing of the resource is about 11,000 mt in the sub-region. Thus, the resource is being over-exploited in western and central Gulf of Guinea.

CECAF Reports of 1990 & 1994 and the results of "Guinea 90" survey (Martos *et al.* 1991) indicate an increase in the abundance of cephalopods in the region, coinciding with the disappearance of the trigger-fish. Martos *et al.* (*ibid.*) assessed that the cephalopod resource is most abundant in deep waters of about 100 m. This stock, together with other demersal stocks that occur in deep waters, remains underexploited for the reason given above.

Development of Fishery Research

The increasing demand for protein foods in the countries in the region after the second world war necessitated the need to develop fishery research in the region (Williams 1968). Several research activities into fisheries were carried-out during the period 1945 to 1960, but only on national basis and restricted to the inshore areas of the Gulf of Guinea (Williams 1968). International expeditions such as Atlantide, Galathea and Belgian Oceanographic worked in the region and the results of these expeditions have made valuable contributions to the basic knowledge of the fauna in the region (Williams 1968).

The need for co-operation in fishery research among the countries in the region led to the establishment in 1952 of the West African Fisheries Research Institute (WAFRI) at Freetown in Sierra Leone by the British Government to deal with fisheries problems in The Gambia, Sierra Leone, Ghana and Nigeria. After independence, these countries established their own research institutions and stopped funding WAFRI, resulting in its disintegration. In the French speaking countries, the Office de la Recherche Scientifique et Technique Outre - Mer (ORSTOM) was responsible for fishery research before independence of these countries. ORSTOM established research stations at Abidjan in Cote d'Ivoire, Point Noire in the Republic of Congo and Dakar in Senegal. Although these countries established their own research institutions after independence, they still cooperate through these ORSTOM centres. The Commission for Technical Cooperation in Africa (CCTA) was set up in 1950 by the UK, France, Belgium, Portugal, South Africa and the Foundation for Mutual Assistance in Africa (FAMA) to promote co-operative planning in scientific matters in Africa, south of the Sahara. The decision to execute research programmes in the marine fisheries resources and environmental conditions of the Gulf of Guinea was taken during the CCTA symposium on oceanography and sea fisheries in 1957. In 1961, CCTA formulated the GUINEA YEAR project to execute research activities into oceanographic conditions, demersal, tuna and sardine resources in the Gulf of Guinea. The demersal resources component of the project was executed between August, 1963 and June, 1964 under the Guinean Trawling Survey (GTS) programme. The GTS was funded by the USAID and the participating countries. The oceanographic component of the project which was planned by the Bureau of Commercial Fisheries (BFC) of the U.S. Fish and Wildlife Service was executed between February, 1963 and March, 1964 under EQUALANT I, II and III. The programme was funded by the Intergovernmental Oceanographic Commission of UNESCO under the name International Cooperative Investigations of the Tropical Atlantic (ICITA). The experimental fishing of the tuna component of the project was planned and executed by BFC between early 1963 and 1965.

The fourth phase of the GUINEAN YEAR project was the experimental fishing for sardines. In 1962, an FAO Extended Programme for Technical Assistance (EPTA) project was established in Ghana to consider aspects of the fishery and biology of sardines. In 1966, the EPTA project was succeeded by a United Nations Special Fund (UNDP/SF) fishery project whose activities were broadened to include the fishery of the sardines and other marine fishery resources of economic importance, their biology and oceanography. The project ended in 1976 by which time a fully fledged Fishery Research Unit (now known as Marine Fisheries Research Division of the Fisheries Department of Ghana) had been established. The Unit was taken over and has been operated by the Government of Ghana since 1976.

After independence, the governments of the various countries realized the need to understand, assess and manage fishery resources sustainably. This has necessitated the establishment of fishery research institutions and departments in all the countries in the region. These institutions and departments have been executing research activities on national basis and through co-operative programmes (e.g. the Cote d'Ivoire - Ghana joint sardinella project in early 1990s and the current UNIDO/GEF funded Gulf of Guinea Large Marine Ecosystem Project involving Cote d'Ivoire Ghana, Togo, Benin, Nigeria and Cameroon with technical assistance from NOAA).

In the region, CECAF and ICCAT are the only international bodies responsible for managing fisheries resources. CECAF which was established in 1967 is empowered to give advice that would assist member countries in establishing the scientific basis for management, conservation and improvement of marine fishery resources. It has been doing this through its sub-committees (Management of Resources within the Limits of National Jurisdiction and Fisheries Development), its Working Parties (Resources Evaluation and Statistics) and Working Groups (Small Pelagics in the western Gulf of Guinea and Demersal Resources on the continental shelf of Guinea Bissau, Guinea and Sierra Leone). Scientists from the region have participated in various CECAF resources surveys (e.g. R/V Dr. Fridtjof Nansen, R/V Cornide de Saavedra and M/V Lagoa Pesca surveys) and have acquired training from such surveys. Thus, CECAF has been playing a leading role in fishery research in the region.

ICCAT which was established in 1966 is the international body responsible for tuna research and management in the whole of the Atlantic Ocean. Research institutions in the region have collaborated with ICCAT in the execution of research activities on skipjack (ICCAT International Skipjack Year, ISYP 1979-1982), yellow-fin (ICCAT Yellow-fin Year Program, YYP 1986-87) and billfishes (ICCAT Enhanced Research Program for Billfishes, ERPB 1990 to date)

The Ministerial Conference on Fisheries Cooperation Among African States Bordering the Atlantic which was formed in 1989 sought to play an important role in the development, management and regulation of exploitation of fisheries resources in the West African marine waters.

Current Fishery Research Programmes

Fishery research programmes being currently executed in the area are either regional or national in character.

Regional Programmes
- Gulf of Guinea INCO - DC Project

This project which ran from 1996-8 was executed by the Fisheries Department of Ghana, ORSTOM of France and University of Warwick of United Kingdom and was funded mainly by the European Union (EU). Member states in the region also collaborated with project. The main objectives of the project were as follows (Anon. 1997; McGlade *et al.* this volume):

> to assess the impact of upwelling and other forms of environmental factors on marine biodiversity and the dynanics of artisanal and industrial fisheries
>
> to develop and implement an information and analysis system for the sustainable management and governance of fisheries resources (FIAS), thereby applying the knowledge obtained from objective (a) to real world problems.

At end of the project:

> the spatio-temporal scales over which environmental forcing occurs were identified,
>
> the biophysical habitats along the continental shelf of the Gulf of Guinea critical to the marine fisheries and biodiversity were identified,
>
> and an information net work for researchers in European member states and the Gulf of Guinea region had been established.

The project has strengthened the linkage between research institutions and increased the capacity for networking and technology transfer by individual researchers and the centres in the regions. It also offered training to scientists in the region to upgrade their knowledge in fisheries science. One important objective that was not entirely achieved was the implementation of the Fisheries Information and Assessment System with a Gulf of Guinea focus in research institutes and selected centres in the regions. To date, only Cote d'Ivoire and Ghana are actively collaborating, thereby leaving gaps in regional data collection and exchange.

- Gulf of Guinea Large Marine Ecosystem (LME) Project

The long-term aim of the project is to develop an effective regional approach to prevent pollution of the Gulf of Guinea and conserve biodiversity in its LME. The project which started in 1995 hopes to cover institutional strengthening, water quality and ecological monitoring. It was originally for four-year duration but has been extended for a further four years. The project is being funded by UNIDO/GEF and with assistance from NOAA and have six countries (Cote d'Ivoire, Ghana, Togo, Benin, Nigeria and Cameroon) collaborating with it.

The project's immediate objective, among others, is to establish a comprehensive program for monitoring and assessment of the Gulf of Guinea LME. To achieve this objective, fish community surveys, *inter alia*, would be executed within the region. It is envisaged that the results of these surveys would provide information on the current LME structure, function, productivity and health. Apart from the planning session and cruise leaders meetings, which worked out protocol for the execution of the surveys, and a survey of limited coverage (only Cote d'Ivoire and Ghana were covered) in late 1996, comprehensive surveys over time and space have yet to be undertaken on any scale. Under the project, research vessels from Ghana and Nigeria are to be used to conduct these surveys, but their initial unavailability delayed the implementation of the Fish Trawl Module component of the project, and efforts by the project to charter other vessels to execute the surveys were not successful. It must, however, be noted that execution of other components of the project such as the pollution, primary production and coastal management components have been on schedule and their objectives achieved.

National Fishery Resource Programmes

Countries in the region are engaged in various national research activities on their different fishery resources depending on priorities and availability of funds. For example, Ghana is benefiting from a World Bank loan to implement the Fisheries Sub-sector Capacity Building Project. One component of the project is to formulate management plans for rational and sustainable exploitation of the resources. As a result, the research institute is being funded under the project to execute both pelagic and demersal resource surveys, the results of which will provide inputs for the development of the management plans.

In addition, the countries are also carryingout programmes in line with recommendations of CECAF Working Party on Resource Evaluation on specific species or resources. These recommendations are reviewed and updated periodically and documented in CECAF reports.

Conclusions

The species composition of the different fishery resources and fish communities are the same from one sub-region to the other in the Gulf of Guinea and have not changed over time. However, the dominant species for each resource and community do vary from one area to the other and have changed significantly over time. For example, while *S. aurita* dominates the western stock of the small pelagics, it is replaced by *E. fimbriata* in the central part. Also, *B. capriscus* replaced the sciaenids between the mid-1970s and late 1980s as the most dominant demersal species in shallow waters while cephalopods (mainly *S. officinalis*) has replaced the sparids as the dominant species in deeper waters.

It is also evident that the offshore demersal and skipjack tuna resources are grossly under-exploited and could respond favourably to increase in fishing pressure. However, all other fishery resources are either close to full exploitation or fully exploited and a further increase in fishing pressure would lead to the collapse of the fisheries.

There is evidence that all the countries in the region are involved in various fishery research activities at a national level: the type of research activity undertaken depending on priorities and the availability of funds. However, most of the countries in region are not engaged in joint research programmes for rational exploitation of their common resources.

References

Ajayi, T.O. 1994 The status of the marine fisheries resources of the Gulf of Guinea. A paper presented at the tenth session of CECAF Working Party on Resources Evaluation held in Accra, Ghana, 10 - 13 October, 1994. CECAF/RE/94/2 Add. September, 1994: 16p.

Anon., 1976 Rapport du Groupe de Travail sur la sardinella (*Sardinella aurita*) des cotes d'Ivoire – Ghaneennes. 28 Juin - 3 Juillet, 1976, Abidjan. 66p.

Anon., 1997 The Gulf of Guinea INCO - DC project. Impacts of environmental forcing on marine biodiversity and sustainable management of artisanal and industrial fisheries in the Gulf of Guinea. *Proceedings of the first Management Meeting and Working Group on the project.* 8th - 9th January, 1997, University of Warwick, U.K.

Blackburn, M. and F. Williams 1975 Distribution and ecology of skipjack tuna in an offshore area of the eastern tropical Pacific Ocean. Fish. Bull. 73: 382 - 411.

Binet, O. 1982 Influence des variations climatiques sur la pecherie des *Sardinella aurita* Ivorio - ghaneenes: relation recheresse - surpeche. Oceanol. Acta 5(4): 443 -452.

Cavervière, A. 1982 *Les espèces demersales du plateau continental ivorien. biologie et exploitation.* Thèse Doctorat d'État-Sciences, Fac. Scien. Luminy Univ. d'Axi - Marseille II. 159p.

CECAF (Fishery Committee for the Eastern Central Atlantic) 1990 Report of the ninth session of the Working Party on Resources Evaluation, Lagos, Nigeria, 18 - 23 November, 1990. FAO Fish. Rep. No. 454. 79 p.

CECAF 1994 Report of the tenth session of the Working Party on Resources Evaluation, Accra, Ghana, 10 - 13 October, 1994. FAO Fish. Rep. No. 511. 95 p.

Fager, E.W. and A.R. Longhurst 1968 Recurrent group analysis of species assemblage of demersal fish in the Gulf of Guinea. Journal of the Fisheries Research Board of Canada 25(7): 1045 - 1121.

FAO (UN Food and Agriculture Orgainization) 1992 *Review of the State World Fishery Resources part 1.* FAO Fisheries Circular 710 Rev. 8 Part 1. Rome.114 p.

Fischer, W., G. Bianchi and W.B. Scott 1981 *FAO species identification sheets for fishery purpose. Eastern Central Atlantic Fishing Areas 34, 47 (in part), Vol. I – VI.* FAO.

Garcia, S. and F. Lhomme 1980 Pink Shrimp Resources. pp.121-145. *In*: J.P Troadec and S. Garcia (eds) *The Fish Resources of the Eastern Central Atlantic. Part One; The Resources of the Gulf of Guinea from Angola to Mauritania.* FAO Fish. Tech.Rep. 186 (1).

Gulland, J. A. 1971 The fish resources of the Ocean. FAO Fish. Tech. Paper 197. 425 p.

Houghton, R.W. and M. A. Mensah 1978 Physical aspects and biological consequences of Ghanaian coastal upwelling. pp 167 - 180. *In*: R. Boje and M. Tomczak (eds): *Upwelling Ecosystems.*

ICCAT (International Commission for the Conservation of Atlantic Tunas) 1997a Report for biennial period 1996 - 97. Part 1 (1996) Vol. 1. English version. 187p.

ICCAT 1997b Report for biennial period 1996 - 97. Part 1 (1996) Vol. 2. English version. 204p.

IMR 1989 Surveys of the Small Pelagic Fish Resources of Cote d'Ivoire and Ghana, 12 - 20 October, 1989. Institute of Marine Research, Bergen, Norway. 9p.

Koranteng, K.A. 1984 A trawling survey off Ghana. CECAF/TECH/84/63, Dakar, Senegal. CECAF Project (FAO). 72p.

Koranteng, K.A. 1998 *The impacts of environmental forcing on the dynamics of demersal fishery resources of Ghana.* PhD Thesis, University of Warwick. 377 p.

Koranteng, K.A. (this volume, chapter 14) Fish species assemblages on the continental shelf and upper slope off Ghana.

Koranteng, K.A. (this volume, chapter 19) Status of demersal fishery resources on the inner continental shelf off Ghana.

Koranteng, K.A. and J.M. McGlade (this volume, chapter 8) Physico-chemical changes in continental shelf waters of the Gulf of Guinea and possible impacts on resource variability.

Koranteng, K.A., J.M. McGlade and B. Samb 1996 A review of the Canary Current and Guinea Current Large Marine Ecosystems. pp 61 - 83. *In*: ACP - EU Fisheries Research Report No. 2. ISSN 1025.

Longhurst, A.R. 1962 A review of the oceanography of the Gulf of Guinea. Bull. IFAN XXIV: 633 - 663.

Longhurst, A.R. 1969 Species assemblage in tropical demersal fisheries. pp 147-216. *In*: Proceedings of the Symposium on the Oceanography and Fisheries resources of the UNESCO/FAO/OAU tropical Atlantic, Paris, UNESCO.

Longhurst, A.R. and D. Pauly 1987 *Ecology of Tropical Ocean.* Academic Press Inc. Harcourt Brace Jovanovich, Publishers, San Diego. 407p

Martos, A.R., I.S. Yraola, L.F. Peralta and J.F.G. Jiminez 1991 The "Guinea 90" Survey. CECAF/ECAF SERIES 91/52. 295p.

Mensah, M.A. 1973 The marine environment and fish production in Ghanaian waters. *The Ghanaian Fishing Industry.* Fishery Research Unit, Tema, Ghana.

McGlade, J.M., K.A. Koranteng and P.Cury (this volume, chapter 1) The EU/INCO-DC programme on impacts of environmental forcing on marine biodiversity and sustainable management of artisanal and industrial fisheries in the Gulf of Guinea.

Oliver, P. and J. Miquel 1987a Report on the Survey carried out by the R/V Cornide de Saavedra in the Gulf of Guinea in 1986. Instituto Espanol de Oceanografia, Palma de Mallorca.

Oliver, P., J. Miquel and J. Bruno 1987b Preliminary Report of the Survey carried out by the R/V Cornide de Saavedra in Liberia, Cote d'Ivoire, and Togo, May to June, 1987. Instituto Espanol de Oceanografia, Palma de Mallorca.

Poinsard. F. and S. Garcia 1984 Stocks assessment and Fisheries Management in the CECAF Region: A perspective view. CECAF/ECAF SERIES 84/32. 34 p.

Rijavec, L. 1980 A survey of the demersal fish resources of Ghana. CECAF/TECH/80/25, Dakar, Senegal. CECAF Project (FAO). 28p.

Schneider, W. 1990 *Field Guide to the commercial Marine Resources of the Gulf of Guinea.* FAO Species Identification Sheets for Fishery Purposes. 268p.

Stromme, T. 1984 Report on the R/V Dr. Fridtjof Nansen Fish Resource Surveys off West Africa: Morrocco to Ghana and Cape Verde. CECAF/ECAF SERIES 84/29.190p.

Troadec, J.P. and S. Garcia (eds) 1980 *The Fish Resources of the Eastern Central Atlantic. Part One; The Resources of the Gulf of Guinea from Angola to Mauritania.* FAO Fish. Tech.Rep. 186 (1). 166p.

Van der Knaap, M. 1985 Preliminary Annotated Inventory of the Scientific Expeditions and Resource Surveys carried out in the CECAF Area. CECAF/TECH/85/64.145p.

Villegas, L. and S. Garcia 1983 Demersal Fish Assemblages in Liberia, Ghana, Togo, Benin and Cameroon. CECAF/ECAF SERIES 83/26.16p.

Williams, F. 1968 Report on the Guinean Trawling Survey. Organisation of African Unity Scientific and Technical Research Commission 99.

The Gulf of Guinea Large Marine Ecosystem
J.M. McGlade, P. Cury, K.A. Koranteng and N.J. Hardman-Mountford (Editors)

18

Environmental Forcing and Fisheries Resources in Côte d'Ivoire and Ghana: Did Something Happen?

P.Cury and C. Roy

Abstract

The main environmental characteristics off the coasts of Côte d'Ivoire and Ghana are reviewed. Seasonal variability as well as long-term environmental changes observed during the last three decades are presented using a seasonal-trend decomposition procedure. Striking similarities are observed between changes in the Gulf of Guinea and the Pacific. This suggests that the observed changes are not the result of local processes but rather are induced by inter-oceanic teleconnection via the atmosphere. Fisheries productivity in the Gulf of Guinea appears to be quite important but is limited by the duration of the upwelling process. Thus, using comparative methods between several upwelling systems, relationships are identified between pelagic productivity and several environmental factors (wind, turbulence, and upwelling index) and only a weak relationship is observed when considering the demersal fish productivity and the upwelling intensity. Fisheries as a whole off Ghana and Côte d'Ivoire have been expanding during the last four decades: the total marine fish catch off Côte d'Ivoire and Ghana has been increasing from less than 50 thousand tonnes in the 1950s to around 400 thousand tonnes in the mid-1990s. For the pelagic species, there has been a noticeable increasing trend is observed since the 1950s and the total catch reached about 250 thousand tonnes in the 1990s. By contrast, the demersal fisheries in Côte d'Ivoire and Ghana have fluctuated between 30 and 50 thousands tonnes since the mid-1960s without any trend. Fishing effort by purse-seiners and trawlers has decreased significantly since the 1970s in Côte d'Ivoire. The abundance, estimated from cpue of both pelagic and demersal, does not appear to be related to fishing activity. Trends in several environmental time series are in phase with the observed trends in the cpue data of the demersal fisheries in Côte d'Ivoire. This suggests that in Côte d'Ivoire the abundance of the demersal fish stock is strongly related to large scale environmental changes.

Introduction

Catch statistics for the major pelagic and demersal fisheries have been collected in West-Africa and in the Gulf of Guinea since the 1960s. In Côte d'Ivoire and Ghana a large amount of information on the environment is also available (Arfi *et al.* 1991; Koranteng 1998; Roy *et al.* this volume). Until now, environmental data have been mostly used to relate pelagic fish stock dynamics to environmental variability (Mendelssohn and Cury 1987; Cury and Roy 1989, 1991; Pézennec and Bard 1992; Pézennec and Koranteng 1998; Durand *et al.* 1998) as

241

pelagic resources tend to be highly variable. However the demersal communities which have been studied for several decades in many fisheries (almost forty years in the Gulf of Guinea) sometimes also exhibit patterns of drastic changes. A general concern in fisheries management is now to address ecological problems and fish population dynamics with an ecosystem perspective and to consider the long-term changes in the environmental or exploitation. Studies that link the different components of the trophic web or the spatial and temporal dynamics of the interaction between the environment and the resource are definitively needed as they have important implication for managing the resources. Recently, an effort to analyse the ecosystems off Ghana has been made which relates the environmental variability to demersal fisheries (Koranteng 1998). The present paper constitutes an attempt to link all the available environmental and fisheries information in order to discuss the observed global patterns in the pelagic and demersal fisheries since the 1950s off Côte d'Ivoire and Ghana.

The Côte d'Ivoire-Ghana coastal upwellings

Mean patterns

The Côte d'Ivoire and Ghana ecosystem is located between 1°E and 8°W at a low latitude (5°N). The wide continental shelf (80 km) east of Cape Three Points narrows on the western side of the Cape and is less than 20km wide off Côte d'Ivoire. The East-West orientation of the coast is a singular characteristic of this tropical upwelling ecosystem. However, the eastward flow of the Guinea current and the westward undercurrent make the structure of the surface and subsurface circulation quite similar to other upwelling areas. The depth of the thermocline is shallow, and varies seasonally between 10 to 60 meters. The seasonal amplitude of the sea surface temperature (SST) is large, SST varies between 21°C during the austral winter and more than 29°C in April (Table 18-1). The southwest monsoon wind is the dominant wind regime, wind speed is maximum (around 5 m/s) during the austral winter. On the eastern side of Cape Three points in Ghana and off Cape Palmas in Côte d'Ivoire,wind is upwelling favourable all year round. Maximum values of the upwelling index are observed during the austral winter. Due to the latitude dependency of Ekman transport, upwelling indices can reach large values in equatorial regions. Off the Côte d'Ivoire and Ghana ecosystem, the wind intensity remains below 5 m/s, but the value of the upwelling index is similar to values observed on other eastern boundary currents. Nutrient concentration is high and comparable to what is found in other upwelling areas. Perhaps, the most important feature of this ecosystem is the occurrence of two upwelling seasons: the main one from June to October and a second one of lesser amplitude in February-March.

Table 18-1 summarises the main characteristics of the Côte d'Ivoire and Ghana ecosystem; information on other upwelling areas are also given for comparison. The mechanism responsible for the occurrence of the seasonal upwellings in the Côte d'Ivoire and Ghana ecosystem remains unclear. For the summer upwelling, several processes are proposed (Picaut 1983; Roy 1995). The contribution of the wind induced Ekman transport to the summer upwelling remains unclear (Binet and Servain 1993). The Guinea current is thought to be an important contributor to the upwelling process (Ingham 1970; Bakun 1978); interactions between the coastal topography and the coastal currents may also be an important factor (Marchal and Picaut 1977). The remote forcing hypothesis is well documented and is

	LATITUDE	UPWELLING	SST (°C)	WIND SPEED (m/s)	UPW. INDEX (m3/s/m)	WIND MIXING (m3/s3)
		JAN-FEB	**27.8**	**3.7**	**1.0**	**95**
COTE			21.4 30.0	1.7 6.1	-0.3 2.6	8 359
D'IVOIRE	5°N	JUNE-OCT.	**27.8**	**3.7**	**1.3**	**95**
-GHANA			21.4 29.9	0.5 6.7	-0.5 3.6	1 335
SENEGAL	15°N	DEC.- JUNE	**25.8**	**4.8**	**1.0**	**197**
(Cap Vert)			17.9 29.2	2.6 7.8	-0.6 2.6	50 683
CAP BLANC	21°N	PERMAMENT	**19.3**	**7.2**	**1.7**	**627**
			16.9 24.9	4.0 10.7	0.5 3.9	160 1548
BAJA			**18.7**	**5.9**	**0.7**	**359**
CALIFORNIA	27°N	PERMANENT	15.1 25.4	4.1 8.7	0.1 1.5	137 1017
(Punta Eugenia)						
Trujillo			**20.4**	**4.9**	**1.2**	**188**
PERU :	8°S	PERMANENT	16.3 28.4	1.6 7.9	0.1 3.3	10 613
Callao	12°S		**19.3**	**5.5**	**1.0**	**276**
			15.0 26.7	2.6 8.3	0.2 3.2	28 789
NAMIBIA	23°S	PERMANENT	**15.9**	**6.0**	**1.1**	**432**
			12.8 20.0	1.7 12.0	0.1 4.4	19 2698

Table 18-1 The physical characteristics of several upwelling areas. For each parameter, minimum, median and maximum values are given. Source COADS data set (Roy 1995).

supported by some numerical models and data analyses but the offshore scale of the thermocline oscillations along the coast suggests that local processes should be considered (Moore *et al.* 1978). Due to the equatorial location of the area, a combination of both remote and local forcing is more likely to be at the origin of the upwelling process.

The winter upwelling has received less attention than the summer upwelling. The intensity of this secondary upwelling has its is maximum in the vicinity of Cape Palmas and sharply decreases toward the east to become almost unnoticeable in the SST time series from the Ghanaian coastal stations (Arfi *et al.* 1991). Using satellite SST images, Hardman-Mountford and McGlade (this volume) provide evidence that the extension over the continental shelf of the winter upwelling is much more pronounced off Ghana than Côte d'Ivoire. The intensification of the Guinea current in January and February is thought to contribute to the upward movement of the thermocline associated with this secondary upwelling (Morlière 1970). Local winds are also an important contributor to the winter upwelling off Côte d'Ivoire: increasing wind induced offshore transport enhances coastal upwelling. As a result coastal SST decreases (Roy 1995).

The climatic variability of the Gulf of Guinea and of the tropical Atlantic has been quite extensively studied during the last decade (Arfi *et al.* 1991; Koranteng 1998; Koranteng and Pézennec 1998; Servain 1991; Carton *et al.* 1996; Bojariu 1997). In the following sections, both the seasonal and the long term variability of the Côte d'Ivoire-Ghana upwellings are investigated using SST data extracted from the COADS database (Woodruff *et al.* 1987) using the CODE software (Mendelssohn and Roy 1996). Mean monthly time series of SST were constructed by combining the COADS data from 7°W to 2°E and from 4°N to the coast from 1963 to 1995. A seasonal-trend decomposition procedure consisting of a sequence of smoothing operations that employ locally-weighted regression or loess (STL) of the monthly time series is performed using the S-Plus statistical package (Cleveland *et al.*

1990). This procedure allows decomposition of the different environmental time series into trend, seasonal, and remainder components. It is used in the following sections to investigate the main patterns of the variability of the Côte d'Ivoire-Ghana marine environment over the last 30 years.

Change in the seasonal cycle of COADS SST time series

Figure 18-1 presents the seasonal component given by the decomposition of the SST monthly time series. The predominant pattern is an intensification of the secondary upwelling (January-February) during the last 20 years. The thermal signature of the minor upwelling is almost unnoticeable in the late sixties and becomes more and more apparent during the late seventies. The amplitude of temperature signal associated with the minor upwelling events reaches 0.8°C in the late nineties. The increase in the amplitude of the seasonal component results from an increase of the warming before and after the minor upwelling and from an accentuation of the cooling in January (Figure 18-2). The intensification of the minor upwelling was previously noticed by several authors (Pézennec and Bard 1992; Koranteng 1998). An intensification of the upwelling-favourable wind is thought to have contributed to the intensification of the secondary upwelling (Roy 1995). Changes in the strength of the Guinea current may also have contributed to the strengthening of the minor upwelling (Binet and Servain 1993).

Another pattern highlighted by the SST seasonal decomposition is a slight shift in the phase of the major upwelling. From 1975 to 1995, there is a pronounced increase of the intensity of the SST drop between June and July (1.7°C in 1975 to 2.8 in 1995, Figure 18-2). Simultaneously, the rate of warming between August and September appears to intensify; the maximum cooling resulting from the upwelling remains in August. These patterns in the seasonal component of SST indicate that the timing of the major upwelling season may have changed during the last 30 years and that upwelling now tends to start earlier than in the early 1960s.

Changes in the seasonal behaviour of a time series are difficult to address, as they require advanced statistical techniques. Minor changes in the timing or the intensity of the seasonal cycle of an environmental variable can have an important ecological impact by

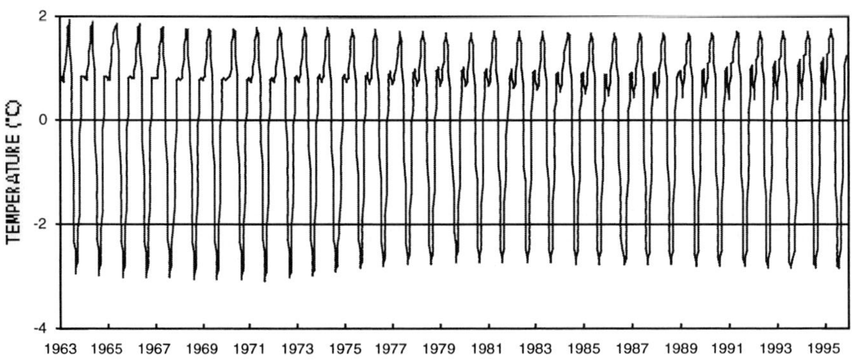

Figure 18-1 Seasonal component of the SST from 1963 to 1995 given by the STL decomposition of the monthly SST time series off Côte d'Ivoire and Ghana (source COADS data set).

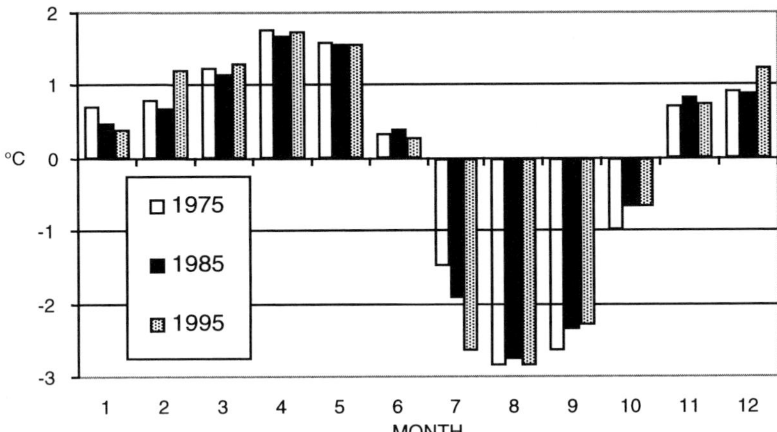

Figure 18-2 SST seasonal component in 1975, 1985 and 1995 given by the STL decomposition of the monthly SST time sereis off Côte d'Ivoire and Ghana (source COADS data set).

modifying the interactions between the physical and biological components of an ecosystem. The intensification of the minor upwelling season is thought to have favoured the increase in abundance of *S. aurita* off Côte d'Ivoire (Pézennec and Bard 1992). This intensification is almost unnoticeable on the raw monthly SST time series but its apparent consequence is, by far, the greatest change observed in the fishery pelagic catch data during the last twenty years.

Long term trend from COADS SST time series

The long-term trend of several environmental time series off Ghana was analysed in detail by Koranteng (1998). The major pattern, common to both surface and sub-surface time series of several environmental variables is a drastic change in the mid-seventies. The long term trend given by the seasonal-trend decomposition performed on the SST COADS data off Côte d'Ivoire and Ghana also shows that the seventies was a period of drastic changes within the ecosystem (Figure 18-3). A relative SST minimum is observed in 1966, then SST peaks in mid-1972 (27.4°C), and sharply decreased until mid-1977 when it reaches an absolute minimum (26.95°C). SST showed a continuous warming trend from 1978 to 1989 and then decreased slightly from 1990 to 1992. The variability of the trend component of the atmospheric pressure monthly time series extracted in the same area from the COADS data set is almost a mirror image of the variability of the SST trend component (Figure 18-4). The similarity between the trend of these two different variables gives strong confidence into the observed pattern of variability.

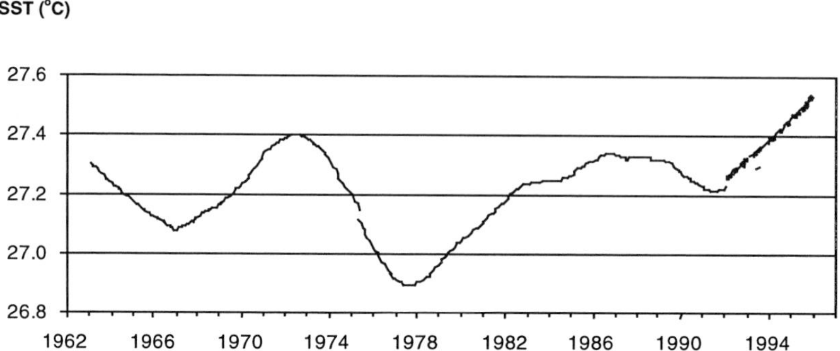

Figure 18-3 Long-term trend of SST off Côte d'Ivoire and Ghana given by the STL decomposition of the monthly SST time series (source COADS data set).

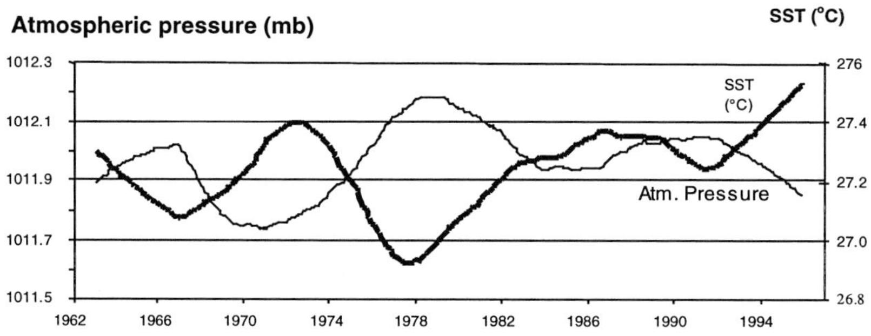

Figure 18-4 Long-term trend of SST and atmospheric pressure off Côte d'Ivoire and Ghana given by the STL decomposition of the monthly SST time series (source COADS data set).

This is also reinforced by the correspondence between the fluctuations of the SST trend and of the anomalies of Atlantic dipole (Figure 18-5). The dipole is an index of the inter-hemispheric gradient of SST anomalies in the tropical Atlantic (Servain 1991). This index is considered to reflect one of the major modes of climatic variability in the tropical Atlantic (Chang *et al.* 1997). The correspondence between the dipole and the SST trend indicates that the long-term climatic changes in the Côte d'Ivoire-Ghana ecosystem are related to basin wide fluctuations over the tropical Atlantic.

Drastic changes in the environment have also been recorded in the Pacific during the seventies (Kerr 1992; Trenberth 1990; Graham 1994). In 1972, the whole tropical Pacific was affected by an intense ENSO event. Later, a major shift in the ocean-atmosphere system occurred during the winter of 1976-1977. In the tropical Pacific, SST and westerly winds suddenly increased after the 1976-1977 winter. These changes were viewed as a shift in the background state of the climate (Graham 1994). The pattern of the climate remained altered up to 1989 when it appeared that pre-1976 climate pattern had been restored. This climate shift has had drastic biological consequences in the Pacific (Hayward 1997); it affected,

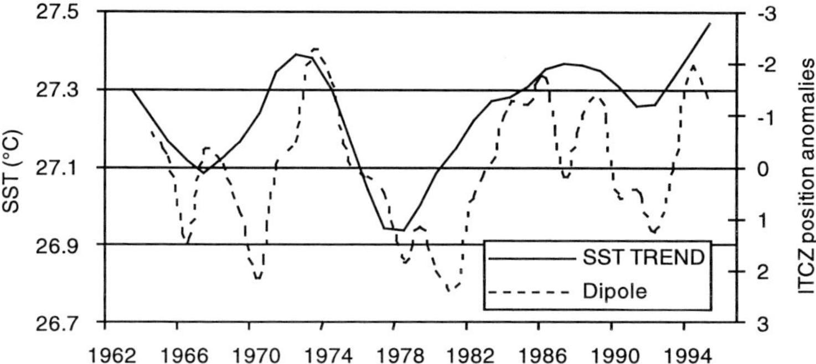

Figure 18-5 Long-term trend of SST off Côte d'Ivoire and Ghana (source COADS data set) and monthly values of the Atlantic dipole index (Servain, 1991) from 1963 to 1995.

sometimes positively and sometimes negatively, zooplankton abundance (Brodeur and Ware 1992) and the production of several commercial species (Ware and McFarlane 1989; Polovina *et al.* 1994; Beamish *et al.*1997).

There are striking similarities between the timing of the major climatic events (1972, winter 1976-1977, 1989) in the Pacific and the timing of environmental changes off Côte d'Ivoire-Ghana as indicated by the SST trend. Such a coincidence between the climatic pattern in the Pacific and the SST trend off Côte d'Ivoire-Ghana indicates that the observed environmental changes are not the results of local processes but rather are induced by inter-oceanic teleconnection through the atmosphere. When considering the interannual SST fluctuations in the tropical Atlantic, Enfield and Mayer (1996) showed that there is a link between these fluctuations and ENSO related anomalies in the Pacific. On the decadal time scale as considered here, the mechanisms involved in the inter-oceans teleconnection are not yet well understood.

Fisheries Productivity

Pelagic fish productivity

Pelagic fish productivity, calculated in different upwelling systems reveals important disparities (Table 18- 2).

Three national fisheries have annual catches of more than one million tonnes: Peru, Chile and Namibia. The Peruvian ecosystem clearly distinguishes itself with a maximum total pelagic catch productivity exceeding the others (Cury *et al.* 2000). Mean total pelagic catch productivity never reaches more than a few hundred thousand tons in Côte d'Ivoire, Senegal or in Morocco. Large ecosystems should *a priori* be able to produce more fish than small ones; however the size of the continental shelf appears to be only one among various important factors (Figure 18-6). Vast upwelling areas might yield relatively unproductive fisheries and vice versa. Thus the pelagic fish productivity in West-Africa and in the Gulf of Guinea appears relatively low compared to that of Peru or Chile (which are both located in the Humboldt Current).

Total pelagic "catch productivity index"

	Upwelling areas	Time period Considered	Mean (tons)	Maximum (tons)	Maximum per unit of surface (tons /km^2)
1	California	1924-1991	200901	609979	6.0
2	Peru	1958-1993	5299183	12286264	142.0
3	Chile	1966-1993	1540109	3708071	59.3
4	Spain-Portugal	1937-1989	331839	368893	6.1
5	Morocco	1950-1991	192885	362023	3.1
6	Senegal	1964-1991	77234	194693	5.9
7	Côte d'Ivoire – Ghana	1966-1993	120414	270570	5.0
8	Namibia	1966-1992	507663	1561300	17.3
9	South-Africa	1950-1992	274312	623200	3.5
10	Venezuela	1957-1989	38032	80079	4.7
11	India	1948-1988	249382	448206	6.4

Table 18-2 Mean and maximum total pelagic fish catch fish catch observed in different upwelling areas and maximum catch per unit of surface of the continental shelf .

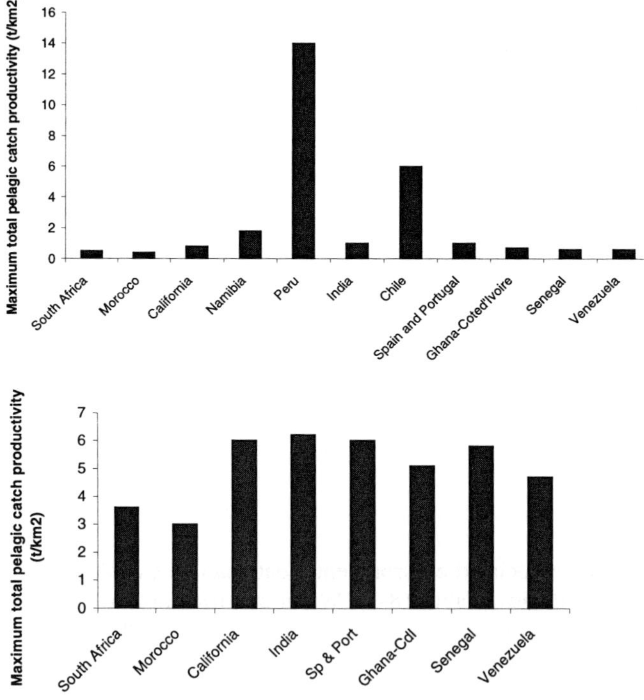

Figure 18-6 Maximum total pelagic catch productivity per unit of surface (t/km^2) with and without the Chilean and Peruvian values (upper and below).

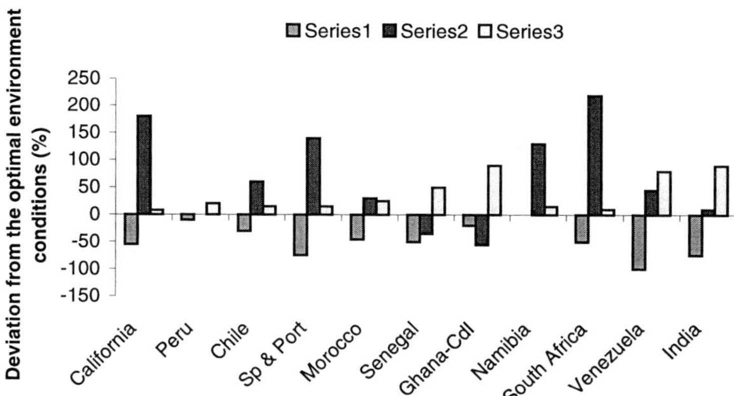

Figure 18-7 Relative deviation of the environmental values (%) from the optimal Environmental values; Series 1 CUI; Series 2 V3; Series 3 SST.

The relationship between estimates of pelagic fish productivity and environmental characteristics in upwelling systems, using non parametric regressive statistical methods, reveals that a combination of several factors are necessary to promote a high productivity:
  - a high upwelling intensity (near to 1.28 $m^3/s^1/m^1$).
  - a moderate turbulence (around 200-250 m^33/s^3).
  - a medium sea surface temperature (15-16°C).
  - a relatively extensive continental shelf (approximately 100 000 km^2).
The Peruvian ecosystem is the only one that combines all the optimal environmental conditions, which is what makes it so productive (Cury *et al.* 2000). A comparison of the environmental values of the upwelling areas with these "optimal environmental values" is presented on Figure 18-7. In Chile and Namibia, the upwelling index is favourable, however it is associated with a negative effect of a high turbulence index. The same high upwelling index is found off Côte d'Ivoire-Ghana and the area appears to be quite productive; in this area the short duration of the upwelling season certainly constitutes a limiting factor for the production. In South Africa, Spain and California, the turbulence index is high and associated with a low upwelling intensity therefore limiting the productivity.

Demersal fish productivity

Similar catch statistics and associated environmental data can be calculated for the demersal fisheries to that used for pelagics (Table 18-3). When one tries to relate them to the size of the continental shelf or the wind intensity or the turbulence or the sea temperature, no relationship can be found.

However it appears that there might be some links between the demersal productivity per unit of surface of the continental shelf and the upwelling index (Figure 18-8). One would expect such relationship to exist, as the global productivity of an ecosystem should be increasing with its global enrichment process. The demersal productivity off Ghana-Côte d'Ivoire appears to be quite high compared to other upwelling systems (with the exception of the very productive areas located in the Humboldt current). In upwelling areas the total pelagic and demersal fish productivity appears to be linked to environmental processes. However these constraints evolve through time, and seasonal as well as inter-annual fluctuations affect the overall resource dynamics.

Country	Surface (km^2)	Catch (t)	Upwelling index	Wind mixing	SST	ratio
Peru	86523	368267	1.20	225	19.1	4.256
Chile	62516	393575	0.93	346	16.5	6.296
Spain-Portugal	59864	81691	0.36	628	16.3	1.365
Morocco	118539	35487	0.66	306	19.6	0.299
Senegal	32887	68517	0.59	150	23.7	2.083
Côte d'Ivoire - Ghana	54647	67727	1.04	103	27.1	1.239
Namibia	90508	136872	1.28	517	16.8	1.512
South-Africa	178315	215818	0.65	723	16.9	1.210
India	70135	153738	0.40	240	28.1	2.192

Table 18-3 Surface of the continental shelf, maximum demersal fish catch, mean upwelling index, turbulence and sea surface and ratio (maximum demersal fish catch/ surface of the continental shelf). (environmental data from COADS and fisheries data from Fishstat95 FAO 1997).

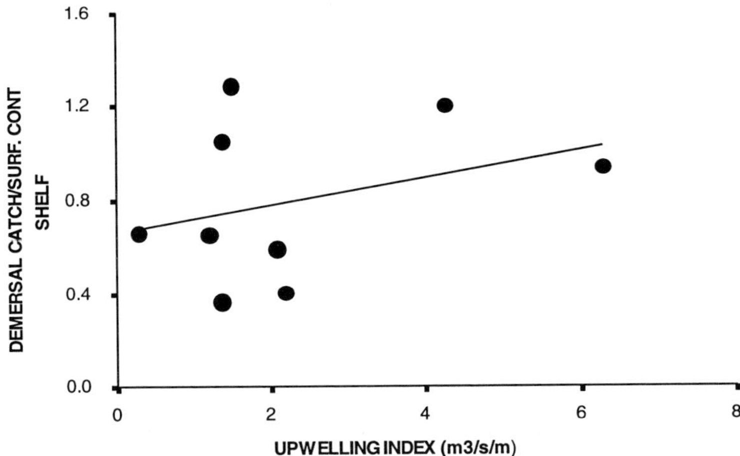

Figure 18-8 Relationship between upwelling index and demersal fish productivity per unit of surface for nine upwelling areas of the world.

Fisheries Variability

In West Africa and in the Gulf of Guinea several oceanographic institutes collect information on biology, fisheries and the environment, in order to support fisheries management. The FAO database (accessible through Fishstat95, FAO1997) also provides an interesting source of information as the fish catch statistics are summarised per year and available since the 1950s, even though the reliability may be sometimes doubtful before the 1970s. However the demersal fish catch from CROA (Centre de Recherches Océanographiques d'Abidjan) and from the FAO data-base appear to be in good agreement.

The global fisheries off Ghana and Côte d'Ivoire have been expanding during the last four decades: the total marine fish catch off Côte d'Ivoire and Ghana has been increasing from less than fifty thousand tonnes in the fifties to around four hundred thousand tonnes in the mid-nineties (Figure 18-9). Ghanaian catches represent 73% of the total catch, and are responsible for the observed trend as the Ivoirian catches have been stable since the 1960s.

For the pelagic species, a noticeable increasing trend has been observed since the 1950s; the total catch reached around 250 thousand tonnes in the 1990s (Figure 18-10). In contrast the demersal fisheries, in Côte d'Ivoire and Ghana have fluctuated between 30 and 50 thousand tonnes since the mid-1960s without any apparent trend.

The fishing effort of both the seiners and the trawlers has consistently decreased since the end of the 1960s in Côte d'Ivoire (Figure 18-11). The diagram of the cpue (catch per unit of effort) of the demersal species versus the fishing effort of the trawlers does not show any relationship which suggests that the link between the fishing activity and the abundance of the resource is not so strong, at least off Côte d'Ivoire.

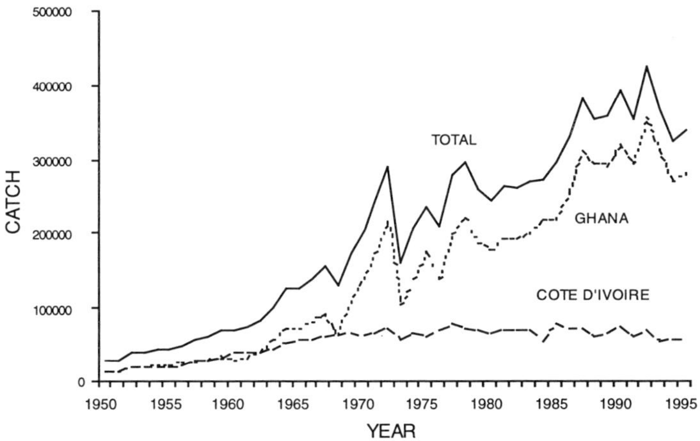

Figure 18-9 Total pelagic fish catch in Ghana and Côte d'Ivoire between 1950 et 1995 (source Fishstat95, FAO 1997).

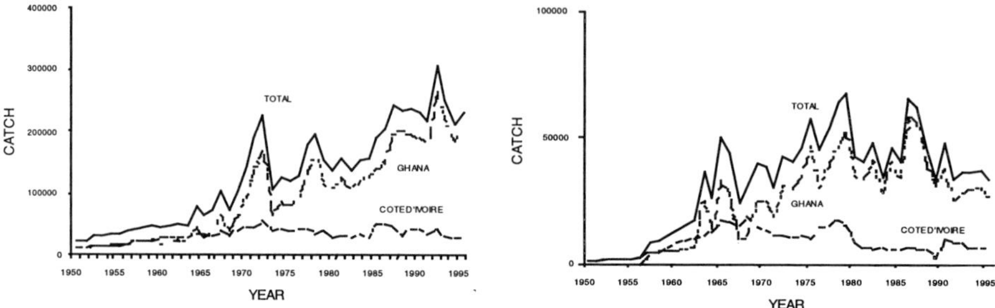

Figure 18-10 Total pelagic (left) and demersal (right) fish catch in Ghana and Côte d'Ivoire between 1950 et 1995 (source Fishstat95, FAO 1997).

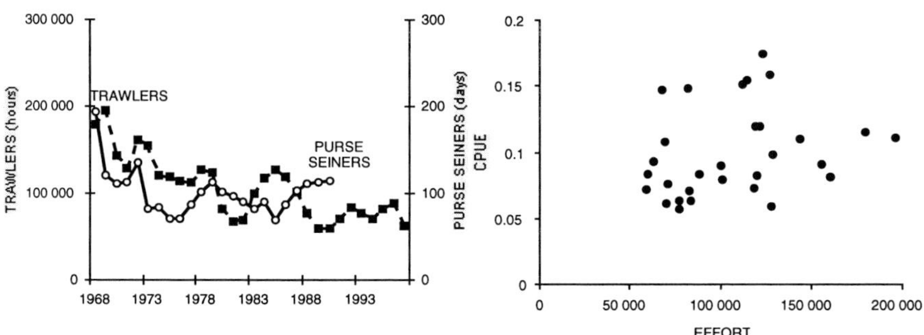

Figure 18-11 Fishing effort (time at sea in hours) between 1968 and 1997 for the trawlers and the purse seiners (source CROA) (left). Catch per unit of effort versus fishing effort for the demersal fishery off Côte d'Ivoire (right).

Ecosystem effects

It may be important to consider interactions between fish stocks, as drastic fluctuations have occurred in both the Ghanaian and Ivoirian ecosystems. *Scomber japonicus* exhibits a similar spatial and temporal pattern as *Sardinella aurita*: they are both found in upwelled waters during the minor and the major upwellings periods. *Scomber japonicus* are also caught seasonally in the purse seine fisheries and migrate into deeper waters during the warm season to join the deeper fauna at 200-300m depth (Longhurst and Pauly 1987), *Scomber japonicus* feed on juveniles as well as larvae of sardinella (Anon. 1976). After the collapse of the *Sardinella aurita* stock in 1973 it appears that two years later the *Scomber* stock collapsed. Despite the recovery of the sardinella in 1976, scomber reappeared only ten years later (Figure 18-12). It suggests that the abundance of prey, which is regulated by the environment, regulates the abundance of the predator. In this case, the *Scomber japonicus* may have collapsed due to the low abundance of sardinella and had, late on, difficulties in recovering from this depleted state.

The ratio of the pelagic predators (such as *Trachurus* sp., *Trichiurus* sp., *Scomber* sp., etc.) to small pelagic prey has decreased substantially since the late sixties, suggesting increasing catch of the small pelagics, which are at lower trophic levels. But this appears to be quite natural as the total small prey pelagic appeared to have increased in biomass over the last two decades.

The total demersal catch appears to have increased consistently, but smoothly through time with a plateau since the 1970s. The appearance of the trigger fish (*Balistes capriscus*) at the beginning of the 1970s and its disappearance at the beginning of the 1990s apparently did not seem to have a

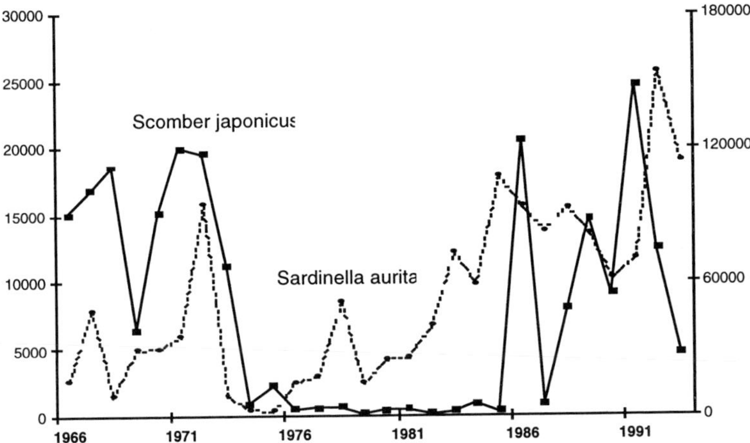

Figure 18-12 Catch of *Scomber japonicus* and *Sardinella aurita* in Côte d'Ivoire and Ghana from 1966 to 1994.

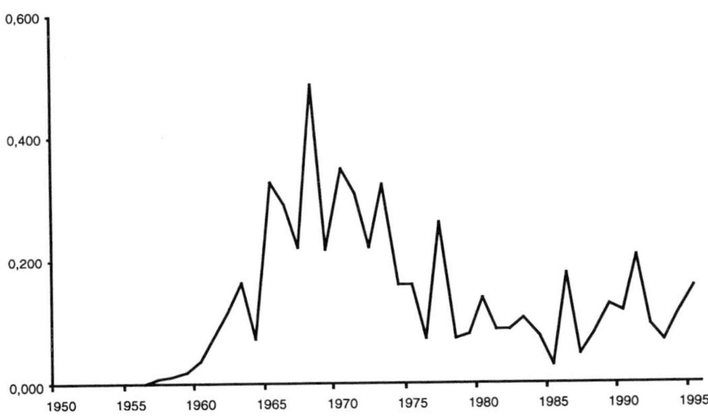

Figure 18-13 Ratio of the catch of pelagic predator (*Trachurus*, *Trichiurus*, *Scomber*, etc) on pelagic prey (sardinellas, anchovies, etc.) from 1960 to 1995.

major impact to the total demersal catch (Figure 18-14), even though it had some major effects on some demersal fish communities (Koranteng 1998, this volume)

Sardinella aurita alternates between periods spent as demersal schools at or near the edge of the continental shelf in the sub-thermocline, and as surface shoals closer to the coast during upwelling seasons (Longhurst and Pauly 1987). It may be worth noticing that the demersal fish abundance was high when the catch of *Sardinella aurita* was low (Figure 18-15). This may suggest a relationship between the two communities, or an environmental factor that favoured one and not the other.

In the ecosystem, strong patterns of fish variability have emerged in the last three decades and appear to be connected in some way, through interactions between species or communities or through environmental forcing.

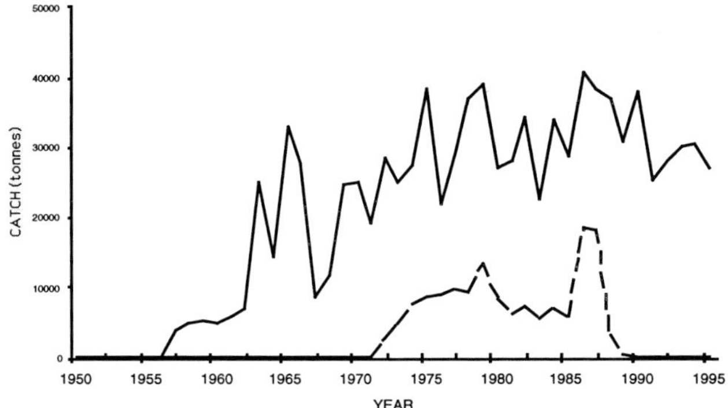

Figure 18-14 Total demersal catch of Ghana (without the trigger fish) and catch of the trigger fish (dashed line) from 1950 to 1995 (source Fishstat95, FAO 1997).

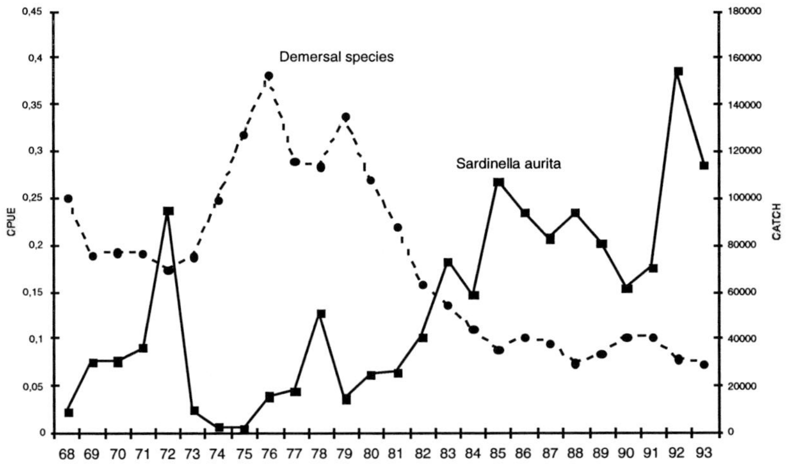

Figure 18-15 Catch of *Sardinella aurita* and CPUE (catch per time at sea) of the demersal species from 1968 to 1993 off Côte d'Ivoire.

Long-term variability and demersal fish productivity

Multi-annual changes were observed in the catch per unit of effort in Côte d'Ivoire for the demersal species. At the beginning of the 1970s a sharp increase was observed that lasted for about ten years, after which the cpue returned to a relatively low level (Figure 18-16). These changes had major effects on the results of the fisheries. In fact, the abundance did increase for many species during the 1970s (for example the sparids, *Pseudotolithus spp.*, rays...). It is also notable that the trigger fish appeared at the beginning of the 1970s (even though it was not fished by the Ivoirian fleet).

When the trends observed in the cpue and the SST (inverse plot) times series are plotted simultaneously, it appears that the two series are fluctuating in phase (a high cpue is observed for a low temperature and conversely) (Figure 18-17). This synchrony between the total demersal production and the SST trend time series suggests that the drop in temperature during the 1970s may have favoured the demersal fish communities. These environmental changes particularly affect the first six months of the year, which are not within the most productive major upwelling season. However they may act as an environmental forcing function for the resources. An enhancement of the productivity during the less favourable season certainly has a drastic effect on the whole sustainable production of fish. This "bottle neck" hypothesis was formulated for the pelagics by Pézennec and Bard (1992) and could perhaps hold also for the demersal community.

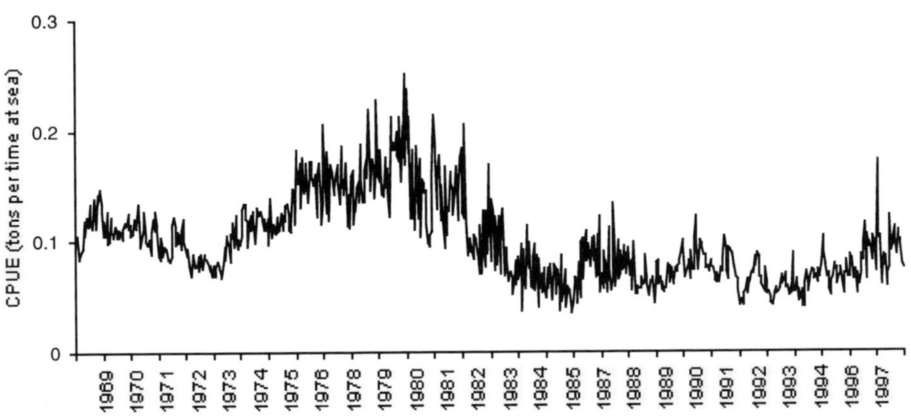

Figure 18-16 Catch per unit of effort for the demersal species off Côte d'Ivoire per fortnight from 1968 to 1995 (source CROA).

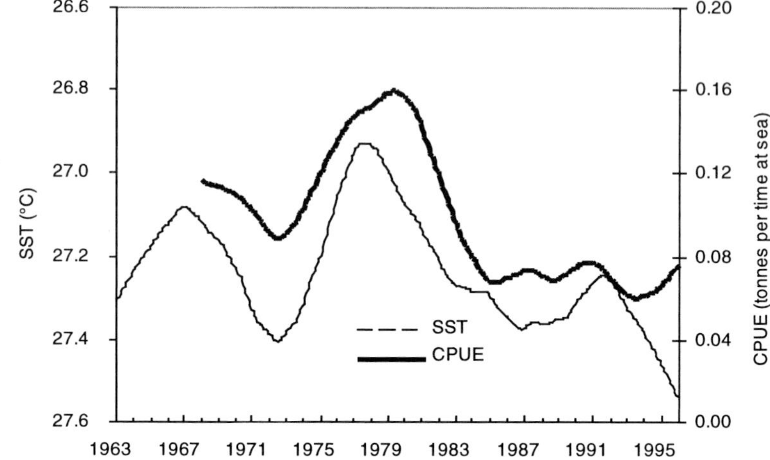

Figure 18-17 Trends in total CPUE of the demersal species (source CROA) and SST (inverse plot) off Côte d'Ivoire from 1964 to 1995 (COADS) calculated using STL.

Discussion: Environmental forcing and resources dynamics

In the Gulf of Guinea drastic changes in the demersal fish community have been observed during the last four decades. The abundance of *Sardinella aurita*, both in Ghana and Côte d'Ivoire, dramatically increased during the last two decades; the biomass of the trigger fish experienced an incredible expansion for eighteen years and, then disappeared; the abundance of the demersal fish community was higher during the 1970s and decreased at the beginning of the 1980s. All these changes were reflected in the demersal and in the pelagic fisheries; they may not be related to fishing effort, although the analyses of Koranteng (1998) and Koranteng and McGlade (this volume) suggest otherwise. The major environmental changes are mostly observed during the first semester of the year (and apparently disconnected from the major upwelling season). These environmental changes appear not to be the result of local processes but rather are induced by inter-oceanic teleconnection through the atmosphere as they are observed in different and independent environmental time series both in the Atlantic and the Pacific oceans. Relationships between the environment and pelagic fish populations have been documented in the Gulf of Guinea (Cury and Roy 1991; Binet and Marchal 1993; Koranteng 1998); however the impact on demersal communities has not been reported. It appears that the low frequency environmental signal is in phase with the demersal fish production. It is reasonable to think that these changes had a major impact in the Gulf of Guinea and consequently affect the total productivity of the ecosystem. According to these changes, cooler temperatures (stronger upwelling) during the 1970s would have increased the productivity in the area, particularly in the Gulf of Guinea which is quite stable from one year to another particularly during the first semester. Moreover, and as mentioned by Pézennec and Bard (1992) this time period constitutes an environmental bottleneck for fish production; consequently any particular favourable environmental events will have a major impact on fish production. Something did happen during the last thirty years in the environment of the Gulf of Guinea and had apparently major consequences on both pelagic and demersal fish abundance. Some drastic changes have sometimes been reported in several ecosystems and related to environmental changes (Arfi 1985; Beamish and Bouillon 1995; Rijnsdorp 1996), mostly in the North Pacific (Beamish 1995). Hallowed *et al.* (1987) examined recruitment variability in 59 stocks in the north-east Pacific, and found strong patterns at all time scales; they concluded that these long-term patterns were due to low frequency climate variability. Whether cyclic temperatures cause cyclic fisheries is a major issue when exploring low frequency environmental changes and their effects (Muter *et al.* 1995).

Relationships between environmental changes and demersal fisheries are not frequently observed. There might have several reasons for that, such as the number of age-classes within the populations which buffer any fluctuation, or the fact that most of the time, these resources are intensively or over-exploited and consequently the observed variability in fish population is mainly driven by the fishing effort. As suggested by Steele (1995), decadal changes in the physical environment are likely to have impacts on marine communities rather than on individual fish stocks. As a whole the dynamics of the demersal fisheries off Côte d'Ivoire may serve as a case studies as its resources were not intensively exploited during the last thirty years and located in an area subject to remote environmental effects. As stated by Longhurst and Pauly (1987) " The Gulf of Guinea upwelling serves as an excellent demonstration of what is frequently forgotten or ignored: that processes important for understanding coastal fishery dynamics may be driven by events a whole ocean away".

References

Anon., 1976 Rapport du groupe de travail sur la sardinelle (*S. aurita*) des côte ivoiro-ghanéennes. FRU Tema, CRO Abidjan, Orstom. 62p.

Arfi, R. 1985 Variabilité inter-annuelle d'un indice d'intensité des remontées d'eaux dans le secteur du Cap-Blanc (Mauritanie). Journal of Fisheries and Aquatic Sciences 42: 1969-1978.

Arfi, R., O. Pézennec, S. Cissoko and M. Mensah 1991 Variations spatiales et temporelles de la résurgence ivoiro-ghanéenne. pp 162-172. *In*: P. Cury et C. Roy (eds) *Pêcheries ouest-africaines. Variabilité, instabilité et changement.* ORSTOM, Paris.

Bakun, A. 1978 Guinea current upwelling. Nature 271: 147-150.

Beamish, R.J. 1995 *Climate change and the northern fish populations.* Canadian special publication of fisheries and aquatic sciences/Publication spéciale canadienne des sciences halieutiques et aquatiques. 121. Ottawa. 739p.

Beamish, R.J. and D.R. Bouillon 1995 Marine fish production trends off the Pacific coast of Canada and the United States. pp. 585-591. *In*: R.J. Beamish (ed.) *Climate change and northern fish populations.* Canadian Special Publication of Fisheries and Aquatic Sciences/Publication spéciale canadienne des sciences halieutiques et aquatiques. 121. Ottawa.

Beamish, R.J., C.M. Ellen and A.J. Cass 1997 Production of Fraser River sockeye salmon (*Oncorhynchus nerka*) in relation to decadal-scale changes in the climate and the ocean. Canadian Journal of Fisheries and Aquatic Sciences 54: 543-554.

Binet, D. and E. Marchal 1993 The large marine ecosystem of shelf areas in the Gulf of Guinea: long term variability induced by climatic changes. pp104-118. *In*: K.

Sherman, L.M. Alexander and B.D. Gold (eds.). *Large marine ecosystems: stress, mitigation, and sustainability.* AAAS Publication, Washington DC.

Binet, D. and J. Servain 1993 Have the recent hydrological changes in the northern Gulf of Guinea induced the *Sardinella aurita* outburst. Oceanological Acta 16: 247-260.

Bojariu, R. 1997 Climate variability modes due to ocean-atmosphere interaction in the central Atlantic. Tellus 49A: 362-370.

Brodeur, R.D. and D.M. Ware 1992 Long term variability in zooplankton biomass in the subarctic Pacific Ocean. Fish. Oceanogr. 1 :32-38.

Carton, J.A., X. Cao, B.S. Giese and A.M. Da Sylva 1996 Decadal and interannual variability in the tropical Atlantic ocean. J. Phys. Oceanogr. 26: 1165-1175.

Chang, P. L. Ji and H. Li 1997 A decadal climate variation in the tropical Atlantic Ocean from thermodynamic air-sea interactions. Nature 385: 516-518.

Cleveland, R.B., Cleveland W.S., J.E. McRae and I. Terpenning 1990 STL: A seasonal-trend Decomposition procedure based on Loess. Journal of Official Statistics 6: 3-73.

Cury, P. and C. Roy 1989 Optimal environmental window and pelagic fish recruitment success in upwelling areas. Canadian Journal of Fisheries and Aquatic Sciences 46: 670-680.

Cury, P. and C. Roy 1991 *Pêcheries ouest-africaines: variabilité, instabilité et cha*ngement. Orstom, Paris. 525p.

Cury, P., A. Bakun, R.J.M. Crawford, A. Jarre, R. A. Quiñones L. J. Shannon, H. M. Verheye 2000 Small pelagics in upwelling systems : patterns of interaction and structural changes in 'wasp-waist' ecosystems. ICES Journal of Marine Science 57 :603-618.

Durand, M.H., P. Cury, R. Mendelssohn, C. Roy, A. Bakun and D. Pauly (eds) 1998 *Global versus local changes in upwelling systems.* Editions ORSTOM Paris

Enfield, D.B. and D.A. Mayer 1996 Tropical Atlantic SST variability and its relation to El-Nino-Southern Oscillation. J. Geophys. Res. 102: 929-945.

FAO 1997 FISHSTAT PC. Release 5097/97 2 disks.

Graham, N.E. 1994 Decadal-scale climate variability in the tropical and North Pacific during the 1970s and 1980s : observations and model results. Clim. Dynam. 10 :135-162.

Hallowed, A.B., K.M. Bailey and W.S. Wooster 1987 Patterns in recruitment of marine fishes in the northeast Pacific Ocean. Biol. Oceanog 5: 99-131.

Hayward, T.L. 1997 Pacific Ocean climate change: atmospheric forcing, ocean circulation and ecosystem response. TREE 12: 150-154.

Ingham, M.C. 1970 Coastal upwelling in the northwestern of Gulf of Guinea. Bull. Mar. Sci. 20: 2-34.

Kerr, R.A. 1992 Unmasking a shifty climate system. Science 255: 1508-1509.

Koranteng, K.A. 1998 *The impacts of environmental forcing on the dynamics of demersal fishery resources of Ghana.* PhD These, University of Warwick. 377p.

Koranteng, K.A. and J.M. McGlade (this volume, chapter 8) Physico-chemical hanges in continental shelf waters of the Gulf of Guinea and possible impacts on resource variability.

Koranteng, K. A. and O. Pézennec 1998 Variability and trends in some environmental time series along the Ivorian and the Ghanaian coasts. pp167-178. *In*: M.H. Durand, P. Cury, R. Mendelssohn, C. Roy, A. Bakun and D. Pauly (eds). *Global versus local changes in upwelling systems.* Editions ORSTOM Paris.

Longhurst, A.R. and D. Pauly 1987 *Ecology of tropical oceans.* Academic Press, London. 407p.

Marchal, E. and J. Picaut 1977 Répartition et abondance évaluées par échointégration des poissons du plateau continental ivoiro-ghanéen en relation avec les upwellings locaux. J. Res. Océanogr. 2: 39-57.

Mendelssohn, R. and P. Cury 1987 Fluctuations of a fortnightly abundance index of the Ivoirian coastal pelagic species and associated environmental conditions. Canadian Journal of Fisheries and Aquatic Sciences 44: 408-421.

Mendelssohn, R. and C. Roy 1996 Comprehensive Ocean Data Extraction Users Guide. U.S. Dep. Comm., NOAA Tech. Memo. NOAA-TM-NMFS-SWFSC-228, La Jolla, CA.. 67pp.

Moore, D. W., P. Hisard, J.P. McCreary, J. Merle, J.J. O'Brien, J. Picaut, J.M. Verstraete and C. Wunsch 1978 Equatorial adjustment in the eastern Atlantic ocean. Geophys. Res. Lett. 5: 637-640.

Morlière, A. 1970 Les saisons marines devant Abidjan. Doc. Sci. Centre Rech. Oceanogr. Abidjan 1: 1-15.

Muter, F.J., Norcross B.L. and T.C. Royer 1995 Do cyclic temperatures cause cyclic fisheries. pp585-591. *In*: R.J. Beamish (ed.) *Climate change and northern fish populations*. Canadian Special Publication of Fisheries and Aquatic Sciences/Publication spéciale canadienne des sciences halieutiques et aquatiques. 121. Ottawa.

Pézennec, O. and F.X. Bard 1992 Importance écologique de la petite saison d'upwelling ivoiro-ghanénne et changements dans la pêcherie de Sardinella aurita. Aquatic Living Resources 5: 249-259.

Pézennec, O. and K.A. Koranteng 1998 Changes in the dynamics and biology of small pelagic fisheries off Côte d'Ivoire and Ghana: the ecological puzzle. pp329-343. *In*: M.H. Durand, P. Cury, R. Mendelssohn, C. Roy, A. Bakun and D. Pauly (eds). *Global versus local changes in upwelling systems*. Editions ORSTOM Paris.

Picaut, J. 1983 Propagation of the seasonal upwelling in the eastern equatorial Atlantic. J. Phys. Oceanogr. 13: 18-37.

Polovina, J.F., G.T. Mitchum, N.E. Graham, M. P. Craig, E.E. Demartini and E. N. Flint 1994 Physical and biological consequences of a climate event in the Central North Pacific. Fisheries Oceanography 3: 15-21.

Rijnsdorp, A.D., P.I. Van Leewen, N. Daan, H.J.L. Heessen 1996 Changes in abundance of demersal fish species in the North Sea between 1906-1909 and 1990-1995. ICES Journal of Marine Science 53:1054-1062.

Roy, C. 1995 The Côte d'Ivoire and Ghana coastal upwellings: dynamics and Change. pp346-361. *In*: F.X. Bard and K.A. Koranteng (eds). Dynamique et usage des ressources en sardinelles de l'upwelling côtier du Ghana et de la Côte d'Ivoire. ORSTOM éditions Paris.

Roy, C., P. Cury, P. Fréon and H. Demarq (this volume, chapter 10) Environmental and resource variability off northwest Africa.

Servain, J. 1991 Simple climatic indices for the tropical Atlantic ocean and some application. J. Geophys. Res. 96:15137-15146.

Steele, J.H. 1995 Climate change and community structure. pp5-10. *In*: R.J. Beamish (ed.) *Climate change and northern fish populations*. Canadian Special Publication of Fisheries and Aquatic Sciences/Publication spéciale canadienne des sciences halieutiques et aquatiques. 121. Ottawa.

Trenberth, K. E. 1990 Recent observed interdecadal climate changes in the Northern Hemisphere. Bull. Am. Meteorol. Soc. 71: 988-993.

Ware, D.M. and G.A. McFarlane 1989 Fisheries production domains in the northeast Pacific Ocean. pp359-379. *In*: R.J. Beamish and G.A. McFarlane (eds). *Effects of ocean variability on recruitment and an evaluation of parameters used in stock assessment models*. Canadian Special Publication of Fisheries Aquatic Sciences 108.

Woodruff, S.D., R.J. Slutz, R.L. Jenne and P.M. Steurer 1987 A Comprehensive Ocean-Atmosphere Data-Set. Bull. Amer. Meteor. Soc. 68: 1239-1250.

The Gulf of Guinea Large Marine Ecosystem
J.M. McGlade, P. Cury, K.A. Koranteng and N.J. Hardman-Mountford (Editors)
© 2002 Elsevier Science B.V. All rights reserved.

19

Status of Demersal Fishery Resources on the Inner Continental Shelf off Ghana

K. A. Koranteng

Abstract

Between 1963 and 1990 the relative importance of major species changed in every trawl survey conducted in continental shelf waters off Ghana. In the 1970s and 1980s, there was a phenomenal increase in the biomass of triggerfish (*Balistes capriscus*) in the area. The species declined in abundance in the late 1980s. Only a few specimens were encountered in trawl surveys conducted in the area in the 1990s.

From the results of bottom trawl surveys, the potential yield of total demersal fish (excluding triggerfish) in Ghana's shelf waters is estimated to be of the order of 43,000 mt per annum. From commercial catch and effort data, the maximum sustainable yield is put at about 49,000 mt per annum. The resource appears to be over-exploited as landed catches usually exceed or are too close to the upper limit of the potential yield. This is further supported by a decline in the trend of time series of catch per unit effort of small-sized inshore trawlers operating in Ghanaian waters.

Introduction

The principal objective of this paper is to assess the status of the total demersal fishery resource in Ghana's continental shelf waters. Natural factors such as pollution and anthropogenic activities, e.g. fishing, induce changes in the status of fishery resources over time. To assess the extent of change and to evaluate the state of the resources, stock assessment surveys are usually conducted. These surveys are able to detect the extent of change and also provide the necessary information for regulating the fisheries. Such surveys have been conducted in Ghanaian waters since the middle part of the twentieth century. Both Ghanaian and foreign fishery research vessels were used in the surveys.

In this paper, most of the data collected during the trawl surveys conducted between 1963 and 1990 are re-analysed. Landed catches of commercial fishing vessels obtained through catch assessment surveys are examined in relation to the biomass estimated in the trawl surveys and the potential yield calculated therefrom. The performance of the commercial trawl fishery in terms of catch per unit of effort exerted in their operations is also examined using time series analysis.

Demersal fishery resources in Ghanaian marine waters and their exploitation

Demersal fishery resources exploited in Ghanaian waters include those of the families Sparidae, Sciaenidae, Lutjanidae and Penaeidae (Table 19-1). In 1972-1982, the Ghanaian shelf experienced a dramatic increase in the biomass of an otherwise reef-dwelling, solitary species *Balistes capriscus* (triggerfish) (Ansa-Emmim 1979; Koranteng 1984, Caverivière 1991). The fish dominated this ecosystem for nearly twenty years (Koranteng 1998).

In Ghana, demersal fishery resources are exploited by artisanal, semi-industrial and industrial fishing vessels. Artisanal fishers use bottom set nets, hook-and-line and beach seines to catch demersal fish and the trawlers use bottom trawl nets. All vessels operate in about the same area and target similar species. Total annual landings of demersal fish species and triggerfish by Ghanaian fishing fleets are shown in Figure 19-1. Landings of other important demersal fish species and species groups are shown in Table 19-2 and the trends depicted in Figure 19-2. These are fish caught in Ghanaian waters only.

The landings rose from under 10,000 mt at the beginning of the1960s to over 60,000 mt in the 1980s. Landed catches of triggerfish increased steadily from under 1 mt in 1970-71 to 13,000 mt in 1979. These declined a bit and stabilised around 6,000 – 8,000 mt per annum between 1980 and 1985. In 1986 and 1987 the landed catch was in excess of 18,000 mt each year. The annual landed catch dropped significantly thereafter and, but for the nearly 200 mt landed in 1992, it never exceeded 20 mt per annum in the 1990s (Table 19-2).

| **Family** | | |
Scientific name	**Common English name**	**Major species**
Sparidae	Seabreams	*Pagellus bellottii*
		Sparus caeruleostictus
		Dentex canariensis
Haemulidae	Grunts	*Pomadasys incisus*
		Pomadasys jubelini
		Brachydeuterus auritus
Sciaenidae	Croakers	*Pseudotolitus* spp.
		Umbrina spp.
Lutjanidae and Lethrinidae	Snappers	*Lutjanus fulgens*
		Lutjanus agennes
		Lethrinus atlanticus
Mullidae	Red Mullet	*Pseudupeneus prayensis*
Serranidae	Groupers	*Epinephelus* spp.
Polynemidae	Threadfins	*Galeoides* spp.

Table 19-1 The most important demersal fish species exploited in Ghanaian marine waters.

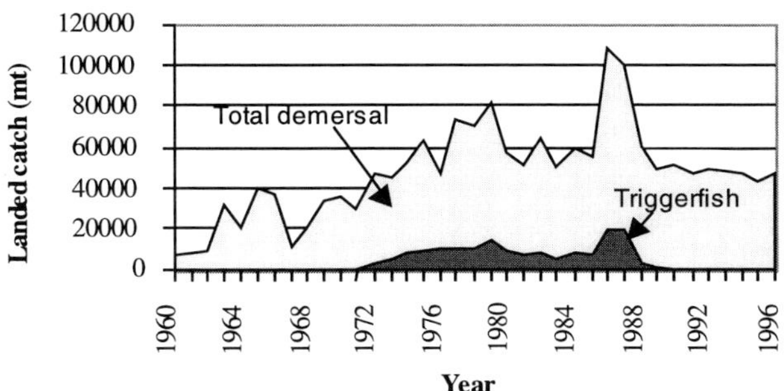

Figure 19-1 Total landings of demersal fish and triggerfish from Ghana's coastal waters, 1960–1996.

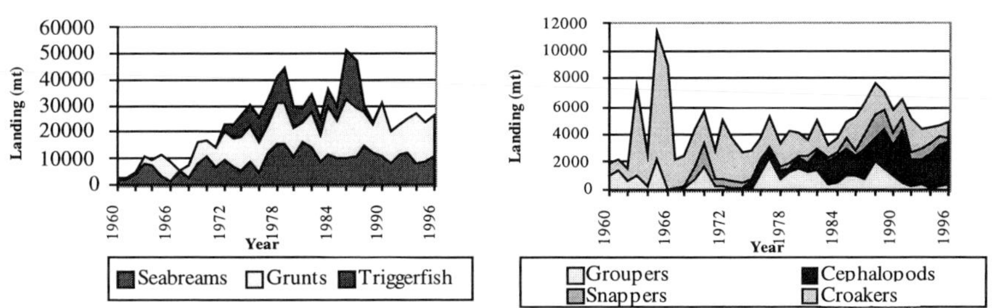

Figure 19-2 Total landed catches of important demersal fish groups from Ghana's coastal waters, 1960 - 1996.

Figure 19-2 shows rising trend in the landings of seabreams, grunts and cephalopods and a decline in importance of croakers between 1960 and 1996. There are pulses in the landings of snappers and groupers.

Assessment of the stocks of demersal fishery resources in Ghana's shelf waters

The first documented fish stock assessment survey conducted in continental shelf waters of Ghana was in May-June 1956 (Salzen 1957). Between August 1963 and June 1964, the

Year	Sparidae (Sea-breams)	Haemulidae (Grunts)	Serranidae (Groupers)	Cephalo-pods	Snappers	Trigger-fish	Sciaenidae (Croakers)	Total
1960	1700	600	1000				800	5800
1961	1900	600	1300				800	7800
1962	3600	700	600				900	8400
1963	7700	2800	1000				6400	31800
1964	7100	2200	300				2300	19800
1965	3500	8000	2100				9300	40100
1966	1200	7500	0				9000	36500
1967	5100	400	100				2000	10600
1968	2800	4400	100		100		2100	19900
1969	8000	8200	700		900		2700	33200
1970	11300	5700	1600		1700	<0.5	2300	35000
1971	6500	7800	300		400	<0.5	2100	29300
1972	9200	10700	200		600	3000	4200	43800
1973	6700	11000	100		500	4900	3100	39500
1974	4896	13621	116		408	7688	2195	44650
1975	8555	13419	17	584	161	8773	2073	54213
1976	4298	11942	1058	795	194	8940	1621	37506
1977	12304	10620	2176	685	196	9685	2116	63842
1978	15639	15866	700	700	212	9389	1684	60007
1979	15468	15605	1186	255	361	13326	2453	67607
1980	11370	10132	1516	705	145	8225	1688	49489
1981	16273	6924	1276	545	526	6198	1214	45165
1982	14384	13185	1312	1355	284	7121	1956	56790
1983	8360	10764	367	1642	194	5536	1039	44126
1984	11587	18092	463	1844	373	6992	1072	52124
1985	9813	14246	1014	2014	644	6104	1199	49347
1986	9939	22964	1002	1355	488	18610	2370	89599
1987	11336	17912	800	2176	948	18283	2379	81919
1988	15256	12109	2030	1726	1507	3033	2369	57442
1989	12503	10308	1496	2852	1298	612	1414	48641
1990	10510	20804	1035	2120	975	0	1602	50658
1991	8058	12213	470	3713	842	8	1557	46712
1992	11438	11473	225	1875	635	198	2340	49336
1993	12204	12599	385	1716	895	9	1301	47943
1994	7742	18791	169	2469	716	11	1104	46590
1995	8214	15330	306	2946	626	2	698	42460
1996	11384	14743	426	2979	328	17	1128	47370

Table 19-2 Total landings of important demersal fish species groups caught in Ghanaian marine waters, 1972 – 1996: Sources: Fisheries Department, Ghana and FAO (1998).

Guinean Trawling Survey (GTS) which was the first large scale survey in the waters of West Africa was conducted through the efforts of the Organization of African Unity (Williams 1968). The Marine Fisheries Research Division of the Fisheries Department has also conducted a number of trawl surveys since 1969. Similar surveys have been conducted in West African waters (including Ghana) under the auspices of the Food and Agriculture Organisation (FAO). Koranteng (1998) gives an elaborate description of all trawl surveys conducted in Ghanaian waters between 1963 and 1990 involving the use of five research vessels

The objectives of these surveys included the following:

i.   exploration of the continental shelf for potential development of a trawl fishery (Salzen 1957; Williams 1968),
ii.   estimation of total biomass and catch rates (Rijavec 1980; Koranteng 1981, 1984),
iii.   monitoring the biomass of fish stocks (Koranteng 1981; 1984, Ramos *et al.* 1990).

In this paper, most of the data collected during the trawl surveys conducted between 1963 and 1990 were analysed with a view to re-assessing the biomass and potential yield of the demersal resource. The data were inputted into the NAN-SIS computer program for survey data logging and analysis (Stromme 1992). Fish names were compared with entries in FishBase (1996). The stock biomass (B) was estimated by the 'swept area method' with catch per haul as the index of abundance. In this method, B is estimated from:

$$B = \frac{A}{a} \cdot \frac{\overline{X}}{q}$$

Eq. 19-1

where A is the total area surveyed, a is the swept area of the net per haul, $\overline{X}$ is the average catch per haul (the index of abundance) and q is the catchability coefficient. Using the relationship between the headline of the trawl net and its horizontal opening (Pauly 1980a), the area swept per haul in each of the surveys was calculated and q was taken to be 0.75 (Koranteng 1998). The potential yield was calculated from:

$$P_y \approx 0.5 \, (MB' + Y') \quad \text{Cadima (1977)}$$

Eq. 19-2

where B' and Y' are the estimated biomass and commercial fishery catch respectively, at the time of the survey. M is the instantaneous rate of natural mortality estimated from:

$$M = \frac{0.985. \, K^{0.654}. \, T^{0.463}}{L_\infty^{\,0.279}} \quad \text{(Pauly 1980b)}$$

Eq. 19-3

where L_∞ (using total length in cm) and K (annual value) are parameters of the von Bertallanfy growth function (VBGF) and T is the mean annual temperature (in °C) of the water body inhabited by the stock. T was taken to be 20 °C (Koranteng 1998). Various estimates of K and L_∞ for fish species in the eastern central Atlantic area, particularly from the Gulf of Guinea, were obtained from literature. M was then calculated for each set of VBGF parameters and mean and standard deviation obtained from these as 0.76 $\pm$ 0.12 yr^{-1} (Koranteng 1998).

	1963/ 1964	1969/ 1970	1973- 1977	1979/ 1980	1981/ 1982	1987/ 1988	1989	1990
Brachydeuterus auritus	1	1	18	2	2	2	1	1
Pagellus bellottii	2	3	2	3	3	3	2	2
Dentex congoensis	3	23	3	11	11	14	11	-
Priacanthus arenatus	4	16	21	9	9	5	10	4
Sparus caeruleostictus	5	6	4	4	5	6	3	6
Epinephelus aeneus	6	10	11	7	7	11	4	11
Pseudupeneus prayensis	7	8	7	5	4	4	4	3
Dentex angolensis	8	-	19	10	19	18	27	-
Galeoides decadactylus	9	11	-	13	30	26	-	26
Pseudotolithus senegalensis	10	-	-	18	-	-	-	-
Loligo sp.	11	-	-	-	-	-	-	-
Paracubiceps ledanoisi	12	29	-	-	22	25	-	-
Dentex canariensis	13	4	6	6	6	7	6	8
Boops boops	14	13	23	22	14	31	19	20
Raja miraletus	15	-	-	-	29	21	24	17
Sphyraena sp.	16	12	-	21	28	32	22	-
Dactylopterus volitans	17	7	8	-	26	8	9	13
Drepane africana	18	-	-	-	-	-	-	33
Dentex gibbossus	19	26	15	16	13	23	-	33
Pseudotolithus brachygnathus	20	-	-	-	-	-	-	-
Balistes capriscus	-	9	1	1	1	1	17	-

Table 19-3 Top 20 species (or genus)(by weight) in the Guinean Trawling Survey (stations of depth ≤ 100 m only) and their ranks in subsequent surveys.

Relative importance of species

For the GTS survey, the total quantity of each species (or genus where species were not clearly identified) caught during the survey was calculated by summing catches from all trawl hauls. The species were ranked according to the weight caught and the 20 most abundant species were listed as shown in column 2 of Table 19-3. The relative importance of these 20 species in subsequent surveys was similarly obtained; these are also shown in the table. For the results to be comparable, only stations of depth less than 100 m in the GTS were considered.

In surveys in 1963/64 and 1969/70, the most abundant species was *Brachydeuterus auritus*. The table clearly shows how from virtually nowhere in the GTS ranking (1963/64) *Balistes capriscus* moved into the ninth position in 1969/70 and then dominated this ecosystem for nearly twenty years.

Biomass and potential yield

Table 19-4 shows calculated biomass of the demersal resource and that of triggerfish in the 9 surveys conducted between 1963 and 1990. Triggerfish has been given special treatment here

Survey years	Demersal species (excluding triggerfish)		Mean biomass of triggerfish (mt)
	Mean catch per haul (kg h^{-1})	Mean biomass (mt)	
1963-64	242.8 - 332.6	78,800	330
1969-70	94.3 - 132.0	48,000	1,370
1973-77	67.6 - 114.1	36,000	14,220
1979-80	189.6 - 290.1	76,700	40,080
1981-82	214.2 - 307.7	83,700	20,180
1987-88	179.5 - 232.8	72,100	5,950
1989	132.7 - 168.8	50,200	1,130
1990a	151.4 - 199.6	58,500	440
1990b	85.0 - 203.6	41,200	80

Table 19-4 Mean catch per haul and estimated mean biomass of demersal species and triggerfish, survey years between 1963 and1990.

because of the significant impact that its proliferation and decline had on the total demersal biomass and the trawl fishery in Ghana (Ansa-Emmim 1979; Caverivière 1991; Koranteng 1998). The 95% confidence interval of the average mean catch per haul calculated from the survey data is also shown for each survey. The biomass estimated from the 1969-70 and 1973-77 surveys appear rather low when compared with the others.

For each survey, the calculated potential yield is shown in Figure 19-3 in relation to the total landed catches. It is important to note that in 1963-64, the landed catch was much lower than the estimated potential yield. A similar situation is evident for the 1979-80 and 1981-82 surveys. In the 1969-70, 1973-77, 1989 and 1990 surveys however, the situation is different whereby landed catches exceeded or were almost equal to the potential yield.

The density (i.e. biomass per unit area) of selected fish species and families and also the total stock, was calculated for each survey. The use of density is to facilitate comparison of results as the various surveys were conducted with different vessels and sizes of trawl. The species and families are grunts (Haemulidae), seabreams (Sparidae), snappers (Lutjanidae and Lethrinidae), groupers (Serranidae), croakers (Sciaenidae), rays (mainly Dasyatidae, Myliobatidae, Rhinobatidae, Torpedinidae), sharks (mainly Carcharhinidae and Squatinidae), cephalopods (Sepiidae, Octopodidae and *Loligo* sp.) and soles (Soleidae, Citharidae and Bothidae). Others are red mullet (*Pseudupeneus prayensis*), bigeye grunt (*Brachydeuterus auritus*), triggerfish (*Balistes capriscus*) and Atlantic bigeye (*Priacanthus arenatus*). These fishes are extremely important in the Ghanaian demersal fishery both in terms of total quantity landed and in value.

In Figures 19-4a,b mean density calculated from the results of the Guinean Trawl Survey (GTS) and Guinea-90 survey are compared. Figure 19-4a for selected species and families, and 4B for the total demersal biomass. GTS was conducted at the beginning of the observational period and Guinea-90 was at the end. These surveys also cover much wider depth ranges (10 – 400/600 m) than the surveys conducted by the Marine Fisheries Research Division. Very clear differences in status of the selected fish species and families are portrayed in the first figure. In the second figure, the difference in density in selected depth zones are shown.

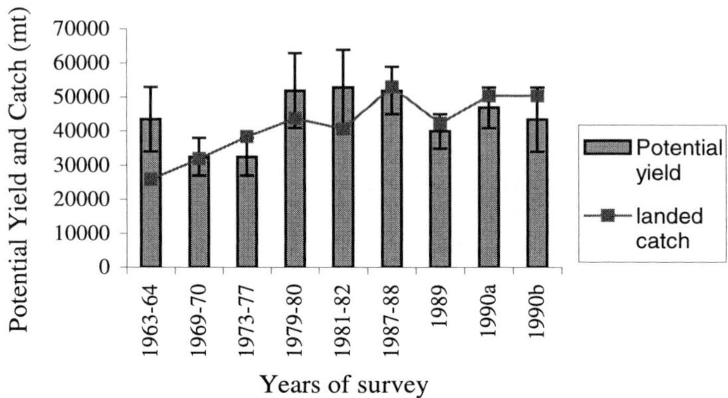

Figure 19-3 Potential yield (with confidence limits) and landed catches.

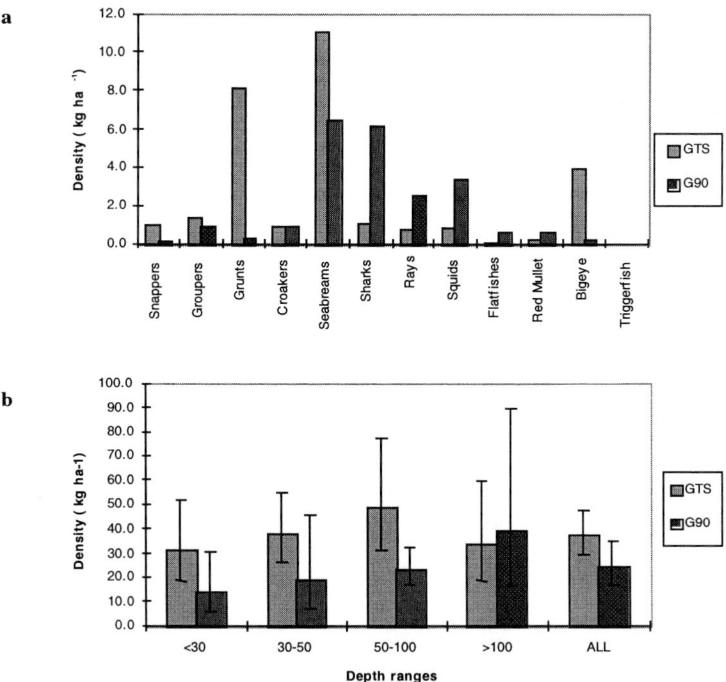

Figures 19-4 a Estimated density of selected species and species groups and **b** total demersal biomass in GTS and Guinea 90 (G90) (B).

Trawling effort and Maximum Sustainable Yield

Figure 19-5 shows evolution of the fishing effort of small-sized inshore trawlers and industrial trawlers operating in Ghanaian waters. This figure portrays a consistent rise in industrial trawling effort and a decline in that of inshore trawlers. From catch and effort data on commercial fishing vessel operations, the maximum sustainable yield (MSY) was calculated using the Schaefer surplus production model.   In this model the equilibrium yield (Y_E) is given by:

$$Y_E = af - bf^2 \qquad\qquad \textbf{Eq. 19-4}$$

where f is fishing effort and a and b are constants obtained by fitting a polynomial regression of Y_E (equilibrium catch) on f and f^2. From this the maximum sustainable yield is calculated as:

$$MSY = -\frac{a^2}{4b} \qquad\qquad \textbf{Eq. 19-5}$$

Two standardised (expected) efforts (TEF1 and TEF2) were defined. TEF1 was calculated from total landed catch in the FISHSTAT Plus database of FAO (FAO 1998) and catch and effort data compiled by the Fisheries Department of Ghana. TEF2 is calculated from FISHSTAT Plus and catch and effort data of industrial trawlers obtained from the Fisheries Department of Ghana. The catch data on Ghana included in the FAO database are also provided by the Fisheries Department of Ghana.

Figure 19-6 is a plot of CPUE of demersal species (excluding triggerfish) against expected effort (TEF2) and Figure 19-7 shows catch versus TEF2. By fitting the surplus production model with TEF2, the following were obtained :

a = 3.405 (s.e. = 0.252),   b = - 6.079 x 10^{-5} (s.e. < 0.001), and

MSY is calculated as 47,700 mt per annum.

When TEF1 is used MSY is calculated to be 51,200 mt. From these two estimates, the maximum sustainable yield is put at around 49,500 mt per annum.

Time series analysis of catch per unit effort of inshore trawlers

The behaviour of catch per unit effort (CPUE) of the exploited total demersal biomass was examined through time series analysis. Monthly CPUEs of small-sized inshore trawlers (i.e. vessels of length 8-12 m) were calculated. Because of the large variation that characterised the landings of triggerfish the species was excluded from the analysis. The resultant time series of CPUEs shows large inter-annual variation (Figure 19-8). The series was decomposed into trend, seasonal variation and remainder; the trend and seasonal variation are also shown in Figure 19-8.

Figure 19-8 shows an increasing trend in CPUE between 1972 and about 1977 and a general downward trend since then. From the plot of seasonal variation of CPUEs, it is evident that the high variability that characterised landings in the 1970s and early 1980s (when triggerfish was in abundance) reduced towards the end of the series.

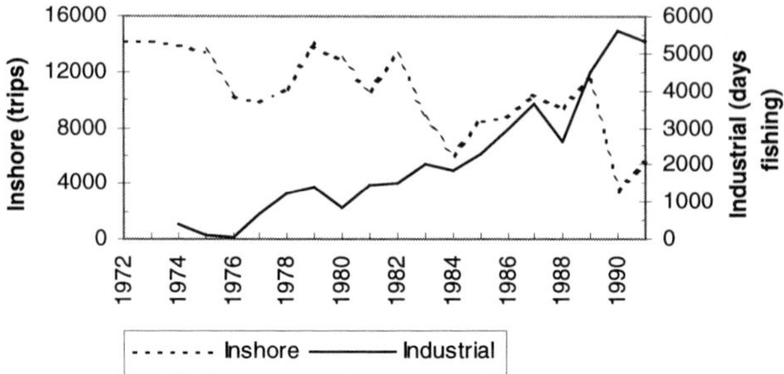

Figure 19-5 Evolution of trawling effort in the Ghanaian demersal fisheries, 1972 – 1990.

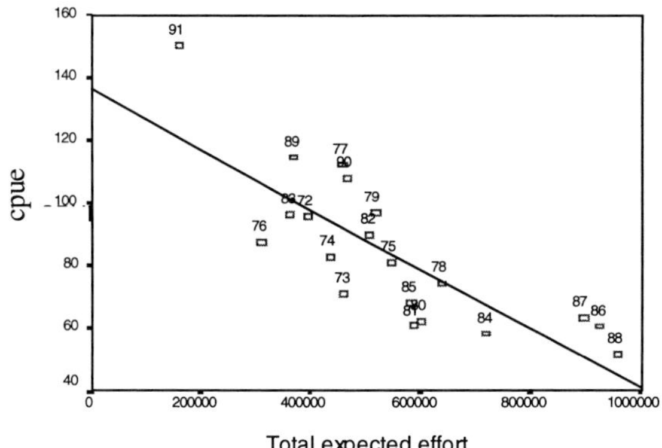

Figure 19-6 Plot of cpue (kg per trip) versus expected effort (TEF1).

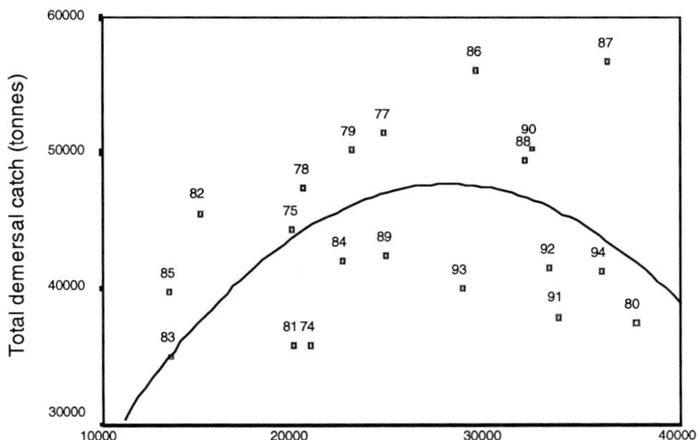

Figure 19-7 Plot of total landed catches versus expected effort TEF2.

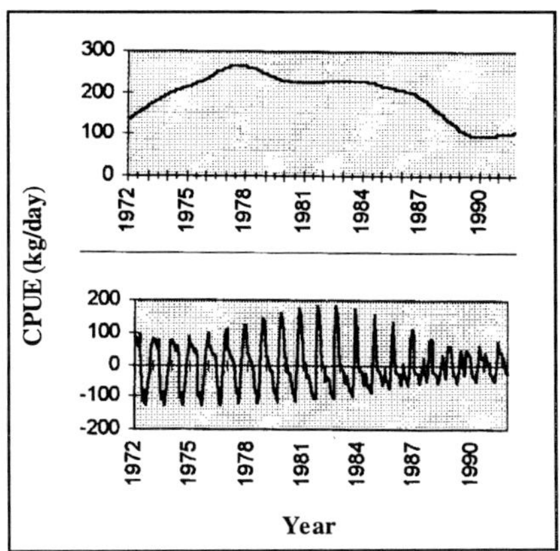

Figure 19-8 Trend and seasonal variation of CPUE of 8 - 12 m long inshore trawlers.

Discussion

The demersal fishery off Ghana saw a steady rise in landed catches in the 1960s to 1980s. Landings rose from under 10,000 mt in the early 1960s to over 60,000 mt in the 1980s. The rise was due partly to an increase in the number of locally-built inshore trawlers (Mensah and Koranteng 1998) but mainly to the deployment of large foreign-built Ghanaian vessels in home waters. These vessels used to fish in more productive foreign waters but had to come home as a result of the EEZ provisions under the 1982 UN Convention on the Law of the Sea (UNCLOS III). This resulted in phenomenal increases in trawling effort (Figure 19-5) and landings.

Landed catches of triggerfish were recorded from about 1970-71 (at less than 1 mt per annum). Following a high value of about 13,000 mt in 1979 the catches declined and stabilised around 6,000 to 8,000 mt per annum between 1980 and 1985. In 1986 and 1987 over 18,000 mt was caught each year. The production dropped to a little over 3,000 mt in 1988 and the resource bore signs of depletion in the late 1980s with extremely low catches. These global changes were reflected in the relative importance of species recorded in the trawl surveys conducted between 1963 and 1990.

At the onset of the proliferation of triggerfish in the Ghanaian shelf ecosystem, the species did not have any commercial value, consequently, it was discarded and no records were kept of the catches. Therefore, the importance of the species is not truly reflected in the catch statistics from the early years of its proliferation. It is also important to note that being a semi-pelagic fish, only the demersal component of the triggerfish was assessed in the bottom trawl surveys. The true magnitude of the abundance of the species is only reflected in the results of acoustic surveys conducted in the area. For example, whereas Koranteng (1989) put the demersal biomass of the species at about 52,000 mt in bottom trawl surveys conducted in 1981/82, Stromme (1983) gave the pelagic biomass of the species assessed in an acoustic survey conducted around the same period as 310,000 mt.

In spite of the impact of triggerfish on the relative importance of species, some species such as *Brachydeuterus auritus, Pagellus bellottii, Sparus caeruleostictus* and *Pseudupeneus prayensis* maintained their relative importance throughout the period of investigation (Table 19-3). For example, except during 1973-1977 when it moved down to the eighteenth position, *B. auritus* maintained its position at the top of the list giving way only to *B. capriscus* in the years that the latter species dominated this ecosystem. *Pseudotolithus brachygnathus and Loligo* sp. which were important during GTS, became completely unimportant for the rest of the time.

The total demersal biomass (excluding triggerfish) in waters of 100 m deep or less, reduced from about 78,000 mt evaluated during GTS to less than half this quantity in the years between 1973 and 1977 (about 36,000mt Table 19-4). In the survey conducted in 1979-80, a higher average biomass of 75,000 mt was obtained. In successive surveys, the biomass increased further to a peak of about 83,000 mt in 1981-82 but declined thereafter. The low value of the estimated biomass between 1969 and 1977 is rather significant. Several factors may have contributed to this situation the most obvious being the vessel's performance and the catchability of the survey gear. It is essential to note that all surveys conducted during the period in question were done with *R/V Research* which belonged to the Marine Fisheries Research Division.

Nevertheless, the density of the total demersal biomass underwent significant change over the 28-year period of investigation; starting at about 50 kg ha^{-1} in 1963, and ending at 32.4 kg ha^{-1} in 1990 (Koranteng 1998). The average density over this period was 35.6 kg ha^{-1}. Koranteng (*ibid.*) has shown that the difference in density of the total demersal biomass estimated in the GTS and Guinea 90 surveys is statistically significant (p = 0.02) with higher density recorded during the GTS than during Guinea 90. The reduction in density in shallow waters reflects the overexploitation of resources in this part of the continental shelf. On individual species and family basis (Figure 19-4), one can see a reduction in density of commercially important groups like snappers, groupers and seabreams.

The maximum sustainable yield estimated from the commercial catch data (about 49,000 mt per annum) agrees very well with the potential yield estimated from the trawl survey data (36,000 - 55,000 mt per annum, with an average of about 43,000 mt). From these results, it appears that the potential yield of the total demersal biomass on Ghana's continental shelf is of the order of 50,000 mt. per annum. In the period that the stocks were assessed through trawl surveys, landings from the fishery averaged between 11,000 and 60,000 mt per annum except in a few years (Figure 19-1). It is essential to note that whereas the biomass has been assessed with survey nets having minimum mesh size of 40 mm in their codends, several nets in use in the industry in Ghana, only have 25 mm codend mesh (Fisheries Department of Ghana, unpublished information). This means that the exploited biomass is much more than estimated in these surveys although much of it would be juvenile fish. For example, a large proportion of landings is made by beach seines, which exploit mainly juvenile fish (Mensah and Koranteng 1988).

The decomposition of the time series of catch per unit effort shows an underlying increasing trend between 1972 and 1977 and a decrease since then (Figure 19-8). This is further evidence of a fishery that has reached its limit of exploitation and showing signs of over-exploitation. The decline in CPUE, especially in the last ten years is worrying and seems to suggest that the effort on the fishery needs to be reduced.

Conclusion

The relative importance of species constituting the demersal fish stock of Ghana, as measured in trawl surveys, underwent considerable changes between the Guinean Trawling Survey in 1963 and the Guinea 90 survey in 1990. For nearly 20 years, *Balistes capriscus* replaced *Brachydeuterus auritus* as the most abundant species in the area. However, some sparid species, notably *Pagellus bellottii* and *Sparus caeruleostictus* and also the west African goatfish *Pseudupeneus prayensis* maintained their relative importance in the stocks over the 28 year period.

The total density of demersal fish species declined between 1964 and 1990, especially in shallow areas. This is a reflection of the intensity of exploitation in the inshore part of the continental shelf of Ghana. From the results of the surveys and the analysis with commercial catch and effort data it appears that the demersal fish stocks in Ghana's continental shelf waters are overexploited. The situation is aggravated by the taking of young immature fish by artisanal fishers using beach seines. The results of time series analysis showed that the fishery underwent two phases, namely a period of increasing catch per unit effort between 1972 and 1977 followed by a decline. This is a further indication of an over-exploited fishery.

References

Ansa-Emmim, M. 1979 Occurrence of the triggerfish, *Balistes capriscus* (Gmel), on the continental shelf of Ghana. pp20-27. *In*: *Report of the special working group on the evaluation of demersal stocks of the Ivory Coast - Zaire sector*, CECAF/ECAF SERIES/79/14(En) FAO, Rome.

Cadima, E.L. 1977 In Méthodes semi-quantitatives d'évaluations. pp131-141. *In*: J.-P. Troadec (ed.) *FAO Circular des Pêches* No. 701.

Caverivière, A. 1991 L'explosion démographique du baliste (*Balistes carolinensis*) en Afrique de l'ouest et son évolution en relation avec les tendances climatiques. pp354-367. *In*: P. Cury and C. Roy (eds). *Pêcheries ouest-africaines. Variabilité, instabilité et changement dans.* ORSTOM Editions, Paris.

FAO (UN Food and Agriculture Organisation), 1998 *FISHSTAT Plus. A PC system for the extended time series of global catches.* Prepared by FAO Fisheries Department, FAO, Rome.

FishBase, 1996 FishBase 96 CD-ROM. International Centre for Living Aquatic Resources Management (ICLARM), Manila.

Koranteng, K.A. 1981 Preliminary report on the 1979/80 trawling survey of demersal fish stocks in Ghanaian waters. Fisheries Research and Utilization Branch, Tema, Ghana (mimeograph).

Koranteng, K.A. 1984 A trawling Survey off Ghana, CECAF/TECH/84/63. Dakar, Senegal: CECAF Project (FAO).72 p.

Koranteng, K.A. 1989 A trawling Survey off Ghana, Supplementary Report No.1: Revised estimates. Marine Fisheries Research Technical Paper No. 3. Fisheries Research and Utilization Branch, Tema, Ghana.

Koranteng, K.A. 1998 *The impacts of environmental forcing on the dynamics of demersal fishery resources of Ghana.* PhD Thesis, University of Warwick. 377 p.

Mensah, M.A. and K.A. Koranteng 1988 A review of the oceanography and fisheries resources in the coastal waters of Ghana, 1981-1986. Marine Fisheries Research Report No. 8. Fisheries Research and Utilization Branch, Tema, Ghana.

Pauly, D. 1980a A selection of simple methods for the assessment of tropical fish stocks. *FAO Fisheries Circular* No. 729, FAO, Rome. 54 pp.

Pauly, D. 1980b On the interrelationship between natural mortality, growth parameters and mean environmental temperature in 175 fish stocks. Journal du Conseil International pour l' Exploration de la Mer 39(3): 175 - 192.

Ramos, A.R., I.S. Yraola, L.F. Peralta and J.F.G. Jimenez 1990 Informe de la Campana Guinea 90, Instituto Español de Oceanografia, Malaga, Spain.

Rijavec, L. 1980 A survey of the demersal fish resources of Ghana. CECAF/TECH/80/25, Dakar, Senegal: CECAF Project (FAO). 28 pp.

Salzen, E.A. 1957 A Trawling Survey off the Gold Coast. Journal du Conseil International pour l' Exploration de la Mer 23(1): 72 - 82.

Stromme, T. 1983 Final report of the R/V Dr. Fridtjof Nansen fish resource surveys off West Africa from Agadir to Ghana. May 1981 - March 1982. CECAF/ECAF Series 84/29 (En). FAO, Rome.

Stromme, T. 1992 NAN-SIS, Software for fisheries survey data logging and analysis. User's manual. FAO Computerized Information Series (Fisheries). No. 4. Rome: FAO.103p.

Williams, F. 1968 Report on the Guinean Trawling Survey. Organisation of African Unity Scientific and Technical Research Commission 99.

The Gulf of Guinea Large Marine Ecosystem
J.M. McGlade, P. Cury, K.A. Koranteng and N.J. Hardman-Mountford (Editors)

A Database for the Trawl Fisheries of Côte d'Ivoire: Structure and Use

F. Ménard, V. Nordström, J. Hoepffner and J. Konan

Abstract

The industrial trawl fishery of Côte d'Ivoire, based in Abidjan, started effectively in 1950. The fishery statistics have been well documented since 1966. However, the data processing has changed significantly during this time and some of the data have been lost. The purpose of this study was to retrieve and tabulate all trawl data available from different sources. Data have then been reformatted and transformed into a database in order to obtain a user-friendly system enabling data-requests and reports. The software chosen was MS ACCESS® because of its accessibility and wide dissemination. The trawl database covers the period from 1968 to 1997. The problems encountered, the solutions used and the structure of the database are presented, and its use illustrated by an application involving the extraction of all the catch and effort data from the continental shelf of Côte d'Ivoire since 1968. Effort data were standardized using a linear model applied to the catch per unit effort data, and a surplus production model was applied using the group of demersal species as a whole. The time series of the mean trophic levels of landings for the period 1968 to 1997, based on the catches of the main commercial species was also calculated.

Introduction

The industrial trawl fishery of Côte d'Ivoire is based in Abidjan. The fishery began in 1950 with the construction of the Vridi Canal (Caverivière 1979a, 1993; Bernard 1989a). A tenth of the vessels were active until 1965. Fishing areas were near Abidjan during this period. From 1965, more powerful vessels arrived in the fishery and the durations of the trips increased. Moreover, fishing zones spread out beyond Côte d'Ivoire: Ghana, Liberia, Sierra Leone and Guinea-Gambia were common fishing areas from 1968 to 1980, with the most significant part of the total catch from the area of Guinea-Gambia (Figure 20-1). From 1980, free access to the foreign countries was difficult and the most powerful vessels left the fishery: 1989 was the last year with catches from foreign countries. A specialized trawl shrimp fishery has existed since the mid-1960s: in 1970 there were about 20 vessels, since when the number decreased, albeit with some vessels always present in Abidjan. Even if they target shrimp, landed catches of demersal fish are significant and discards very important. Trawlers fish mainly in the 10-75m bathymetric strata. Total landed catches are dominated by 3 families: Sciaenidae, Sparidae and Pomadasyidae. Among the species, croakers (*Pseudotolithus spp*, Sciaenidae), red pandora (*Pagellus bellotti*, Sparidae), and small fish, dominated by bigeye

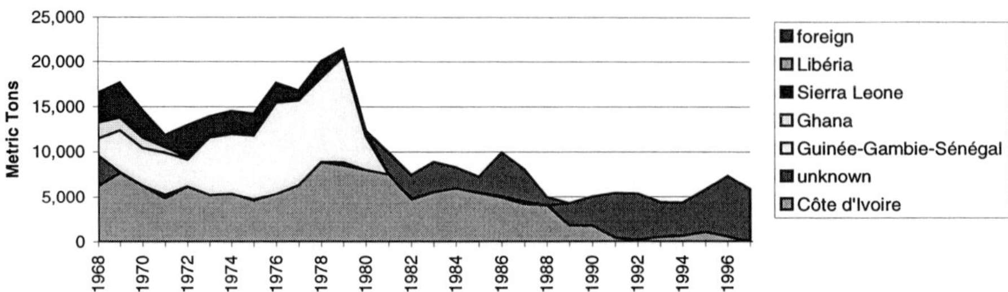

Figure 20-1 Total trawling landings (in MT) of the industrial fishery of Côte d'Ivoire by fishing area from 1968 to 1997.

grunt (*Brachydeuterus auritus*, Pomadasyidae), are the most frequent.

From 1966, the CRO (Centre de Recherches Océanologiques), the fishery research institute in Côte d'Ivoire, collected fishery statistics; these have been well documented since 1968. Unfortunately, most of the data were subsequently lost or dispersed. The purpose of this study was first to retrieve and process all the trawling data collected from the Ivorian trawler fleet available from different data sources, then to reformat the data and finally to transform them into a database format in order to obtain a user-friendly system covering data entry, queries and reports. The software chosen was MS ACCESS® and the trawl database covers the period from 1968 to 1997.

The next section presents the problems encountered, the solutions used and the structure of the database. The last section illustrates the use of the database, involving extraction of all the effort and catch data from the continental shelf of Côte d'Ivoire since 1968; fitting of a surplus production model using the group of demersal species; and estimating the trend in the mean trophic levels of landings for the period 1968 to 1997, based on the catches of the main commercial species.

Data analysis on the Côte d'Ivoire trawling fleet

Data overview and problems encountered

Trawl data from different sources are available from 1968 to 1997. They mainly cover the information from the landings in Abidjan (where all landings take place) on a trip by trip basis. The landed catches are put into boxes before the fish auction takes place, and the basic statistics are thus mainly measured in number of boxes by commercial species groups (some individual species). Sometimes the catches are measured in weight or 'batches' of fishes or in number of fish. The landings cover all the fish auction data, but certain 'noble' species (mainly lobsters, squids, groupers,...) go through other commercial channels, and are thus not included in the statistics. For certain years, some sampling of box weights, size and detailed catch areas/depths have been carried out. Unfortunately, most of the data had been lost due to a series of computer problems amongst others. It was thus necessary to retrieve the data from the different sources available; several types of problems were encountered during this stage and as a result some data have had to be re-entered in order to complete the data series.

- Computer-related problems

The data series corresponded to an extended time period, during which time data processing changed significantly (Fonteneau and Troadec 1969; Caverivière and Barbe 1977; Lhomme 1983; Bernard 1989b). In the beginning, the data were entered on punched cards and processed on a centralized computer using programs written in Fortran. Later the data were entered on diskette (of different types) and the processing was decentralized at the CRO. The data entry and the programs were finally recoded into Dbase, and the data dispersed on several PCs. The lack of permanent personnel assigned to work in this area also caused some supplementary problems.

i) Data coding and sampling problems

Several file formats were used by the CRO, partly due to data processing limitations, see above, but also due to changes in the collection of the parameters considered important.

Species - The most difficult problem to solve related to the changes in the species codes, since the data are entered according to commercial species groups rather than individual scientific species codes (due to marketing reasons and to the high number of species in the catches). A special list of 'species' was thus established, where a certain number of 'species' corresponded to commercial species groups.

Categories and box weights - The catches are coded according to different categories and are expressed in number of boxes, weight, number of fish or 'batches' of fishes. In order to find a uniform measure (number of boxes for example), conversion formulas should then be applied. This represents another problem, since the weight of the boxes has fluctuated over the time, and is not always well known from sampling due to lack of data (the box weights are only sampled and controlled on a very irregular basis). The 'lots' are considered to have 10 fishes and the mean weight by species has been established from different sources. An extensive study on the mean weight of the boxes and the categories found in separate documents was carried out. Because catches in weight for the period 1968-1985 occur by month and area they were used to improve the mean box weights before data-entry.

Logbook coverage - The sampling system of the logbooks was initiated at the CRO in 1966. On the whole, the catch series was complete, but the percentage of trips covered by logbooks has decreased during the last few years and at present there is a very low logbook coverage.

Areas/depths - The fishing areas are known for trips covered by the logbook sampling system run by the CRO. For the data from 1968 to 1974, only one area per trip was considered. From 1975 onwards, seven possible different fishing areas and depths for a trip were considered. In the case of no sampling, an indicator local/foreign has been created showing whether the trip took place in the Côte d'Ivoire or abroad. As the codes used varied over the years they were recoded to obtain a uniform series.

The vessels - Some problems were encountered when studying the vessel file, which is still not complete. Some vessel numbers used in the files were not included in the corresponding vessel file and have not yet been identified. Certain vessel characteristics were missing, which constituted a problem for the standardization of effort. Some vessels fish using a pelagic trawl net between two vessels: these vessels will have different species compositions in their catches and their effort will not always be clearly expressed.

Shrimp catch files - A specialized shrimp fishery has existed since the mid-1960s. The data corresponding to this fishery have to be recovered. In general, the vessels targeting shrimps have a special vessel code. This is however not always the case, and it is thus not easy to decide whether a trip/vessel has targeted shrimps or not. Shrimps also occur as by-catches in the landings of the generic trawlers. Some shrimp trips were also found mixed within the general files; these data were thus separated out into another file.

A separate file covering the global shrimp catches exists from 1985 to 1995. It also includes the total global catches of fish without species specification for these vessels.

Data availability

- Catch, price and size data

The data sampling started in 1966, but the data covering the two first years were only exist on paper. An attempt to scan these data was unsuccessful. There were four major periods with different data formats and types of data.

- Box mean weights

1968-1974	No data.
1975-1983	Mean annual weights are available by species or species groups.
1984	Detailed box weights available on paper.
	The means have been recovered by species or species groups.
1985	Mean annual weights are available by species or species groups
1986-1987	Detailed data available from box weight sampling by species
1988	Detailed file available with weights by species
1989-1994	Detailed data available by month
1995-1996	No data
1997	Some samples are available for June, July and December

- Total catch file

The mean box weights covering the period from 1968 to 1985 exist, but they have not yet been recovered. In order to retrieve the mean box weights from catches in number of boxes and catches in weight, the catches in mt coming from the annual statistics (which exist on paper) have been input via scanner.

Data processing: reformatting of the basic files and creation of intermediate files

This part of the data processing was carried out using SAS and Fortran programs. Initially SAS was not available, and the Dbase files used were converted into ASCII files before use in the Fortran programs. A detailed description of the programs as well as the flowcharts corresponding to the data processing of the basic files and the file formats are available on request from the authors.

- Trip data, detailed catch/price data and effort

The data corresponding to these series (ASCII or dbf files) were transformed into three groups of files – trip files, detailed catch/price files and effort files. Comparative tables of catches in # of boxes and weight have been established and through use of mean weights, when available, the outputs have been verified.

- Mean box weights

First, the different data files were put together in order to obtain a file by year and species when available. For the years 1968-1975 catch data were available (with some problems concerning the species groups) in weight and number of boxes, and the mean weights were then calculated from these two sets. The tables were then analyzed and some adjustments and interpolations carried out. Figure 20- 2 illustrates the fluctuations of the yearly mean box weights for all the species or group of species.

Description of the database

The database was created using Microsoft Access 97. A database enables data to be organized coherently, modified and augmented in a simple way. It can also optimize data-storage and makes it relatively easy to compute statistics. Microsoft Access 97 was chosen because of its ease of accessibility, both in terms of the software itself as well as its operation and data processing.

In order to access the database a user interface was developed. The purpose of this interface was to provide a simple and very intuitive tool for the user, that would require a minimum learning period. The interface was designed around certain types of queries but it is evolutionary and can easily be extended when new analyses become necessary.

- Description of the interface

The interface covers two major topics : calculation of catches and the associated efforts. The following parameters are used to define the queries :

i.   Time period, in years. The user should select beginning and end years for the data to be included in the query;
ii.   Fishing area;
iii.   Time interval by which the results will be output : year, quarter, month or fortnight;
iv.   And, only for the catches, the species required.

The results of a query are given in metric tons. Concerning effort, the outputs are shown in time fishing and time at sea, both expressed in number of hours.

For the effort data, there is a problem concerning missing data. When showing the results of any effort query the user is given the possibility to check the percentage of data present in the database for the present query compared to the total number of trips. In this way the user can get an idea of the quality level of the data and the possible adjustments that should be applied to the data in order to obtain the best results

The next stage is to develop a data entry interface in order to directly input the landing and logbook data which are collected each day in the port of Abidjan. This interface will facilitate the data entry and will also provide validation procedures.

- Database organization

Ten tables make up the database, each table covering a group of coherent information. The tables are inter-related via certain variables (date and vessel number, which uniquely define a trip) allowing the reconstruction of the initial information. The database is composed of :

i.   Three tables covering the catch and effort data
    fish auction table with the general data from the fish auctions (by date and vessel);
    catch table with the detailed information of the trips (species, category, catch and price);
    effort table contains the information on the areas and duration of the trips;
ii.   Five parameter tables, that should be transparent to the user (fishing areas, depths, species, categories, fishery type)
iii.   Two information tables : mean box weights by year and species and the vessel file (vessel type, vessel characteristics, age, ...). Figure 20-3 shows the organization of the database.

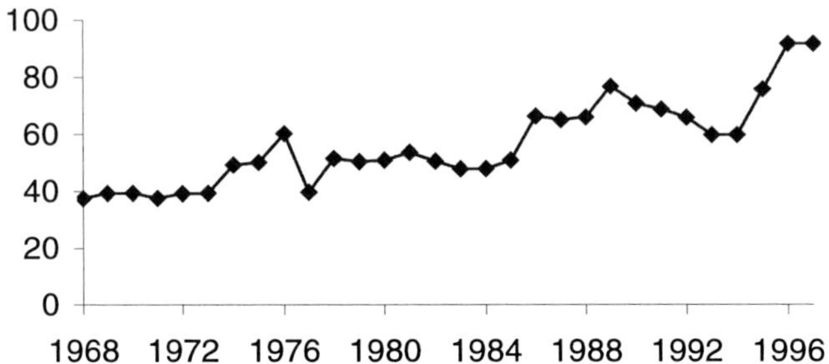

Figure 20-2 Yearly mean box weight (in kg) estimated for all the species or group of commercial species from 1968 to 1997.

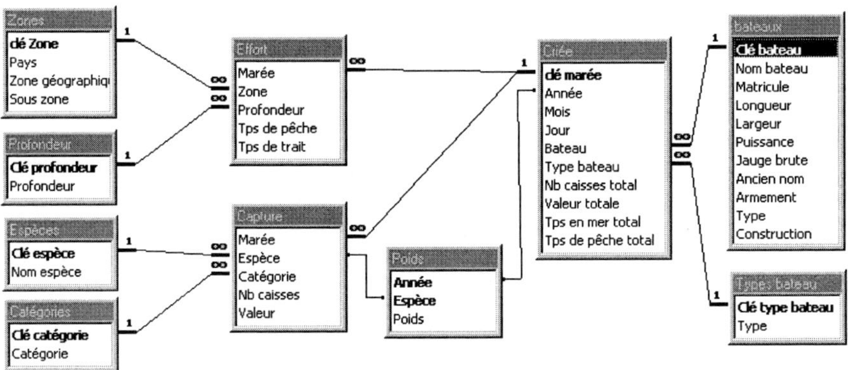

Figure 20-3 Organization of the database: composition of tables and relations.

An example of the use of the database: application of a surplus production model

Extraction of the data

To illustrate the use of the database, two preliminary analyses was undertaken: a surplus production model using the group of demersal species, and an estimation of the fluctuations of the mean trophic level of landings for the period 1968 to 1997.

Yearly catch data were extracted from the data base for the period 1968-97, considering the continental shelf of Côte d'Ivoire only. Catch data from unknown areas since 1969 were assumed to come from the continental shelf of Côte d'Ivoire. For the fishing effort of the trawlers, it was not possible to extract a fishing time for the complete series. Thus, the yearly fishing effort computed was the sum of the time at sea of all the active fishing vessels in the selected area.

Table 20-1 gives the results of the extraction: total yearly catch (in mt) and effort data (in hours).

- Standardization of fishing data

The fishing effort has to be standardized, because trawlers are of varying sizes and types. The fishing power correction factor by vessel has to be extracted These factors are applied to the fishing effort of individual vessels.

However a lot of vessel characteristics are missing from the data files. A mutiplicative model for the CPUE of the group of demersal species has been developed. The model includes a factor for the year (α_{year}), a factor for the season ($\alpha_{quarter}$), considered here as the quarter, and a factor for vessel (α_{ship}). Interactions between these main effects have been investigated, but the interaction between year and quarter is only kept in the final model, including a transformation of the cpue that normalizes the distribution :

$$\log(\text{cpue}+.005) = \mu + \alpha_{year} + \alpha_{quarter} + \alpha_{ship} + \alpha_{year*quarter} + \varepsilon \qquad \textbf{Eq. 20-1}$$

where α_{ship} is the fishing power factor.

Table 20-2 presents the ANOVA table (R-square = 0.759). Figure 20-4 illustrates the results of the model estimations, the fishing power correction factor of individual ships (Figure 20-4a) and the pattern of the residuals (Figure 20-4b). Another way to evaluate the model estimations is to plot the fishing power factor versus the engine power of the individual ships, when available (Figure 20-4c). The R-square is not very high and several outliers need to be analyzed more closely.

- Surplus production model estimation

The standardized effort (Table 20-1) was related to year 1997, leading to a series of multiplicators of effort (noted mf_n, n for the year, with $mf_{1997} = 1$). Two surplus production models were fitted to the catch data and the series of fishing efforts using the effort averaging method of Fox (1975):

i. a generalized surplus production model (the Pella and Tomlinson 1969 form written $\text{cpue} = (a + b \times mf)^{1/(m-1)}$) leading to the estimation of 3 parameters a, b and m ;

ii. a Fox 1970 model (written $\text{cpue} = a \times e^{b \times mf}$) because the estimation of m was very close to 1.

Both models give the same equilibrium relation for the cpue versus the multiplicator of effort. Estimated parameters (Table 20-3) are used finally to compute equilibrium yields for different multiplicators of fishing effort. Figure 20-5 illustrates the equilibrium and observed yields versus the multiplicator of effort, and Table 20-3 gives the estimation of MSY and the corresponding multiplicator of effort.

Year	catch (in mt)	effort	std effort
1968	6240	78345	4699
1969	7761	116803	7092
1970	6282	80551	5135
1971	5274	70472	4603
1972	6165	92345	5634
1973	5253	81696	5099
1974	5364	67032	4590
1975	4546	56196	3628
1976	5254	61792	3955
1977	6293	65817	5313
1978	8900	86090	7203
1979	8785	73280	6365
1980	7952	62432	6645
1981	7414	54721	5352
1982	4502	47267	4186
1983	5487	77964	5978
1984	5946	94323	7017
1985	5205	94121	6896
1986	5039	67750	4857
1987	4425	66581	4847
1988	4082	66493	4673
1989	4258	59068	4127
1990	5022	59988	4507
1991	5394	70514	5001
1992	5319	83862	6280
1993	4409	77458	6286
1994	4320	69590	5782
1995	5779	83865	7854
1996	7331	90474	9658
1997	5819	63359	6134

Table 20-1 Catch (in mt), effort (in hours) and standardized effort (see text) data of the industrial demersal fishery of Côte d'Ivoire from 1968 to 1997.

Source	DF	Mean Square	Pr > F
year	29	2.27	0.0001
quarter	3	0.83	0.0001
ship	124	1.86	0.0001
year*quarter	87	0.15	0.0001

Table 20-2 Analysis of variance table of the multiplicative model for the CPUE.

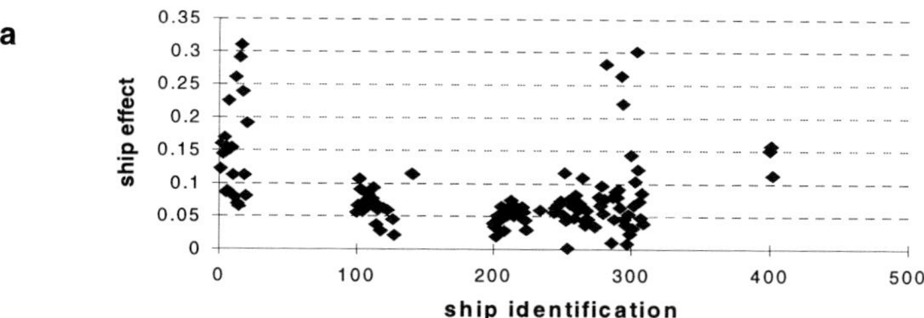

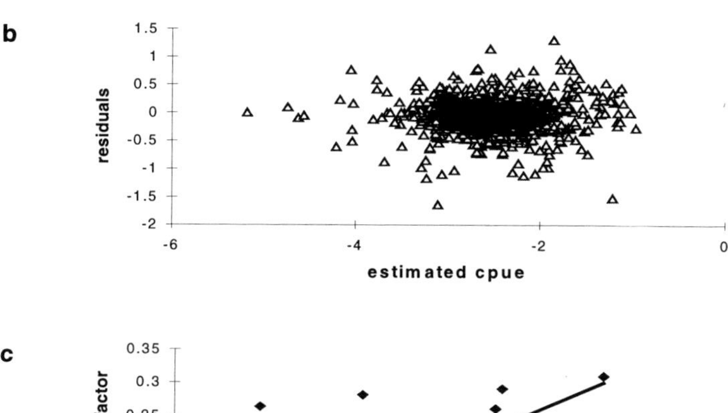

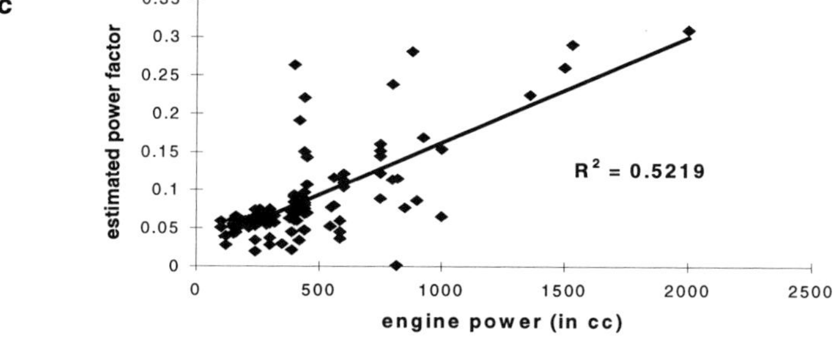

Figure 20-4 Results of the estimations of the multiplicative model for the CPUE : **a** fishing power factor of individual ships; **b** residuals versus estimated CPUE; **c** fishing power factor versus engine power.

Model	mf_{MSY}	MSY	Y_{97}/MSY
$\text{cpue} = (-0.0173 + 1.263 \times mf)^{1/(1.025-1)}$	1.8	6940	0.782
$\text{cpue} = 10437.1 \times e^{-0.552 \times mf}$	1.8	6958	0.780

Table 20-3 Estimated parameters of the surplus production models, maximum sustainable yields (MSY) and corresponding multiplicators of effort related to year 1997.

Mean trophic level of landings

Following the methodology of Pauly *et al.* (1998), the yearly mean trophic level of landings of the industrial demersal fishery was estimated. The landings of the main commercial species caught from 1968 to 1997 (except 1972-1974) were extracted from the database. Sixteen commercial groups were selected. Their trophic levels (see Pauly *et al.* 1998 for details) come from the FishBase 97 CD-ROM (Froese and Pauly 1997) (Table 20-4). Mean trophic level tl_i for year i is computed as the mean of the trophic level of each commercial species j weighted by the catches (Y_{ij}) :

$$tl_i = \sum_{ij} tl_{ij} \frac{Y_{ij}}{\sum_j Y_{ij}}$$

Eq. 20-2

Figure 20-6 exhibits no particular trend of the mean trophic level.

Discussion

The demersal stock of Côte d'Ivoire, when considered as a whole, seems to be under-exploited. Caverivière (1979b) estimated a maximum sustainable yield of 9950 MT for the period 1959-1970. Bard *et al.* (1988) gave an estimation of 5800 MT for the period 1971-1985. The purpose here is not to discuss the validity of these estimations, nor to give a diagnosis for the demersal fishery of Côte d'Ivoire from these estimations. Indeed all the assumptions of such an assessment that could not clearly be considered as a result of the complexity of such a tropical fishery need to be kept in mind.

The use of a surplus production model does not allow a drift in the exploitation pattern to be taken into account. Assessment by age-structured models are the only ones able to take account of changes in the exploitation pattern. But the quality of the size data does not allow the use of such models. Surplus production models need fishing effort estimations. Here, fishing powers are assumed to be stable in time and standardized effort were estimated from effort computed from time at sea for all the vessels. Effective trawling time were not available because of the poor return of the log books. Moreover, the standardization does not take into account any spatial effects; this is important as the sciaenid and sparid fishery, the main group of species exploited by the trawl fishery, is carried out over different bathymetric strata. The standardized fishing effort should thus be considered as a proxy of the actual effort applied to the stocks.

Data from the artisanal fleet are not available, and catches from high value/quality fish are not incorporated in the statistics. Furthermore, the demersal catches taken by the purse seiners targeting *Sardinella* resources, nor the catches carried out by trawlers targeting shrimps were not considered, even though the amounts are significant in some years. All the demersal species were considered as a whole in this assessment. Fishing mortalities were maintained in the same proportions for all the species or group of species. The relative species abundance was allowed to fluctuate, whilst the global level was held stable.

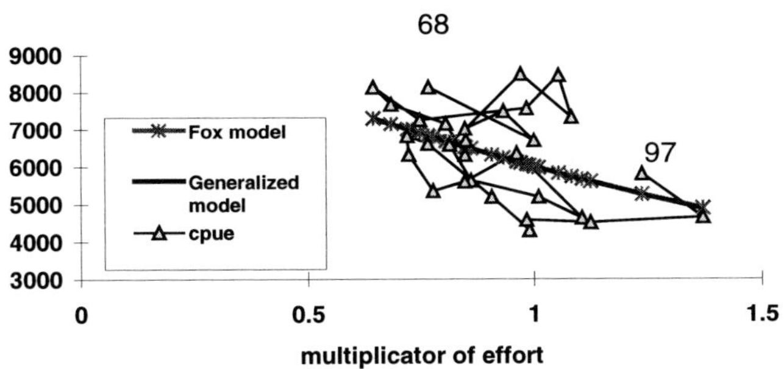

Figure 20-5 Observed and equilibrium CPUE and yields estimated by surplus production models fitted to the demersal commercial species against the multiplicator of effort.

Species or taxa	Common name	Trophic level
Pseudotolithus spp.	Croakers, canary drum	3.30
Brachydeuterus auritus, Pteroscion peli, Penteroscion mbizi	Bigeye grunt, boe grunt	3.10
Polydactylus quadrifilis, Galeoides decadactylus	African threadfins	3.30
Pagellus bellottii, Dentex congoensis, Dentex angolensis	red pandora	3.20
Ilisha africana	west African ilisha	2.80
Sphyraena spp.	Barracudas	4.50
Chloroscombrus chrysurus, Vomer setapinis	Carangidae	3.30
Brotula barbata	Bearbed brotule	3.40
Trichiurus lepturus	Largehead hairtail	3.80
Pomadasys spp.	Spotted grunts	3.50
Epinephelus sp, Mycteroperca rubra, Cephalopholis spp.	Groupers	3.80
Pseudupeneus prayensis	Goatfish	3.30
Cynoglossus spp, *Synaptura sp,Dicologoglossa hexophthalma*	Tonguesoles	3.00
Many species	Squales	3.80
Rajidae spp., *Dasyatis centroura, Dasyatis marmorata*	Rays	3.80
Many species	others	3.50

Table 20-4 Trophic levels of the main commercial species caught from 1968 to 1997.

The lack of a decreasing trend in the yearly mean trophic level over the past 30 years shows that there are no any major changes in the structure of the demersal food web of the continental shelf of Côte d'Ivoire. However, as mentioned previously, only the landings of the industrial fishery has been taken into account.

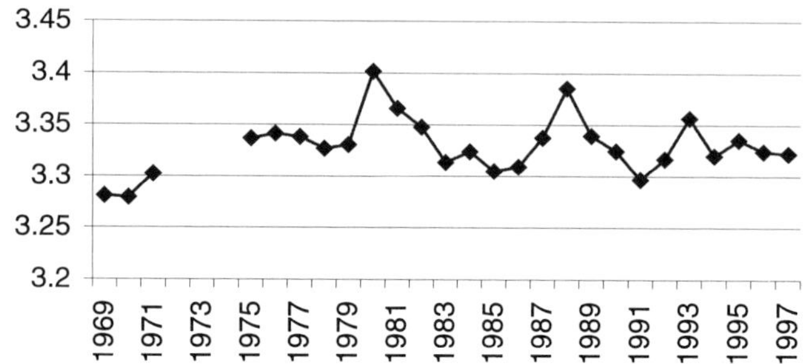

Figure 20-6 Time series of mean trophic level.

Conclusion

The database developed in this study is a powerful tool to analyze the dynamics of the industrial trawl fishery. To analyze the data in more detail, the next step will be to take into account the species or the group of commercial species using a multispecies approach. Such an approach has been applied to the main demersal resources exploited off Senegal (Gascuel and Ménard 1997). Similar tools should now be developed for other fisheries, in order to have data to allow a more extensive assessment of the health of the exploited fish communities. However, the availability and the quality of the data could be a limiting factor.

The database should also be incorporated into a regional database, although the confidentiality of some of the data will make it necessary to aggregate the data into a suitable format.

References

Bard, F.X., P. Vendeville and J. Konan 1988 Analyse des pêches chalutières ivoiriennes par l'usage du modèle de production global. *FAO/COPACE/TECH/88/88*. 9p.

Bernard, Y. 1989a La pêche chalutière ivoirienne de 1981 à 1988. Comparaison avec la période1959-1980. Rapport interne Centre Rech. Océanogr. Abidjan. 40p.

Bernard, Y. 1989b STATCHAL Chaîne de statistiques chalutières – Notice d'utilisation. Rapport interne Centre Rech. Océanogr. Abidjan. 29p.

Caverivière, A. 1979a La pêche industrielle des poissons démersaux en Côte d'Ivoire. Généralités–localisation–homogénéité relatives des zones de pêche ivoiriennes. Doc. Scient. Centre Rech. Océanogr. Abidjan 10(2): 43-93

Caverivière, A. 1979b Estimations des potentiels de pêche des stocks démersaux ivoiriens par les modèles globaux. Effet de la prolifération du baliste (*Balistes capriscus*). Doc. Scient. Centre Rech. Océanogr. Abidjan 10(2): 95-164

Caverivière, A. 1993 Les ressources en poissons démersaux et leur exploitation. pp427-488. *In*: P. Le Loeuff, E. Marchal and J-B. Amon-Kothias (eds). *Environnement et ressources aquatiques de Côte d'Ivoire. I: Le milieu marin*. ORSTOM, Paris.

Caverivière, A., F. Barbe 1977 Traitements statistiques des données de la pêche au chalut des poissons démersaux en Côte d'Ivoire. Prises-efforts-distributions de fréquence. Arch. Scient. Centre Rech. Océanogr. Abidjan, 2(1). 16p.

Fonteneau, A. and J.P. Troadec 1969 Récolte, stockage et traitement des données statistiques relatives à la pêche au chalut en Côte d'Ivoire. Doc. Scient. Prov. Centre Rech. Océanogr. Abidjan 38.

Fox, W.W. 1970 An exponential surplus-yield model for optimizing exploited fish populations. Trans Am. Fish. Soc. 99: 80-88.

Fox, W.W. 1975 Fitting the generalized stock production model by least-squares and equilibrium approximations. U.S. Fish Bull. 73: 23-37.

Froese, R. and D. Pauly 1997 FishBase 97, concepts, design and data sources. International Center for Living Aquatic Resources Management (ICLARM), Manila, Philippines.

Gascuel, D. and F. Ménard 1997 Assessment of multispecies fishery in Senegal, using production models and diversity indices. Aquatic Living Resources 10(5): 281-288.

Lhomme, F. 1983 Méthodes de codage et traitement informatique utilisés par le programme "exploitation rationnelle des stocks du plateau continental ivoirien". Note à diffusion restreinte Centre Rech. Océanogr. Abidjan. 63p.

Pauly, D., V. Christensen, J. Dalsgaard, R. Froese and F. Torres 1998 Fishing down marine food webs. Science 279: 860-863.

Pella, J.J. and P.K. Tomlinson 1969 A generalized stock production model. Bull. Inter-Am. Trop. Tuna Comm. 13: 421-458.

The Gulf of Guinea Large Marine Ecosystem
J.M. McGlade, P. Cury, K.A. Koranteng and N.J. Hardman-Mountford (Editors)

21

Spatial Modelling of Trophic Interactions and Fisheries Impacts in Coastal Ecosystems: A Case Study of Sakumo Lagoon, Ghana

D. Pauly

Abstract

A trophic model of a small West African coastal lagoon (Sakumo Lagoon, near Tema, Ghana), consisting of 14 interacting functional groups, was constructed using the Ecopath approach and software, based on field data gathered in the early 1970s, and reflecting an early stage in the development of this now much-modified ecosystem. This trophic model, after some test runs with Ecosim, was used as basis for an application of the 'Ecospace' routine of Ecopath, recently developed by Carl Walters, Fisheries Centre, University of British Columbia. This generates plots of the relative biomasses of functional groups onto a grid map, while accounting for their trophic interactions, their movements towards preferred habitats and the distribution of fishing effort. This example illustrates an approach that may be widely used to represent and analyse the coastal and marine ecosystems of the Gulf of Guinea, and other areas where 'ecosystem-based management' is required, while data are sparse.

Introduction

There is, among decision-makers and scientists, wide agreement that present practices in fisheries and coastal development are unsustainable, and should be replaced by 'ecosystem-based' approaches. There is, on the other hand, much less of a consensus as to the integrative conceptual and/or software tools that may be used to implement such approaches. Also, many scientists claim that "there is not enough data" for integrative approaches of any sort, and this is one of the reasons why many decision-makers have lost patience with us.

This loss of patience is understandable, given that making this claim is, with many scientists, a reflex that kicks in every time they are confronted with the task of integrating the immense amount of largely un-analyzed data that publicly-funded research has been allowed to accumulate over the last decades.

This author and his colleagues have argued elsewhere that, at least as far as aquatic ecosystems are concerned, models explicitly accounting for trophic interactions, and structured around the principle of mass-balance, allow for rigorous integration of much of what we know about aquatic ecosystems. This approach also allows the embedding of fisheries, and their resources, base into the ecosystems that sustain them (Christensen and Pauly 1992, 1993; Pauly and Christensen 1995).

Critics of this approach, and of Ecopath, the software developed for its implementation (Christensen and Pauly 1992; Pauly and Christensen 1995, and see www.ecopath.org), have mainly concentrated on three points:

a) it did not include an explicit procedure for considering uncertainty in input parameters;

b) it did not consider temporal dynamics, and thus did not allow asking 'what if questions?'

c) it did not include a spatial structure, though spatial processes are crucial to trophic interactions.

Item (a) was accommodated by including a semi-Bayesian resampling routine (Ecoranger) into Ecopath, which allows consideration of prior distributions of Ecopath inputs, and the output of their posterior distributions, besides distributions of estimates, a procedure found suitable by practitioners (see contributions in Pauly 1998).

Item (b) was accommodated through the development of Ecosim (Walters *et al.* 1997), whose latest version, incorporated into Ecopath with Ecosim (available from www.ecopath.org) includes, for split pools, a version of the delay-difference model of Deriso (1980).

Item (c) was recently resolved, again by Carl Walters, through the development of Ecospace, a spatially-structured version of Ecopath documented in Walters *et al.* (1999), and whose first tests by a wide range of users are documented in Pauly (1998).

This contribution presents an application of Ecospace, now a routine of Ecopath with Ecosim (see above), to Sakumo Lagoon, a small (1 km^2) water body near Tema, Ghana, West Africa, which the author studied in 1971, and which may be seen as representing many other coastal lagoons along the coast of the Gulf of Guinea. Since it was originally studied, Sakumo Lagoon has been much modified by overfishing, pollution and other injuries (Ntimoa-Baidu 1991; Koranteng *et al.* 1998), a topic not followed upon here.

Materials and methods

The data used for this study stem mainly from three earlier contributions describing the ecology of Sakumo Lagoon, and the adaptations of its dominant fish species, the cichlid *Sarotherodon melanotheron* (Pauly 1975, 1976; Pauly *et al.* 1988). This was complemented with data from (models of) similar systems, mainly adapted from contributions in Christensen and Pauly (1993).

The data requirements of an Ecopath model are expressed by its mass-balance equation, which for each of its functional group i demands that

$$B_i * (P/B)_i * EE_i = Y_i + \Sigma B_j * (Q/B)_j * DC_{ij}$$ **Eq. 21-1**

where B_i and B_j are biomasses (the latter pertaining to j, the consumers of i); P/B_i their production/biomass ratio, equivalent to total mortality under most circumstances (Mertz and Myers 1998); EE_i the fraction of production ($P = B * (P/B)$) that is consumed within, or caught from the system Y_l is equal the fisheries catch (i.e., $Y = F*B$); Q/B_j the food consumption per unit biomass of j; and DC_{ij} the contribution of i to the diet of j.

Fourteen functional groups were found appropriate, given the data at hand, to represent the interacting species of Sakumo Lagoon. They are defined in Table 21-1, which also gives the values of B, P/B, Q/B, and EE either entered into (with sources), or output by the Ecopath parameterization routine, which solves (1) through a robust matrix inversion (MacKay 1981).

Table 21-2 presents the diet matrix (the DC_{ij}) used, based on field data for the fish, on the known ecology of the invertebrates groups, on the mass-balance requirement implied in (Eq. 21-1) and on several runs of Ecosim (Walters *et al.* 1997) to verify that the system was dynamically stable.

Group Name	TL	Biomass (t·km²)	P/B (year⁻¹)	Q/B (year⁻¹)	EE	Preferred habitat
Adult tilapias	2.00	3.50	4.0	35.1	**0.986**	I,R.
Juv. Tilapias	2.80	0.20	6.0	20	**0.840**	I,R
Penaeids	2.31	0.03	3.5	15	**0.960**	C
Ethmalosa	2.30	0.06	2.0	16	**0.423**	All
Mar. carn.	3.36	0.05	2.0	15	**0.508**	C
Callinectes	2.81	0.05	2.0	10	**0.500**	C
Clibanarius	2.35	0.01	2.0	15	**0.500**	C
Tympanotonus	2.21	0.05	2.0	10	**0.500**	I,R
Oysters	2.10	5.00	1.0	20	**0.500**	I,R
Misc. benthos	2.05	0.15	5.0	50	**0.908**	All
Zooplankton	2.00	1.00	50.0	250	**0.271**	All
Phytoplankton	1.00	1.00	400.0	0	**0.740**	All
Detritus	1.00	10.00	-	-	**0.894**	All

Table 21-1 Basic inputs of Ecopath model of Sakumo Lagoon (see text for sources), and outputs (in bold characters). Habitats are C (Coastal); I (Interior), R (Rivers), or All.

Prey No.	Prey\Predator	1	2	3	4	5	6	7	8	9	10	11
1	Adult tilapias	-	-	-	-	-	0.6	-	-	-	-	-
2	Juv. Tilapias	-	-	-	-	0.010	-	-	-	-	-	-
3	Penaeids	-	-	-	-	0.001	-	-	-	-	-	-
4	*Ethmalosa*	-	-	-	-	0.001	-	-	-	-	-	-
5	Mar. carnivores	-	-	-	-	0.001	-	-	-	-	-	-
6	*Callinectes*	-	-	-	-	-	-	-	-	-	-	-
7	*Clibanarius*	-	-	-	-	-	-	-	-	-	-	-
8	*Tympanotonus*	-	-	-	-	-	-	-	-	-	-	-
9	Oysters	-	-	-	-	-	-	-	-	-	-	-
10	Misc. benthos	-	-	0.2	-	0.001	0.2	0.1	0.2	-	0.05	-
11	Zooplankton	-	0.8	0.1	0.3	0.001	-	-	-	0.1	-	-
12	Phytoplankton	-	0.2	0.1	0.1	0.001	-	-	-	0.7	-	0.9
13	Detritus	1	-	0.6	0.6	-	0.2	0.2	0.8	0.2	0.95	0.1
14	Import	-	-	-	-	0.984	-	0.7	-	-	-	-

Table 21-2 Diet matrix used for construction of Sakumo Lagoon ecosystem model. Most inputs are from Pauly (1975, 1976); other inputs are based on basic biology, then refined by back-calibration.

Ecosim is structured around the reinterpretation of Equation (1) as a system of coupled differential equations:

$$dB/dt_i = [B_i * (P/B)_i * EE_i] - F_i * B_i + \Sigma a_{ij} \qquad \textbf{Eq. 21-2}$$

where dB/dt_i is the growth rate of groups i in terms of its biomass, $a_{ij} = Q_{ij}/(B_i * B_j)$ is the Lotka-Volterra mass-action term, and all other terms are defined as in (Eq.. 21-1). However, the biomasses, in Ecosim do not strictly follow Lotka-Volterra dynamics:

a) the B_i are divided into vulnerable and invulnerable components (Walters *et al.* 1997), and it is the transfer rate between these two components (adjustable by the user) which determines if control is top-down (i.e., Lotka-Volterra), bottom-up (i.e., donor-driven), or intermediate (as used here);

b) in case of split pools (juveniles vs. adults of the same species), account is kept of the numbers that recruits from the juvenile to the adult stages (using a delay-difference model, which allows stock-recruitment relationships (not further discussed here) to be among the Ecosim outputs. (Split pools are used here only for the tilapia *Sarotherodon melanotheron*, which dominates the biomass of Sakumo Lagoon, and which undergoes at about 4 cm a strong ontogenic diet shift, from plankton to detritus (Pauly 1976, and see Table 21-2).

Ecospace is a dynamic version of Ecopath, incorporating key elements of Ecosim (including a & b above), and which allocates biomass across a grid map (usually 20 x 20 cells, as used here, but more are possible), while accounting, among other things, for:

1) symmetrical movements from a cell to its four adjacent cells, of rate m, modified by whether a cell is defined as 'preferred habitat' or not (running means over sets of five cells allows for smooth transitions between habitat types);

2) increased predation risk (here by a factor of 1.5 for all groups) and reduced feeding rate (here by a factor of 0.5 for all groups) in non-preferred habitat;

3) a level of fishing effort that is proportional, in each cell, to the biomass of the target group in that cell, and whose distribution can also be made sensitive to costs (e.g., of sailing to certain areas).

Based on Pauly (1975), three habitat types can be distinguished within Sakumo lagoon:

1) *Coastal*, near the mouth of the lagoon, which is kept permanently open by a sluice, and in which therefore allows coastal organisms to swim, or to be washed into the lagoon, especially at high tide;

2) *Interior*, referring to the much larger, interior part of the lagoon, in which the dominant process is the deposition at a mean depth of 50 cm, of the terrigenous and planktonic detritus that is the food of adult *S. melanotheron*; and

3) *Rivers*, referring to the small areas covered by the mouths of small rivers connected to the *Interior*, and which contain, at least following the raining season, residues of a freshwater fauna.

The assignment of the 14 functional groups to either of these three preferred habitat, or to '*all*' for groups with uniform distribution, are given in Table 21-1. The defaults movement rates (m) assumed for the various groups were used. These values are tentative, but similar results (i.e., distribution maps, see below) were obtained using different values.
Based on field observation (Pauly 1975), the fishery was assumed to be distributed uniformly with regards to the habitat types, i.e., it is only the distribution of the organisms (mainly

tilapia) which determined where the fishers operate, and not costs (which can also be accommodated by Ecospace).

Results and Discussion

Figure 21-1, representing the trophic interactions implied in Tables 21-1 and 21-2, complements a previous qualitative representation of these same interactions (Pauly 1975; Figure 12). As might be seen, *S. melanotheron* entirely dominates the biomass of Sakumo Lagoon. Also, its food consumption is higher than that between all other groups, except those between phytoplankton and detritus.

This is emphasized by the matrix of mixed trophic impacts (not shown), from which one can infer that only changes in tilapia or in detritus biomass could markedly impact the fishery, or the other groups in the system. Consequently, Ecosim suggests that increased fishing on tilapia would directly or indirectly affect other groups in the systems (Figure 21-2), while increased fishing on other groups would not affect the tilapias (simulations not shown). Overall, the robust results obtained here with Ecosim suggest that the original description of Sakumo lagoon as an ecosystem driven by the detritus consumption of *Sarotherodon melanotheron* still holds.

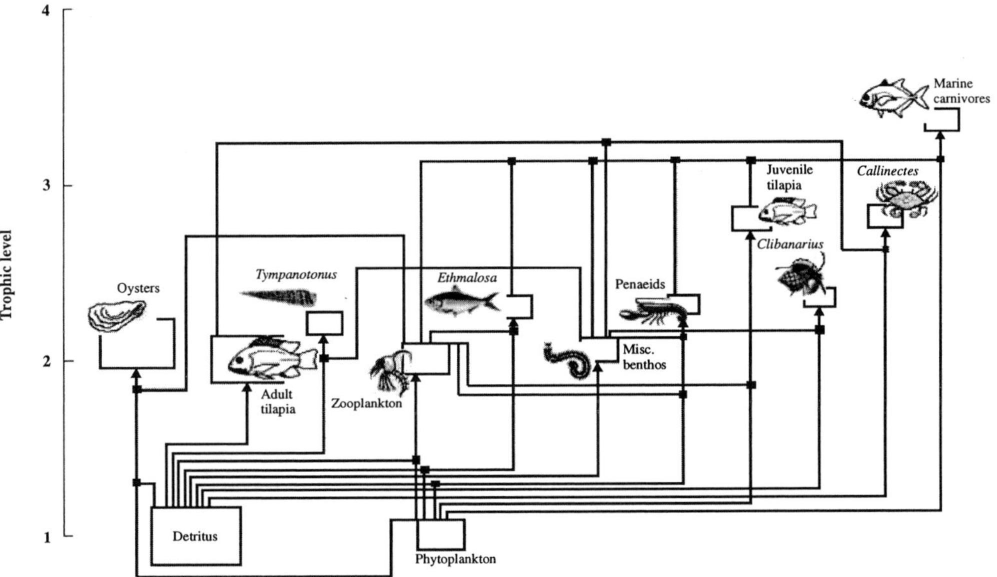

Figure 21-1 Flowchart of trophic interaction in the Sakumo Lagoon ecosystem in the early 1970s, as constructed using Ecopath based on the data in Tables 21-1 and 21-2. Backflows to the detritus (ungrazed phytoplankton, fecal matter, etc.) are not shown.

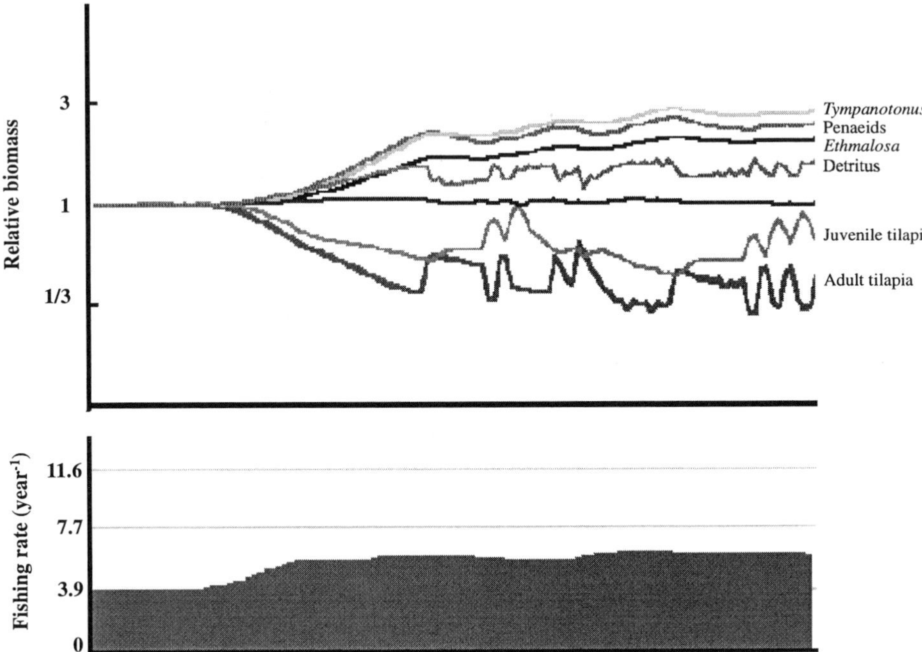

Figure 21-2 Result of a run with Ecosim, the dynamic version of Ecopath (Walters *et al.* 1997) of the file representing the Sakumo Lagoon ecosystem. The scenario run here assumes an increase of about 30 % of the fishing mortality on adult tilapia (baseline). Note resulting increase of various competitors, and of detritus. Note also the oscillations induced by the delay-difference model, wherein the juvenile and adult tilapia alternate in abundance. This behavior can be suppressed, or amplified by changing the setting of Ecosim, and is presented here only for information, as it has so far not been shown to really occur in *S. melanotheron*, or any other tilapia.

The different panels of Figure 21-3 (colour plate) illustrate the key results of the present exercise, i.e., that Ecospace can reproduce the observed distribution of different functional groups in Sakumo Lagoon, while simultaneously accounting for their movements, and their interactions with their preys and predators (including the fishery). As might be seen, the distribution of all groups (except phytoplankton) reflects our definition of three habitat types through their relative equilibrium biomass (as attained after a simulation period of 10 years; see insert on lower right corner of graph).

Overall, these distributions correspond to those observed by the author (see Pauly 1975, especially Figure 11). More importantly, a few interesting predictions are made that suggests realism. Notably, the miscellaneous carnivores, of entirely marine origin, were very hard to accommodate in term of their food requirements, within or close to that small part of the lagoon here defined has having a 'coastal' habitat. Indeed, the miscellaneous carnivores declined under most scenarios (as indicated in the insert), reflecting their nature as occasional visitors (Pauly 1975). However, we shall abstain here from 'tweaking' the model until it closely matches the observed distributions. This would be pointless, given that the model, sophisticated as it is, cannot capture the key change that appears to have occurred in the lagoon: the miniaturization (as a genetic response to constant, strong fishing pressure?) of *Sarothorodon melanotheron*, which now appears to reach 5 cm at most, and to mature at an even smaller size (pers. obs., July 1998).

Rather, we shall conclude by proposing that the tools presented here – Ecopath for food web construction and analysis, Ecosim for temporal simulations, and Ecospace for spatial simulations – are well suited for studying the various subsystems of the Large Marine Ecosystem represented by the Gulf of Guinea, just as they have been shown to be useful when applied to other ecosytem types (Longhurst and Pauly 1987; Christensen and Pauly 1993; Walters *et al.* 1997, 1999). Further, the author offers to assist interested colleagues in constructing models such as presented here (see also www.ecopath.org.). The time has come to make sense of the data we have been gathering all these years.

Acknowledgments

I thank Dr. Kwame Koranteng for enabling me to re-visit Sakumo Lagoon, and the town of Tema, and Mr. Alasdair Beattie for preparing Figures 21-1 - 213.

References

Christensen, V. and D. Pauly 1992 The ECOPATH II - a software for balancing steady-state ecosystem models and calculating network characteristics. Ecological Modelling 61: 169-185.

Christensen, V. and D. Pauly (eds) 1993 Trophic models of aquatic ecosystems. ICLARM Conference Proceedings No. 26. 390p.

Deriso, R.B. 1980 Harvesting strategies and parameter estimation for an age-structured model. Canadian Journal of Fisheries and Aquatic Sciences 37: 268-282.

Koranteng, K, M. Entsua-Mensah and P.K. Ofori-Danson. 1998 Comparative study of the fish and fisheries of three coastal lagoons in west Africa. International Journal of Ecology and Environmental Science 24: 371-382.

Longhurst, A. and D. Pauly 1987 *Ecology of Tropical Oceans.* Academic Press, San Diego, 407p.

MacKay, A. 1981 The generalized inverse. Practical Computing, September 1981: 108-110.

Mertz, G. and R.A. Myers. 1998. A simplified formulation for fish production. Canadian Journal of Fisheries and Aquatic Sciences 55(2): 478-484.

Ntiamoa-Baidu, Y. 1991 Conservation of coastal lagoons in Ghana: the traditional approach. Landscape and Urban Planning 20: 41-46.

Pauly, D. 1975 On the ecology of a small West African lagoon. Meeresforschung 24(1): 46-62.

Pauly, D. 1976 The biology, fishery and potential for aquaculture of *Tilapia melanotheron* in a small West African lagoon. Aquaculture 7(1): 33-49.

Pauly. D. (ed.) 1998 *Use of Ecopath with Ecosim to evaluate strategies for sustainable exploitation of multi-species resources.* Fisheries Centre Research Reports 6(2). 49p.

Pauly, D. and V. Christensen 1995 Primary production required to sustain global fisheries. Nature 374: 255-257.
Pauly, D., J. Moreau and M.L. Palomares 1988 Detritus and energy consumption and conversion efficiency of *Sarotherodon melanotheron* (Cichlidae) in a West African lagoon. J. Appl. Ichthyol. 4: 150-153.
Walters, C., V. Christensen and D. Pauly 1997 Structuring dynamic models of exploited ecosystems from trophic mass-balance assessments. Reviews in Fish Biology and Fisheries 7: 139-172.
Walters, C., V. Christensen and D. Pauly 1999 Ecospace: prediction of mesoscale spatial patterns in trophic relationships of exploited ecosystems with emphasis on the impacts of marine protected areas. Ecosystems 2: 539-554.

IV

Ecosystem Health &
The Human Dimension

Pollution

The Gulf of Guinea Large Marine Ecosystem
J.M. McGlade, P. Cury, K.A. Koranteng and N.J. Hardman-Mountford (Editors)

22

Environmental Pollution in the Gulf of Guinea: A Regional Approach

P.A.G.M. Scheren and A.C Ibe

Abstract

This chapter reports on the results of the pilot project Water Pollution Control and Biodiversity Conservation in the Gulf of Guinea Large Marine Ecosystem (LME), funded by the Global Environment Facility (GEF). It is divided into three sections. The first section gives the results of an assessment of the socio-economic circumstances that are the root causes of the environmental pollution problems. The second section presents the results of the assessment of the state of pollution in the natural ecosystems of the coastal zone, as well as a qualitative and quantitative assessment of the land-based sources of pollution. The final section assesses the institutional constraints related to control of environmental pollution, and suggests management options for resolving them.

Introduction

The importance of the coastal ocean, including lagoons, estuaries, bays, creeks, etc., to the socio-economic development of countries bordering the Gulf of Guinea is largely recognised because of its vast potential resources. It is thus critical that the marine environment is maintained in a state that is supporting this enormous productivity. Regional monitoring of pollution and ecosystem health has allowed for better quantification of the problem, providing a more accurate overall picture of the conditions of waters and the effects of pollution from various sources. Through this it is now known that the health of the coastal ocean in this region is increasingly in jeopardy as a result of the rapid intensification of human activities on or near the coast (Ibe 1986).

Deriving mainly from the history of the region's early contact with European "seafarers", nearly all the major cities, harbours, airports, industries, extensive agricultural plantations and other socio-economic infrastructure in the region are located in coastal areas. Pollution from these various sources affects the waters of the Gulf of Guinea and the natural living resources therein, which depend on clean water for their survival. Environmental degradation, including habitat destruction, loss of biological diversity and degenerating human health are among the major impacts.

Recognising that marine pollution and living marine resources are not constrained by political boundaries or even many geographical ones, the countries in the region have recognised that only through a concerted, regional approach will the ongoing degradation of their marine environment be counteracted. Under a pilot project, "Water Pollution Control and Biodiversity Conservation in the Gulf of Guinea Large Marine Ecosystem (LME)", funded by the Global Environment Facility (GEF), five countries in the region (Benin, Cameroon, Côte d'Ivoire, Ghana and Nigeria) agreed to adopt common strategies and mechanisms in the battle to restore and protect the health of the Gulf of Guinea LME; Togo joined subsequently, making up the six partners (Figure 22-1). The approach adopted by the countries was based on the concept of the Large Marine Ecosystem (LME), which recognises that to be successful, successful coastal ecosystem management must be linked with a Marine Catchment Basin (MCBS) i.e. those areas adjacent to LMEs and from where many of the impacts due to human activities originate (Sherman 1993, 1998; Sherman and Solow 1992; Sherman *et al.* 1993).

Assessment of Socio-economic Circumstances Related to Environmental Pollution

According to GESAMP (1990), "Marine Pollution", refers to the introduction by man, directly or indirectly, of substances or energy into the marine environment (including estuaries) resulting in such deleterious effects as harm to living resources, hazards to human health, hindrance to marine activities including fishing, impairment of quality of sea water and reduction of amenities. In line with this definition, it is apparent that the socio-economic circumstances prevailing in the region, including the rapid rate of urbanisation and industrialisation in the coastal areas, are at the core of the pollution problem and must be appreciated in any effort to counter the problem.

Demography
The coastal zone is the economic nerve centre of the states bordering the Gulf of Guinea. In Nigeria, 22.5%, of the total population, or 20 million people, live in the coastal zone, as narrowly defined, while in the other countries in the region, proportions vary from 30 to 90% (World Resources Institute 1991). Of the nearly 180 million inhabitants of the six countries 41.2 % live in rapidly expanding urban settlements, with 29.1 % living in coastal urban settlements alone. Population growth, at an average annual rate of 2.8%, is just slightly higher than the average for Sub-Saharan Africa, with the probability of a doubling every 25 years.

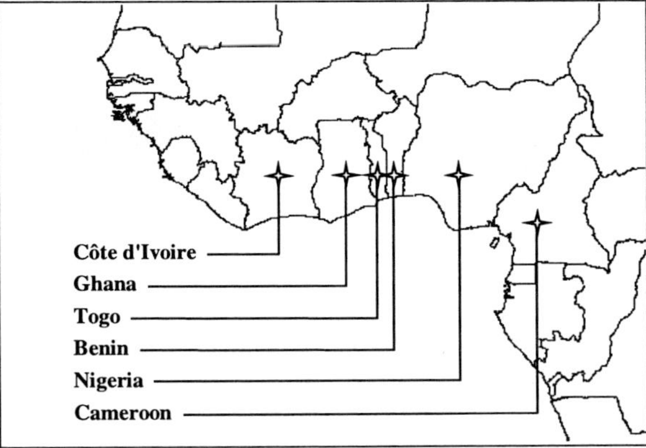

Figure 22-1 Map highlighting the six countries in the Gulf of Guinea in the GEF-LME project.

This is substantially higher in the larger coastal settlements (e.g. 4% in Abidjan). The latter is merely a result of migration from rural areas to urban centres. Immigration from inland neighbouring countries, an important factor in some of the countries, adds to the problem. For example, in Côte d'Ivoire, immigration has been estimated to account for an additional 1.3% annual population growth (Métongo *et al.* 1993). Table 22-1 provides an overview of the demographic characteristics of the region.

The increasing pressure from a growing population on the region's natural resources is exerting a toll. More and more resources are exploited to support the growing need for food, basic consumables such as fuelwood and materials for building and construction. Despite growing industrialisation in the 6 countries, poverty rates have been declining only

	Ivory Coast	Ghana	Togo	Benin	Nigeria	Cameroon
DEMOGRAPHY AND POVERTY						
Population (millions)	14.7	18.0	4.3	2.8	117.9	14.1
Annual growth 1991-1997 (%)[1]	2.9	2.7	2.7	2.9	2.9	2.9
Urban population (% of total population)[2]	51.0	37.0	32.0	40.0	41.0	46.0
Coastal urban population (% of total urban population)	84.0	72.0	100.0	100.0	20.0	54.0
Poverty: national headcount index (% of population)	--	31.0	--	33.0	34.0	--
GNP per capita (US$)[3]	690.0	370.0	330.0	380.0	260.0	650.0
Projected annual GNP per capita growth 1998-2002 (%)	3.4	3.1	3.1	2.5	--	2.4
ECONOMY						
GDP (US$ billions)	4.7	6.8	1.5	2.1	52.2	9.1
Projected annual GDP growth 1998-2002 (%)	5.7	5.0	5.0	5.4	--	5.3
Agriculture (% of GDP)	27.3	47.4	41.8	41.0	32.7	44.7
Industry (% of GDP)	21.2	16.6	20.8	14.8	46.9	21.5
Manufacturing (% of GDP)	17.6	9.5	8.7	8.8	4.8	10.5
Services (% of GDP)	51.5	36.0	37.4	44.2	20.4	33.8

Table 22-1 Socio-economic parameters for the Gulf of Guinea Region (World Bank 1998, apart from data on coastal and urban population, World Resources Institute 1991). [1]Average annual population growth for Sub-Saharan Africa: 2.7 %. [2]Average % urban population for Sub-Saharan Africa: 32 %. [3]Average GNP per capita for Sub-Saharan Africa: 500 US$.

slowly, with an average rate still over 30%, according to the National Headcount Index. Consequently, natural resources are being over-exploited at a rapid pace with little attention being paid to their long-term sustainability.

Industry

Many of the major cities located in the coastal zone are important industrial and commercial centres. Major coastal harbours are located near the shoreline in all of the project countries: Abidjan and San Pedro in Côte d'Ivoire; Tema and Takoradi in Ghana; Lomé in Togo; Cotonou in Benin; Lagos, Warri, Port Harcourt and Calabar in Nigeria; and Douala, Limbé and Kribi in Cameroon. The main industries are found on or near the coast. Abidjan, for example, hosts 70% of all industrial enterprises in Côte d'Ivoire, accounting for 80% of the total industrial production (Ministère du Logement, du Cadre de Vie et de l'Environnement 1997). In Cameroon, industries in the coastal zone account for 60% of the national production (Ministry of Environment and Forestry 1999), while in Nigeria, the greater Lagos region alone hosts 80% (Federal Environmental Protection Agency, 1998) of industrial production nation-wide. The rates are similar for Togo (85%), Benin (80%) and Ghana (65%).

Industry accounts for 21.5 % of GDP in the region, but the economy is substantially more industry driven in Nigeria (Table 22-1), where oil exploitation forms the core of the country's economy (Chikwendu 1998; Federal Environmental Protection Agency 1998). The main industrial sectors in the project countries are either related to the direct needs of the population (agro-food processing, beverage, garment, textile and leather industries, soap and detergent manufacturing, wood processing and cement production) or to the processing of raw materials exploited in the region (petroleum refining in Nigeria, Cameroon, and to a lesser extent in Côte d'Ivoire and Ghana, aluminium in Ghana, Cameroon and Nigeria, iron and steel in Nigeria and Côte d'Ivoire, and phosphate in Togo). Fertilizer and pesticide plants, pharmaceutical industry, paper pulp industry and plastic product manufacturers exist in some of the countries.

The first industrial plants in the region were built in the 1950s and 1960s, mostly as "green-field projects" at the perimeter of the then existing coastal communities. Employment opportunities have led to the coastal population growing by a factor of four to five since the 1960s. Urban development plans meanwhile have failed to keep pace with the increasing industrial development and population expansion. The problem of waste collection and disposal has assumed proportions that are beyond control, leading to the degradation of the living environment and depriving the populace of certain critical resources (e.g. clean water, fish) and causing diseases which sometimes take on dramatic forms (Duchasin *et al.* 1973; Dosso *et al.* 1984; Ibe 1993, 1996; Kouassi *et al.* 1995).

It is only in the most recent decade that awareness of the consequences of the uncontrolled industrial development has been growing; and that in most of the countries in the region policies for environmental regulation of industrial pollution have been formulated. However, regulatory systems are still far from ideal, and presently most industries are still under-regulated with regard to the discharge of effluents and solid wastes.

Exploitation of Mineral Resources

Mineral resources mining in the coastal region takes place in most of the countries. Extraction of sand, gravel, rocks and other materials for construction widely exacerbates land

and coastal erosion. In Togo, dramatic levels of land-erosion have resulted from the extraction of soil for the production of phosphate. The residues of this extraction are furthermore disposed of in coastal and lagoon waters, contaminating the waters. Gold mining takes place in Côte d'Ivoire, near Aboisso. The availability of other minerals for exploitation, such as titanium in Ghana, iron ore in Cameroon and limestone in several of the countries, has been determined for future exploitation. The soil degradation resulting from these mining activities, result in increased runoff and leaching of nutrients from land, and the disposal of its residues in coastal and lagoon waters contribute an important trigger mechanism in the eutrophication of these waters.

Oil and gas exploration in the region is a special case, having impacted the regional and global environment since its inception in 1956, when the first commercial oil field at Oloibiri, in present day Rivers State of Nigeria, was discovered. From an initial 5,000 barrels per day (BPD) in 1958, crude oil production in Nigeria rose to a height of 2.4 million BPD in 1979, and currently averages around 2 million BPD. Today, the industry accounts for over 85 % of Nigerian's annual export (Chikwendu 1998; Federal Environmental Protection Agency 1998). This dependence has caused a situation where the enormous environmental harm related to oil exploration and exploitation has been and still is largely being neglected.

The environmental pollution effects from oil exploration are related to the actual exploration and production activity, product pipeline operations, marine and terminal operations up to the actual petroleum refining process. Blowouts and oil spills have caused disastrous situations. According to official estimates from the National Nigerian Petroleum Company, in Nigeria, approximately 2,300 m^3 of oil is spilled in about 300 separate accidents annually (Awobajo 1981). The environmental impacts of these spills has been devastating and is responsible for mass fish mortality and destruction of critical habitats such as mangroves (Ibe *et al.* 1985; Odu and Imevbore 1985; Powell 1986; Saenger *et al.* 1996).

Nigeria, and especially the Niger Delta, may be the most dramatic case of the effects of oil exploration and exploitation; but, to a lesser extent, exploitation takes place in Benin, Cameroon, Côte d'Ivoire and Ghana. New reserves are continuously being discovered, including more and more off-shore wells. Although environmental awareness is rising, and under national and international pressure the oil exploiting companies are more and more called upon to account, progress is still slow and the harms and threats of large scale pollution remain.

Agriculture

Apart from Nigeria, agriculture is the dominant economic activity in the countries, in terms of contribution to GDP (Table 22-1). In terms of manpower involved in agricultural activity, however, it is by far the largest in all countries. Especially along the coastline and the fertile areas around rivers and wetlands, cultivation of both food crops and cash crops is widespread. The major food crops are maize, plantain, cassava, rice, and vegetables. Major cash crops include oil palm, rubber, cotton, pineapples and coconut; while slightly inland, cacao and coffee are cultivated.

With the increasing population pressure in the coastal region and the growing industrialisation of agriculture, the pressure on the environment is enormous. Although most of the characteristic cash crops in the coastal region do not require an extensive use of pesticides, the uncontrolled use of it causes serious threats. The use of internationally banned agrochemicals such as DDT, aldrin and lindane is still widely practised. Moreover, the increasing land cultivation and extensive and unsustainable exploitation of soils causes erosion and increased runoff and leakage of nutrients to surface waters, contributing to the eutrophication of these waters.

Forestry

Under pressure from a rapidly expanding population, little of the original coastal forests now remain. For decades, the forests have been exploited for firewood and construction materials, but also for the lucrative commercial export to industrialised countries. Forests have been turned into agricultural lands. Although large parts of the coastline forests have been deprived of any commercially valuable wood, the majority of households still depends on carbon and fresh firewood for their cooking, putting the little remaining forests under immense pressure. In Cameroon, where much of the forest remains, export of timber remains a major contributor to the economy, accounting for approximately 20% of total export (Ministry of Environment and Forestry 1999).

Apart from the direct consequences for biodiversity, such as the disappearance of flora and fauna and the pressure on critical habitats for marine living resources such as the mangrove forests, the destruction of what once where vast extents of forest has caused severe degradation of soils. Increased soil erosion and runoff and leakage of nutrients to surface and ground waters result.

Tourism

Tourism, along the coastline, is a rapidly growing commercial activity in all of the countries. With little regard for the natural values and environmental concerns, the coastline is being exploited. On the beaches, one hotel after the other is being built, and on weekends and holidays, they are now crowded with tourists. With the African continent becoming increasingly a holiday destination for those outside the region, the rate at which tourism is developing is astonishing. For example, in Ghana, earnings from tourism multiplied more than ten-fold between 1987 and 1994.

Litter and marine debris have had their obvious impact on the aesthetics of the coastal beaches, while construction on the coast has accelerated coastal erosion and organic pollution (c.g. untreated sewage), but in terms of overall marine environmental pollution the effects of tourism are probably not yet significant. For the future, however, the uncontrolled exploitation of the coastline for tourism is regarded with concern.

Environmental Pollution Assessment

Although more and more attention is now given to the proper monitoring of the state of the environment in the countries in the region, data on environmental pollution is still patchy. As part of a general collection of baseline data on the health of the Gulf of Guinea Large Marine Ecosystem, the countries embarked on a region-wide exercise to assess and collect existing data on environmental pollution, to monitor principal parameters in their coastal and marine waters, and to assess the sources of pollution. It is important to stress that only those data that seem to meet QA/QC criteria (Topping 1992; Ibe and Kullenberg 1995) are summarised here.

Assessment of the levels of pollution in lagoon and coastal waters

Although, without an in-depth and long-term analysis, it is difficult to differentiate pollution effects or gradients from natural variations, the environmental impacts of several decades of substantial and continuous input of untreated domestic and industrial wastes are clearly manifested by high levels of contamination of the coastal ecosystem. Such effects are most evident in the sensitive semi-enclosed or enclosed coastal lagoon systems, especially around the large coastal urban centres, where pollution levels are highest. Tables 22-2, 22-3 and 22-4 provide an overview of the typical levels of respectively organic, heavy metal and organochlorine pollution found in the coastal zone lagoon systems around the main urban centres.

The urban lagoon systems show high levels of organic pollution, resulting in eutrophication and, as reported for the Korle and Chemu II lagoons in Ghana and for several of the bays of the Ebrié lagoon in Côte d'Ivoire, in near total oxygen depletion (Guiral 1984; Dufour *et al.* 1985,1994; Guiral *et al.* 1989; Biney 1994; Acquah 1998a; Awosika and Ibe 1998; Gordon 1998). As a result, many of the coastal lagoons that formerly provided fish for its riparian population, and also served as breeding grounds for many kinds of marine fish species, hardly support a fishery today. Fish landings in the Lagos Lagoon have dropped significantly in the past 25 years due to overfishing and pollution (Awosika *et al.* 1992; Ajayi 1994). Advanced levels of pollutants and microbial contaminants related to water borne diseases have been found in all of the coastal lagoon systems near the large urban centres of the region (Pagès 1975; Pagès and Citeau 1978; Dufour *et al.* 1985; Kouassi *et al.* 1990; Métongo *et al.* 1993; Acquah 1998a; Ajao and Anurigwo 1998).

The elevated levels of heavy metals and organochlorines can be attributed to the disposal of untreated industrial and domestic effluents, as well as to oils spills and leakage. Studies on the concentrations of heavy metals and hydrocarbons in fish and shellfish have revealed that the levels found in living resources exploited for human consumption are generally below the international consumption standards for fish and fish products. Their increasing concentrations are, however, a major source of concern.

Assessment of Land-based Sources of Pollution

According to a world-wide study carried out by the UN Group of Experts on the Scientific Aspects of Marine Pollution (GESAMP) in 1990, the majority of marine pollution originates from run-off and land-based discharges of pollutants, followed by atmospheric sources. Table 22-5 summarises the results of this study.

In the absence of excessive maritime transportation activities in the region, land-based sources of pollution deserve primary attention in the control of marine environmental pollution. Land-based activities have furthermore been known to have considerable influence on the level of atmospheric deposition of certain substances, with its adverse effect on the aquatic environment. For example, extensive biomass burning, such as through forest fires, is known to cause substantial increases in the atmospheric deposition of nutrients (Reimold and Daiber 1967; Lemasson and Pagès 1982; Crutzen and Andreae 1990; Goldman *et al.* 1990; Bootsma and Hecky 1993; Scheren 1995; Scheren *et al.* 1998). A two-dimensional approach can be applied for classifying the pollution per source of origin (domestic, industry, agriculture) and per media (solid, liquid, atmospheric) (See Table 22-6).

	Korle Lagoon, Accra[1]	Chemu II Lagoon, Tema[1]	Lagos Lagoon, Lagos[2]	Ebrié, Lagoon, Abidjan[3]	Background[4]
DO [mg/l]	0-6.2	0-0.5	2.2-9.5		6.4-6.6
BOD [mg/l]	4.4	71.2-240			3.2-5.5
PO_4-P [mg/l]	0.86	0.59-2.85	<0.01-0.5	0.06-0.27	0.06-0.09
NH_4-N [mg/l]	3.8	1.3-12.6		0.18-1.11	0.2
NO_3-N [mg/l]		0.2-0.35	0.1-0.8	0.01-0.28	
Total coliform (Count/100 ml x 1000)	635-1,604			0-1,735	

Table 22-2 Typical levels of organic pollution in some of the coastal lagoon systems [1] Sources: Biney (1994) and Acquah (1998a); [2] Sources: Ajao (1990), Kusemiju *et al.* (1999) and Oyewo (1999); [3] Source: Affian (1999); [4] Values measures for unpolluted lagoons in Ghana (Laloi and Mokwe lagoons), according to Biney (1994).

Sample	Cd	Cr	Cu	Fe	Hg	Mn	Pb	Zn	Reference
Sediment [µg/g dry wt]									
Lagos Lagoon, Lagos	0.01-15.5	2.9-167	1.5-132	510-85548		98-2757	0.4-483	7.8-831	Okoye *et al.* 1991; Oyewo 1999
Ebrié Lagoon, Abidjan		20.7-465	3.0-76.3	1.3-67.0	0.05-0.49	24.0-534	4.0-88.8	5.5-398	Arfi *et al.* 1994
Unpolluted sediments	0.2-5				0.01-0.08		8-60		GESAMP 1985, 1988
Water [mg/l]									
Korlé Lagoon, Accra (median)	0.24		0.31				0.08	0.08	Acquah 1998a
Lagos Lagoon, Lagos (median)	0.002		0.003	0.086		0.021	0.009		Okoye 1991a
Natural sea water levels	0.005		0.003				0.003	0.02	Acquah 1998a
Shellfish [µg/g fresh wt]									
Lagos Lagoon, Lagos (median)	0.18		23.6				5.1	240	Okoye 1991b
Ebrié Lagoon, Abidjan	0.35-0.95		17.5-33.5		0.065-0.19			608-2115	Métongo 1991
WHO standard	2		30		2		2	1000	Kakulu et al. 1997

Table 22-3 Typical levels of heavy metal pollution in some of the coastal lagoon systems.

Sample	aldrin	dieldrin	endrin	heptac hlore	lindane	DDT	PCB	Reference
Sediment [ppb]								
Ebrié lagoon, Abidjan					0.5-19	1-997	2-213	Métongo 1998
Lagos Lagoon, Lagos	19.3	28	12	85.3				Okonna 1985
Lekki Lagoon, Nigeria	nd-347	190-8460	nd-129	nd-1845	0.11-4.9			Ojo 1991
Shellfish [ppb]								
Ebrié Lagoon, Abidjan	19.9-132	13.5-168	6.1-74.0		13.5-93.0		8.9-43.9	Etien 1998
Ocean at Limbé, Cameroon	nd-12.0				nd-5.3	nd-481	nd-716	Enoh *et al.* 1998
Health standard	100-500	100-500	100-500	100-300	500-2000	2000-5000	1000-5000	

Table 22-4 Typical levels of organochlorine pollution in coastal waters and coastal lagoon systems.

Source	Percentage
Run-offs & Land-based Sources	44
Atmosphere	33
Maritime Transport	12
Ships	10
Offshore Production	1

Table 22-5 Sources of Marine Pollution (after GESAMP 1990).

Under the project, a rapid quantitative assessment of land-based sources of pollution in the region was undertaken with UNEP, on a country-by-country basis. The assessment procedure followed the methodology developed by WHO (World Health Organization, 1982), and concerned primarily a quantification of point sources of pollution of industrial and domestic origin. The effects of increasing land exploitation for agricultural and other uses, as qualitatively described in the previous section, would be difficult to quantify because of their diffuse nature.

Based on the results of this rapid survey, a more detailed survey on industrial pollution was executed, in which manufacturing industries in the various countries co-operated in completing a series of questionnaires and in allowing on-site investigations aimed at obtaining an in-depth knowledge of the nature of the manufacturing process, the types and quantities of wastes generated, and the nature of waste treatment and discharge practices. Furthermore, an extensive review of urban sanitary systems in the region, including their problems and constraints, was executed. Finally, the problem of solid waste management and marine debris was extensively studied, in collaboration with the International Oceanographic Commission (IOC) of UNESCO, including a 5-year monitoring program of marine debris on beaches over the region.

	Solid	**Liquid**	**Atmospheric**
Domestic	Household degradable solid waste. Household synthetic non-degradable solid waste.	Household degradable effluents via sanitary or sewage system. Degradable leaching and runoff from septic tanks and pit latrines. Partly degradable, partly hazardous leaching and runoff from waste dump sites.	Traffic combustion - mainly gaseous carbon, sulphur and nitrogen components. Biomass burning for heating and cooking - particulate and gaseous nitrogen, phosphorous, sulphur components.
Industry	Organic degradable solid waste from spills, by-products and residuals. Synthetic non-degradable solid waste from packaging materials and residuals.	Waste water effluents – partly degradable, partly hazardous substances and heavy metals. Leaching and runoff from local dump sites and storage – partly degradable, partly hazardous substances and heavy metals. Liquid waste (oil spills) from transport by waterways.	Industrial combustion - mainly gaseous carbon, sulphur and nitrogen components. Transport traffic combustion - mainly gaseous carbon, sulphur and nitrogen components.
Agriculture	Organic degradable solid waste from agricultural residuals. Non-degradable solid waste from packaging materials and other synthetic waste material.	Runoff and leaching of agrochemicals and nutrients from fertilisers. Increased 'natural' runoff and leaching of nutrients due to soil erosion.	Forest burning - particulate and gaseous nitrogen, phosphorous, sulphur components. Increased dust due to soil erosion - nutrients transported as particles. Transport traffic and machinery combustion - mainly gaseous carbon, sulphur and nitrogen components.

Table 22-6 Schematic overview of environmental pollution sources.

- Solid waste / Marine debris

The enormous bulk of solid waste daily produced by households and industries in the region forms a serious threat to the environment (Portmann *et al.* 1989). Solid waste generation per head of population for low income areas is approximately 150 kg per year, totalling to 3,8 million tonnes per year, mostly putrescible or non-hazardous waste, for the coastal population in the region. The distribution of solid waste generation per country is displayed in Figure 22-2.

Although domestic solid waste production constitutes the larger bulk, industries also produce considerable amounts of waste, some of which are classified as hazardous or infectious. The total solid waste production from industries is estimated at 435,000 tonnes per year. Figures 22-3 and 22-4 provides an overview of the amounts and types of waste produced, per industrial sector, in the 5 countries reviewed, and shows that the bulk of solid waste is of a putrescible or low-hazard nature. Putrescible waste originates mainly from the food manufacturing industries. Other major contributors are the aluminium and steel industry, mainly in Cameroon and Nigeria, with the petroleum industry as the most important producer of hazardous waste in the form of oily and toxic sludge.

As part of a 5-year regional marine debris monitoring program three types of beaches were monitored in each country: a populated beach, a tourist beach and a reference beach

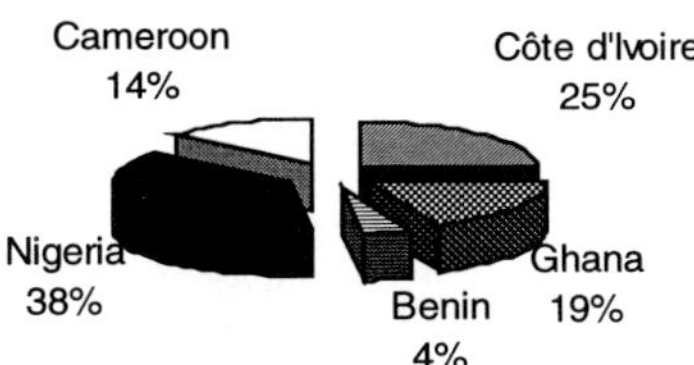

Figure 22-2 Estimated domestic solid waste production per country.

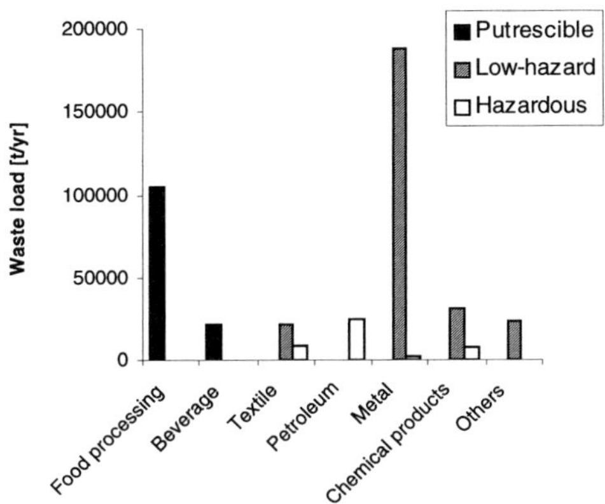

Figure 22-3 Types of solid waste in the coastal zone per industrial sector.

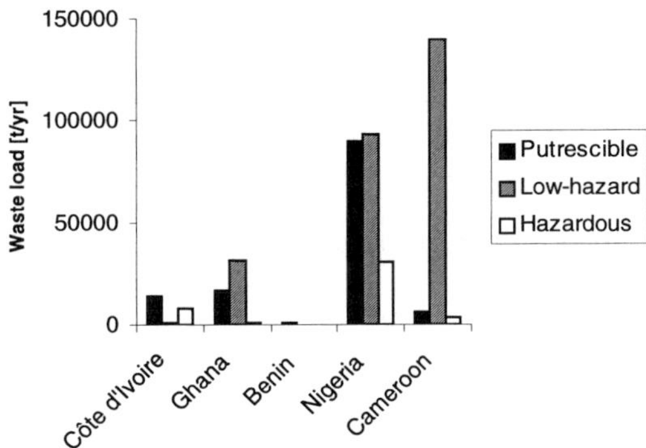

Figure 22-4 Loads and types of waste in the coastal zone per country.

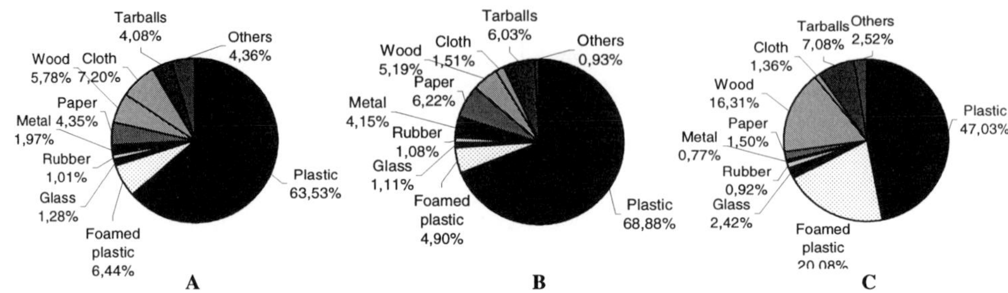

Figure 22-5 Averaged marine debris distribution pattern for **A** populated, **B** recreational and **C** reference beaches in the Gulf of Guinea region.

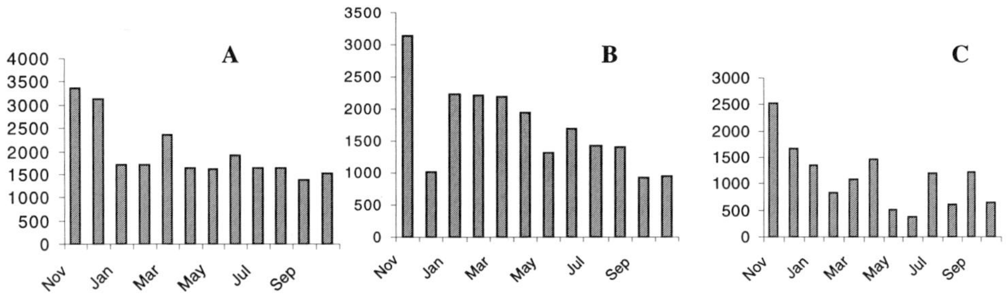

Figure 22-6 Seasonal variation of marine debris on **A** populated, **B** recreational and **C** reference beaches in the Gulf of Guinea region.

with low human activity. The distribution of various debris items collected per 500 m^2 in 1997/1998 for the three different types of beaches in the participating countries are shown in Figure 22-5. The total number of items collected per 500 m^2 was 6,798 for Benin, 6,279 for Cameroon, 19,730 for Ghana, 7,693 for Nigeria and 17,109 for Togo, leading to a region-wide average number of debris items found of 23 per m^2. Dominant items were plastics, with percentages between 42.76% (Cameroon) and 77.90% (Benin) and a mean of 61.60%.

The dominance of plastic items on almost all beaches sampled is related to fishing activity and the use of plastics as carrier bags, packaging materials, etc. With regard to monthly dynamics of total numbers of debris items, peaks observed varied according to the type of beach (Figure 22-6), except for the month of November, where a peak occurred in all the three types of beach. In the populated beach peaks were observed in November, December and March, which corresponds to periods of increased tourism activities within the region.

- Liquid waste

Liquid waste is the most direct source of marine pollution. Daily, large quantities of polluting effluents from industrial and domestic sources are discharged directly into the coastal waters and lagoons, having as effects eutrophication and oxygen depletion of the waters. The assessment of liquid water pollution sources carried out by 5 countries in the region, Côte d'Ivoire, Ghana, Benin, Nigeria and Cameroon, provided a general overview of the enormous amounts of polluting effluent flows concerned.

The studies revealed that most industrial and domestic effluents are untreated. For example, in the Lagos area, home to the highest density of industrial establishments in the region, 80% of industrial effluents are discharged into Lagos lagoon, through ditches, shallow pits, gutters and trenches (Ibe 1986; FMWH 1988; Ajao 1990; Ajao and Anurigwo 1998). The Ebrié lagoon in Côte d'Ivoire, the Lagos Lagoon in Lagos, Korlé and the Chemu Lagoons in Accra/Tema are examples of the levels of pollution that arise from these uncontrolled waste loads (see Table 22 - 4), specifically in terms of organic pollution. Figure 22-7 compares the estimated loads of domestic and industrial effluents on the environment. Depending on the level of industrial pollution and the size of the urban population, waste loads vary (Figure 22-8).

Although in all coastal urban centres sewage systems have been installed to varying degrees, only a small part of the urban population is actually branched to the networks. Rates of connectivity range from as high as 40% in Abidjan to less than 5% in the other major cities, including the 8 million population of Lagos. The majority of the population in the coastal urban settlements depends on septic tanks and pit latrines for their sanitation. Where they exist, central sewage treatment facilities are either out of order or operating badly. An additional problem is posed by the fact that the stormwater drainage systems crossing the cities have often been turned into open sewers, causing not only problems of hygiene but also substantial runoffs into coastal and lagoon waters, specifically during the rainy season.

The main industrial sector responsible for BOD loading in the region is the food manufacturing sector (Figure 22-9), which processes agricultural products for consumption in the respective countries themselves. In terms of organic pollution, the other industrial sectors, by comparison, play a minor role. However, their influence may not be less important, specifically with regard to pollution related to oil exploration/exploitation and processing in Nigeria, where waste water, from the separation of oil and production water during the exploration/exploitation process and oil spills, adds up to enormous amounts of oil going into the environment annually. The World Bank (1998) estimated that waste water accounts for approximately 710 tons of oil discharged by oil producing companies in Nigeria annually. The volume from oil spills is roughly three times as high as that of wastewater.

With few of the industrial establishments having installed any form of treatment at all, industries exceed effluent standards in most, if not all cases. Table 22-7 shows sample effluent discharge data for Nigeria and Ghana: this indicates that only very few industries would pass the test if a proper compliance monitoring system was in place.

	pH		BOD		COD	
	Ghana	**Nigeria**	**Ghana**	**Nigeria**	**Ghana**	**Nigeria**
Chemicals & pharmaceuticals	8.2-10.6	8.9-10.7	1.0-380	6-584	24-6200	56->1500
Food manufacturing & beverage	4.0-11.0	8.7-10.8	17-1318	22-562	700-30200	17-1318
Textile manufacturing	6.5-11.2	9.2		354		13
Metal processing	2.5-11.0	9.1-10.1		38-45		117-439
World Bank Guidelines	6-9		50		250	
Ghanaian standards	6-9		50		250	
Nigerian standards	6-9		50		100	

Table 22-7 Typical physical-chemical characteristics of industrial effluent discharges as measured in the region (Nigeria - Junaid 1999; Ghana - Acquah 1998b).

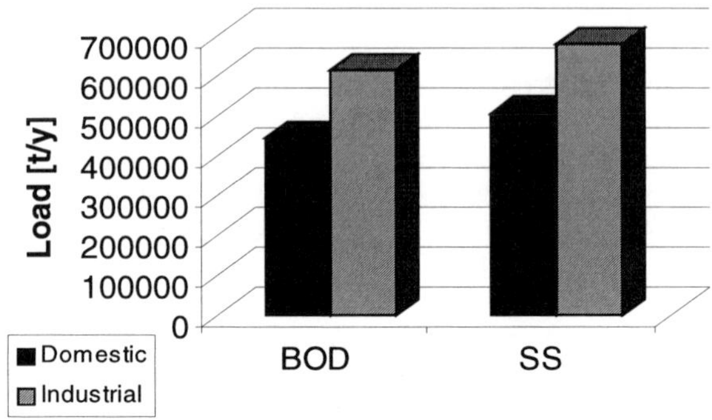

Figure 22-7 Domestic and industrial BOD and Suspended Solids Loads in the Gulf of Guinea coastal zone Region.

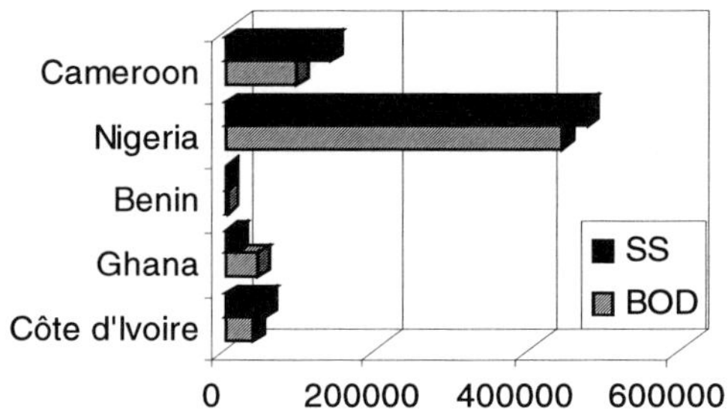

Figure 22-8 Industrial Waste loads in the Gulf of Guinea Coastal Zone Region.

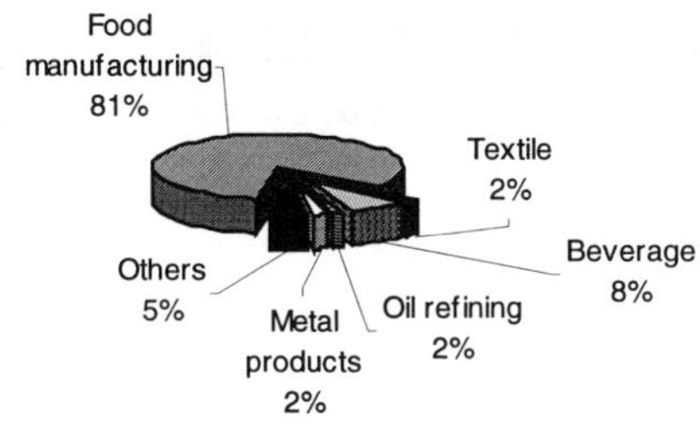

Figure 22-9 Industrial BOD loads per category of industrial production process in the Gulf of Guinea Coastal Zone.

- Air pollution

The main sources of air pollution in the region are traffic (vehicles, aeroplanes, ships), power stations, industrial units and households using charcoal and wood for cooking and kerosene for lighting. A typical distribution of air pollution loads per source is depicted in Figure 22-10.

The distribution excludes Nigeria, where large-scale flaring of natural gas, a by-product from the oil exploitation industry, takes up an important part in the air emission balance. Eighty percent of natural gas production in Nigeria is flared, reaching up to approximately 27000 million m^3 annually (Central Bank of Nigeria 1996), making Nigeria the largest gas-flaring country in the world. The estimated annual emissions of CO_2, nitrogen oxides and SO_2 from gas-flaring alone are respectively 35,000,000 t/yr, 260,000 t/yr and 49,000 t/yr (Ajao and Anurigwo 1998).

Figure 22-10 also shows that, on a regional scale, traffic (road, air and water) accounts for most CO_2, SO_2 and nitrogen oxides emissions, with road traffic being dominant. Furthermore, domestic biomass burning is an important contributor of particulate material, nitrogen oxides, and to some extent, SO_2 and hydrocarbons emissions. Industrial emissions are important factors in terms of particulate material and SO_2, increasing concern in view of their effects, such as acidification of rain water. Control over air emissions is generally absent in the countries concerned.

The effects air emissions on the marine environment in the Gulf of Guinea are difficult to quantify, and the effects of air pollution sources difficult to prove. In Nigeria, for example, studies on the effects of gas-flaring have not found serious levels of acidification in wet and dry deposition. However, it may be expected that the transport of substances via the atmospheric avenue has considerably influences on the state of the Gulf of Guinea LME (Table 22-5).

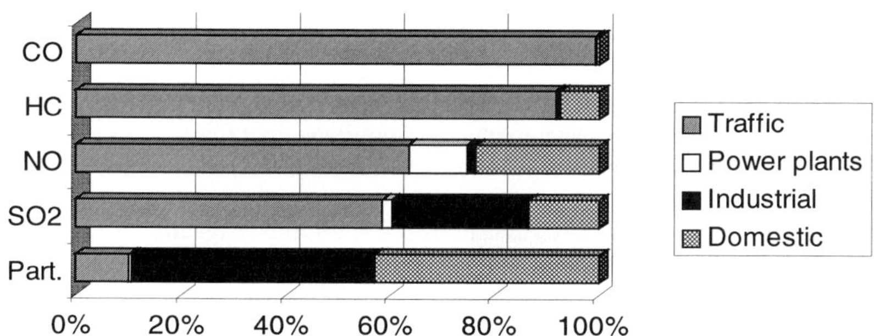

Figure 22-10 Typical distribution of air pollution loads per source for the Gulf of Guinea Region.

Institutional Constraints and Management Options

The following section provides an overview of the policy deficits and institutional constraints relating to effective marine environmental management in the region. These have to be viewed against the background of the continued significant phenomenal population growth and weak economic situation, both of which have induced not only unsustainable exploitation of natural resources, but also negligence of environmental considerations. Finally, measures for cost effective environmental management are suggested.

Institutional Constraints

Although Environmental Action Plans exist for all project countries (except Togo), clearly set targets, guidelines and standards are still lacking, as are comprehensive Urban Management Plans that would include systematic pollution prevention and abatement strategies. Pricing policies for many natural resources, such as water, land and wood, are missing, inviting intensive use of these resources. Few industrial enterprises and investment projects have been subjected to an Environmental Impact Assessment (EIA), Environmental Audit or Environmental Compliance Monitoring. Information on pollution abatement options and technologies at the level of industrial establishments is inadequate, particularly regarding small and medium size enterprises.

The capacity of human resources capacity in government institutions, in most fields of expertise, is still largely inadequate, while public awareness is insufficient, hampering the role of public opinion in self-regulating environmental pollution. Furthermore, even with the definition of effective policies, financial and economic constraints hamper many of the investments to be made in terms of public sector infrastructure, and constrain governments in their implementation of effective regulatory systems, as far as the operational costs of such systems. Financial constraints also prevent the private sector from taking necessary pollution abatement actions, whether this concerns investment costs related to procurement and installation of pollution abatement technologies, or the costs of running such practices.

Management Options

Each of the countries in the sub-region, should review and strengthen the provisions and application of existing National Environmental Action Plans, including:
- development of environmental legislation and regulations, including guidelines and standards;
- strengthening of environmental institutions by building capacity for environmental pollution assessment and control;
- EIAs and Environmental Audits should form integral parts of pollution control;
- establishment of mechanisms for funding environmental protection through the financing of pollution abatement and clean technologies (e.g. tax on polluting products to be paid to a Central Environmental Protection Fund; accelerated depreciation and tax cuts; applying the "polluter pays" principle; etc.); and
- protection of wetlands and nearly stagnant water bodies (e.g. lagoons) with strong prohibitions on releasing untreated wastewater and hazardous wastes into them.

Governments should furthermore upgrade urban management plans to incorporate the creation of centres for waste collection and treatment, as well as the creation of regulated and controlled dump sites. Policies that encourage excessive resource use should be eliminated and replaced by policies that encourage individuals and industry to use natural resources appropriately without adversely affecting productivity. This can be done through the introduction of a system of pricing of resources to reflect as much as possible, not only the direct costs of such resources, but also the cost of environmental damages. Additional resources need to be assigned to the implementation of pollution control investments. Since, under present conditions, governments will be financially restricted, an effective implementation strategy should contain the following elements:

- a prioritisation of investment needs to be borne by national budgets, and a clear strategy for investment projects that need to be put forward for external loans and grants;
- a close examination of low cost options, such as the application of mangroves and wetlands as natural 'purifiers' of putrescible domestic effluents, the use of settling pits for sewage treatment in small communities and the sorting of domestic wastes prior to disposal as a means of encouraging recycling and reuse. Pilot projects on such methods have already been undertaken in several countries in the region; and
- an exploration of "Build, Operate and Transfer (BOT)" options, to involve the private sector in public sector investments.

The Integrated Coastal Areas Management Approach

Coastal areas, including the shallow oceans, are often subject to conflicting uses. The very attraction of one use may be the direct cause of environmental degradation. Attempts in the past at reconciling such conflicts have been sectoral in nature and, given the complexity and intimate relationships of the issues described in this paper, were doomed to failure from the onset. An institutional framework is, therefore, imperative and the concept of Integrated Coastal Areas Management affords a vantage platform for a cross sectoral, multi-disciplinary, multi-stakeholder approach to resolving the seeming contradictions inherent in the proper management (including appropriate pollution prevention and control measures) of coastal areas.

The management options suggested in this paper must be viewed in this context. It is for this reason that a considerable emphasis was placed, in the implementation of this project, on the formulation, through national across-the-board consultative processes, of national Integrated Coastal Areas Management Plans. These Plans, while national in character, derive from common policies and strategies defined through regional consultations, in accordance with the declared determination of the project countries for joint management of their shared international waters and their natural resources. The national accent stems from the fact that it is at the national level that one finds the necessary legislative and administrative structures for monitoring and enforcement.

Today, thanks to the GEF project, draft National Integrated Coastal Areas management Plans that incorporate all of the suggested management measures for cost effective pollution prevention and control in the Gulf of Guinea, have been produced and are at different stages of adoption by governments. As part of this effort, elements of regional and national GIS-based Decision Making Support Systems have been put in place. Hopefully, making choices from options described in this paper will no longer be a leap in the dark. In this regard, the partnership formed through the network of NGOs and CBOs for sustained community based actions will definitely be an asset.

References

Acquah, P. 1998a Control of Industrial Pollution: Ghana's Experience. pp165-175. *In*: A.C. Ibe, A.A. Oteng-Yeboah, S.G. Zabi and O. Afolabi (eds). *Integrated Environmental and Living Resources Management in the Gulf of Guinea: The Large Marine Ecosystem Approach*. Proceedings on the First Regional Symposium on the Gulf of Guinea Large Marine Ecosystem, UNIDO/UNDP/NOAA/UNEP, CSIR Press Accra, Ghana.

Acquah, P. 1998b *Ghana Industrial Pollution Assessment*. UNIDO/GEF Gulf of Guinea Large Marine Ecosystem Project. Accra, Ghana. 40p.

Affian, K. 1999 *Base de données sur la pollution de la lagune Ebrié*. UNIDO/GEF Gulf of Guinea Large Marine Ecosystem Project.

Ajao, E.A. 1990 *The Influence of Domestic and Industrial Effluents on Populations of Sessile and benthic Organisms in Lagos Lagoon*. Ph.D. Thesis. University of Ibadan, Nigeria, 413 p.

Ajao, E.A. 1996 Review of the State of Pollution of Lagos Lagoon. NIOMR Techn. Paper No. 106. 20p.

Ajao, E.A. and S. Anurigwo 1998 *Assessment of Land-based Sources of Pollution in Nigeria*. UNIDO/GEF Gulf of Guinea Large Marine Ecosystem Project, Abidjan, Nigeria. 163p.

Ajayi, T.O. 1994 The Status of Marine Fisheries Resources of the Gulf of Guinea. *In*: *Proc. 10th Session FAO CECAF, 10-13 October 199*. Accra, Ghana.

Arfi, R., J-F. Agnèse, M. Bouvy, N. Kaba, E. Kouassi, M. Pagano, L. Saint-Jean, Y. Sankaré and B.S. Métongo 1994 *Protection de l'environnement d'Abidjan, Programme de suivi de la qualité des eaux lagunaires et marines*. Rapport final. 36p.

Awobajo, S.A. 1981 Analysis of Oil Spill Incidents in Nigeria 1976-1980. *In*: *Proceedings of a Seminar on the Petroleum Industry and the Nigerian Environment, Nov*. Warri, Nigeria.

Awosika, L.F. and A.C. Ibe 1998 Gulf of Guinea Coastal Zone and Strategies for Integrated Management Plan. pp119-127. *In*: A.C. Ibe and S.G. Zabi (eds). *State of the Coastal and Marine Environment of the Gulf of Guinea*, UNIDO/IOC, CEDA, Cotonou.

Awosika, L.F., C.E. Ibe, E.O. Nwankwo, A. Balogun and E.A. Adekoya 1992 Physical and Geological Factors Affecting the Transport, Dispersion and Deposition of Pollutants in the Lagos Lagoon. *A Progress Report for the Intergovernmental Oceanographic Commission (IOC) of UNESCO*, Paris. 45p.

Biney, C.A. 1994 Assessment of Water, Air and Land Pollution Sources in Accra-Tema Metropolitan Area, Ghana. World Health Organization, Regional Office for Africa.

Bootsma, H.A. and R.E. Hecky 1993 Conservation of the African Great Lakes: A Limnological Perspective. Conservation Biology 7(3): 644-655.

Central Bank of Nigeria - CBN 1996 Annual Report and Statement of Accounts for the Year Ended 31st Dec. 1996. 195p.

Chikwendu, C.C. 1998 Oil Production and Environmental Pollution in Nigeria's Coastal Areas. pp195-206. *In*: A.C. Ibe, A.A. Oteng-Yeboah, S.G. Zabi and O. Afolabi (eds). *Integrated Environmental and Living Resources Management in the Gulf of Guinea: The Large Marine Ecosystem Approach*. Proceedings on the First Regional Symposium on the Gulf of Guinea Large Marine Ecosystem, UNIDO/UNDP/NOAA/UNEP, CSIR Press Accra, Ghana.

Crutzen, P.J. and M.O. Andreae 1990 Biomass Burning in the Tropics: Impact on Atmospheric Chemistry and Biogeochemical Cycles. Science 250(4988): 1669-1678.

Dosso, M., M. Duchassin, A. Lombardo, M. Koné and V. Edoli 1984 Cas sporadiques ou début d'une nouvelle épidémie de choléra? Bull. Soc. Path. Exot. 76: 121-125.

Duchassin, M., C. Clerc, A. Bourgeade and M.T.Hosotte 1973 Survie de vibrion cholérique Et Torr dans les eaux de lagune d'Abidjan. Bull. Soc. Path. Exot. 66 : 679-684.

Dufour, P., J.M. Chantraine and J.R. Durand 1985 Impact of man on the Ebrié lagoonal ecosystem (Ivory Coast). International Symposium on the Utilization of Coastal Ecosystems, 22-27 November, 1982 Rio Grande.

Dufour, P., A.M. Kouassi and A. Lanusse 1994 Les Pollutions. pp309-334. *In*: J.R. Durand, P. Dufour, D. Guiral and S.G. Zabi (eds) *Environnement et Ressources Aquatiques de Côte d'Ivoire*, Tome II - Les Milieux Lagunaires. ORSTOM Edittions, France.

Enoh, E.P., T.F. Mbenkum, I.L. Mbome and J. Folack 1998 The State of Cameroon's Coastal and Marine Environment. pp35-41. In: A.C. Ibe and S.G. Zabi eds., *State of the Coastal and Marine Environment of the Gulf of Guinea*, UNIDO/IOC, CEDA, Cotonou.

Etien, N. 1998 *Etat de la Pollution des Eaux Côtières Ivoiriennes*. Expert report. UNIDO/GEF Gulf of Guinea Large Marine Ecosystem Project. 28p.

Federal Environmental Protection Agency - Republic of Nigeria 1998 *Coastal Profile of Nigeria*. Ceda Press - Cotonou, Benin. 94 p.

FMWH - Federal Ministry of Works and Housing 1988 *The State of the Environment in Nigeria*. Monograph Series No. 1. Industrial Waste management. 124p.

GESAMP - IMO/FAO/UNESCO/WMO/WHO/IAEA/UN/UNEP Joint Group of Experts on the Scientific Aspects of Marine Pollution 1985 Review of Potentially Harmful Substances - Cadmium, Lead and Tin. Rep. Stud. GESAMP 22 and UNEP Reg. Seas. Rep. Stud. 56.

GESAMP - IMO/FAO/UNESCO/WMO/WHO/IAEA/UN/UNEP Joint Group of Experts on the Scientific Aspects of Marine Pollution 1988 Review of Potentially Harmful Substances - Arsenic, Mercury and Selenium. Rep. Stud. GESAMP 28 and UNEP Reg. Seas. Rep. Stud. 92.

GESAMP - IMO/FAO/UNESCO/WMO/WHO/IAEA/UN/UNEP Joint Group of Experts on the Scientific Aspects of Marine Pollution 1990 The State of the Marine Environment UNEP Reg. Seas. Rep. Stud. 39.

Goldman, C.R., A.D. Jassby and E. de Amezaga 1990 Forest Fires, Atmospheric Deposition and Primary Productivity at Lake Tahoe, California, Nevada. Verhandlungen der Internationale Vereinigung für Limnologie 24: 499-503.

Gordon, C. 1998 The State of the Coastal and Marine Environment of Ghana. pp57-67. *In*: A.C. Ibe and S.G. Zabi (eds). *State of the Coastal and Marine Environment of the Gulf of Guinea*, UNIDO/IOC, CEDA, Cotonou.

Guiral, D. 1984 Devenir de la matière organique particulière dans un milieu eutrophe tropical. Rev. Hydrobiol. Trop. 17 : 191-206.

Guiral, D., R. Arfi and J-P. Torreton 1989 Mécanismes et incidences écologiques de l'homogénéisation annuelle dans un milieu eutrophe stratifié. Hydrobiologia 183 : 195-210.

Ibe, A.C. 1986 *Experience from Marine Pollution Monitoring in West and Central Africa.* Symposium on Global Marine Pollution, Sixth Session of the IOC Scientific

Committee on the Global Investigation of Pollution in the marine Environment, Paris, 22-24 September 1986.

Ibe, A.C. 1996 The Coastal Zone and Oceanic Problems of Sub-Saharan Africa. pp201-214. *In* G. Benneh, W.B. Morgan and J.I.Uitto (eds). *Sustaining the Future: Economic, Social and Environmental Change in Sub-Saharan Africa.* United Nations University Press.

Ibe, A.C., L.F. Awosika, A.E. Ihenyen, C.E. Ibe and A.I. Tiamiyu 1985 *Coastal Erosion in Awoye and Molume Villages, Ondo State, Nigeria.* A Report for Gulf Oil Co. Nigeria Ltd.

Ibe, A.C. and C.E. Kullenberg 1995 Quality Assurance/Quality Control (QA/QC) Regime in Marine Pollution Monitoring Programmes: The GIPME Perspective. Marine Pollution Bulletin 31 : 209-213.

Junaid, K.S.A. 1999 Industrial Pollution Problems and Effects on Health of the Marine Environment. *In:* P.A. Scheren (ed.) *Proceedings of a Workshop on Environmentally Sustainable Industrial Development in the Gulf of Guinea Regio*n, 18-21 January 1999 Accra, Ghana. UNIDO/GEF Gulf of Guinea Large Marine Ecosystem Project. Abidjan.

Kakulu, S.E., O. Osibanjo and S.O. Ajai 1987 Trace Metal Content of Fish and Shellfishes of the Niger Delta Area of Nigeria. Environ. Int. 13: 247-251.

Kouassi A. M., D. Guiral and M. Dosso 1990 Variations saisonnières de la contamination microbienne de la zone urbaine d'une lagune tropicale estuarienne, Cas de la ville d'Abidjan (Côte d'Ivoire). Rev. Hydrobiol. Trop. 23(3): 181-194.

Kouassi, A.K., N. Kaba and B.S. Métongo 1995 Land-based sources of pollution and environmental quality of the Ebrié lagoon waters. Mar. Poll. Bull. 5: 295-300.

Kusemiju, K., D.I. Nwankwo and M.O. Oyebanji 1999 *A Survey of the Physico-Chemical Parameters and Plancton of Lagos Lagoon, Nigeria.* Report for the Gulf of Guinea Large Marine Ecosystem Project. Lagos, Nigeria.

Lemasson, L. and J. Pages 1982 Apports de Phosphore et d'Azote par la Pluie en Zone Tropicale (Côte d'Ivoire). Rev. Hydrobiol. Trop. 15(1): 9-14.

Métongo, B.S. 1991 Concentrations en métaux toxiques chaz *crassostrea gasar* (huître de mangrove) en zone urbaine lagunaire d'Abidjan (Côte d'Ivoire). J. Ivoir. Océanol. Limnol. Abidjan 1(1): 33-45.

Métongo, B.S. 1998 Evaluation Rapide des Sources de Pollution Atmospherique, Hydrique et Tellurique en Côte d'Ivoire. UNIDO/GEF Gulf of Guinea Large Marine Ecosystem Project. 78p.

Métongo, B.S., N. Kaba and A.M. Kouassi 1993 *Evaluation Quantitative et Qualitative des Effluents et des Polluants.* Centre de Recherches Océanologiques, Abidjan, Côte d'Ivoire.

Ministère de l'Environnement et de la Production Forestière - République de Togo 1999 *Etat de l'Environnement du Littoral du Togo (Profil Côtier).* Université du Bénin - Lomé, Togo. 160 p.

Ministère du Logement, du Cadre de Vie et de l'Environnement - République de Côte d'Ivoire 1997 *Côte d'Ivoire Profil Environnemental de la Zone Côtière.* Ceda Press. - Cotonou, Benin. 87p.

Ministry of the Environment and Forestry - Republic of Cameroon 1999 *Cameroon Coastal Profile.* ADE Graphics Industries - Yaounde, Cameroon. 102p.

Odu, E.A. and A.M. Imevbore 1985 Environmental Pollution in the Niger Delta: The Mangrove Ecosystem of the Niger Delta. pp133-155. *In: Proceedings of the International Seminar on the Petroleum Industry and the Nigerian Environment.*

Ojo, O.O. 1991 *Bottom Sediments as Indicators of Chemical Pollution in Lekki Lagoon.* M.Sc. Thesis, University of Ibadan, Nigeria.

Okonna, S.I. 1985 *A Survey of Pesticide Residues in some Nigerian River Waters and Sediments.* M.Sc. Thesis. University of Ibadan, Nigeria.

Okoye, B.C.O. 1991a Heavy Metals and Organisms in the Lagos Lagoon. Int. J. Environ. Stud. 37: 285-292.

Okoye, B.C.O. 1991b Nutrients and Selected Chemical Analysis in the Lagos Lagoon Surface Waters. Int. J. Environ. Stud. 38: 131-136.

Okoye, B.C.O., A. Afolabi and A. Ajao 1991 Heavy Metals in Lagos Lagoon Sediments. Int. J. Environ. Stud. 37: 35-41.

Oyewo, E.O. 1999 *Pollution Monitoring in the Lagos Lagoon, Nigeria.* Progress Report for the Gulf of Guinea Large Marine Ecosystem project. Lagos, Nigeria.

Pagès, J. 1975 Etude de la pollution bactérienne en lagune Ebrié. Doc. Sci. Centre Rech. Océanogr. Abidjan 6(1): 97-101.

Pagès, J. and J. Citeau 1978 La pollution bactérienne de la lagune et de la mer autour d'Abidjan. Doc. Sci. Centre Rech. Océanogr. Abidjan 9(1) : 43-50.

Portmann, J.E., C. Biney, A.C. Ibe and S. Zabi 1989 *State of the Marine Environment in the West and Central African Region,* UNEP Regional Seas Reports and Studies no. 108.

Powell, C.B. 1986 Ecological Effects of Human Activities on the Value and Resources of Nigerian Wetlands. pp120-129. *In: Proceedings of the Workshop on Nigerian Wetlands.* Man and the Biosphere/UNESCO.

Reimold, R.J. and F.C. Daiber 1967 Eutrophication of Estuarine Areas by Rain Water, Chesapeake Sci. 8: 120-133.

Saenger, P., Y. Sankaré and T. Perry 1996 *Effects of Pollution and Overcutting on Mangroves.* Report for the Gulf of Guinea Large Marine Ecosystem Project. Abidjan, Côte d'Ivoire.

Scheren, P.A.G.M. 1995 A Systematic Approach to Lake Water Pollution Assessment. Water Pollution in Lake Victoria, East Africa. Eindhoven University of Technology.

Scheren, P.A., H.A. Zanting and A.M. Lemmens 1998 Estimation of Water Pollution Sources in Lake Victoria, East Africa: Application and Elaboration of the Rapid Assessment Methodology. Unpublished Manuscript.

Sherman, K. 1993 *Emerging Theoretical Basis for Monitoring the Changing States (Health) of Large Marine Ecosystems.* Summary Report of Two Workshops: 23 April 1992 National Marine Fisheries Service, Narragansett Rhode Island, and 11-12 July 1992 Cornell University, Ithaca, New York. NOAA Tech. Mem. NMFS-F/NEC-100. 27p.

Sherman, K. 1998 The Gulf of Guinea and Global Developments and Sustainability of Large Marine Ecosystems. pp50-75. *In:* A.C. Ibe, A.A. Oteng-Yeboah, S.G. Zabi and O. Afolabi (eds). *Integrated Environmental and Living Resources Management in the Gulf of Guinea: The Large Marine Ecosystem Approach.* Proceedings on the First Regional Symposium on the Gulf of Guinea Large Marine Ecosystem, UNIDO/UNDP/NOAA/UNEP, CSIR Press Accra, Ghana.

Sherman, K. and A.R. Solow 1992 *The Changing States and Health of a Large Marine Ecosystem.* CM/L38 Session V. 31p.

Sherman, K., L.M. Alexander and B.D. Gold 1993 *Large Marine Ecosystems: Stress, Mitigation, and Sustainability*. AAAS Press, Washington DC. 376p.

Topping, G. 1992 The Role and Applications of Quality Assurance in Marine Environmental Protection. Marine Pollution Bulletin 25: 61-65.

World Bank 1998 *World Development Indicators 1998*. World Bank, Washington DC.

Word Health Organisation 1982 *Rapid Assessment of Sources of Air, Water and Land Pollution*. WHO Offset Publication No. 62, Geneva.

World Resources Institute 1991 *World Resources* (4th ed.). World Resources Institute, Washington DC.

Socio-Economics

The Gulf of Guinea Large Marine Ecosystem
J.M. McGlade, P. Cury, K.A. Koranteng and N.J. Hardman-Mountford (Editors)

23

Socio-Economic Aspects of Artisanal Marine Fisheries Management in West Africa

E. Bortei-Doku Aryeetey

Abstract

Artisanal fisheries in West Africa including the Gulf of Guinea are facing serious challenges due to the virtually open access nature of the industry and the fact that the natural resources supporting this industry are beginning to show serious signs of stress, linked to over-exploitation and natural environmental variability. This has been traced to an over-dependence on fishing and allied activities as a means of livelihood in fishing communities and also expansionary policy measures in the past that encouraged more people to enter the fishing sector. The nature of artisanal fisheries in West Africa are described. National and sub-regional structures for traditional and formal management of the sector are reviewed.

Introduction

The need to integrate the socio-economics of fisheries activities into general fisheries management frameworks is now widely acknowledged as a necessary condition for the attainment of sustainable fisheries, and poverty reduction policies in West Africa. By so doing not only will policy makers and development practitioners be more responsive to the management peculiarities of the ecosystem, but they will also recognise the capacity building needs of the labour force in the sector and the whole community. The issue of the promotion of sustainable fisheries has become critical to the over-fished waters of West Africa as it has everywhere. Besides, economic downturn in the sub-region calls for a more judicious use of natural resources. GNP per capita declined in most countries of the West African Region between 1980 and 1994, averaging about US$350.00 (Horemans 1996).

The pressure on fisheries resources is expected to worsen given the rapidly increasing populations of West Africa, which could easily double in the next 25 years. In many countries the population growth rate is about 3 percent per annum (Heinbuch 1992). Some of the signs of what is to come can already be seen there, with the near collapse of the pelagic stocks between Côte d'Ivoire and Benin attributed to high fishing pressure and bad climatic conditions in the 1970s (Charles *et al.* 1994).

The main objective of this paper is to review the general socio-economic aspects of artisanal fisheries and the ways in which these influence the management of the resources. In this regard the paper focuses on the *social and economic organisation of artisanal marine fisheries in the West African sub-region, resource management strategies and prospects for the adoption of sustainable fisheries practices.*

Most of the information presented in the paper has been gathered from secondary sources. Considerable use has been made of the reports published by the Integrated Development of Artisanal Fisheries (IDAF) Project in West Africa, whose area of operation stretches from Mauritania to Angola. The IDAF project covers 10,000 km which represents approximately 30 percent of the African coastline, and includes all of the Gulf of Guinea. Other sources that have been used include the FAO reports on sustainable fisheries, United Nations Human Development Report (1997), as well as country studies where possible.

In general, the existing data on fisheries activities in the sub-region are seriously deficient. Several of the countries do not conduct regular frame surveys. The few countries where such efforts are made include Mauritania, Guinea Bissau, Ghana, Togo and Benin. Fewer still undertake socio-economic studies on fisheries. The problem appears to be compounded by the highly diverse nature of the fisheries with respect to the species that are caught, and types of gear in use. In addition the highly dispersed settlements with numerous landing sites make data collection expensive and time consuming. The issue is further complicated by the high incidence of migration in the sub-region (Charles *et al.* 1994).

In many respects the prevailing norms of relatively easy entry and open access to fishing and allied activities which have provided a haven for labour in fishing communities, impede diversification in the acquisition of occupational skills and livelihoods. In addition, the situation precludes any serious regulation of resource use. The consequences of this has been the overfishing and reduced catches. Under these conditions incomes are said to have declined, and the general poverty of artisanal fishing communities may have worsened.

The rest of the paper first outlines human capital conditions of the fishers in the West African sub-region, including gender issues and the mobilisation of productive assets adopted by artisanal marine fisheries. The next section briefly describes the different types of artisanal marine fisheries practised on the West African Coast including the Gulf of Guinea (Guinea Bissau to Gabon). This is followed by a review of the different approaches that are being used to regulate access to the resource, such as the prevailing institutional arrangements for marine resources management in the sub-region.

Socio-Economic Profile of Fishing Communities

Artisanal fishing is not only a means of income, it is a way of life; men and women are born into it, and it is an integral aspect of the institutional arrangements governing the life of fishing communities. It is rare to find that people have joined artisanal fisheries in their adult life, as found surprisingly by Mino-Kahozi *et al.* (1997). Their frame survey conducted in Zaire revealed that 51 percent of the fishermen were previously civil servants, farmers and traders. All along the West African coast there are reports of serious levels of poverty in fishing communities, demonstrated in inadequate social amenities and poor housing, low levels of human capital in education and health conditions and limited employment opportunities outside fishing and related activities (Kotnik 1981; Seynabou 1996; Cassel and Jallow 1991; Ijff 1990).

Economic Services in Fishing Communities

In most fishing communities the local infrastructure support for fisheries and other economic activities is very limited and so the industry continues to rely on traditional facilities. Financial assistance for example, is still predominantly from local sources including personal savings, fish traders and relatively wealthy fishermen. Rarely are fishermen, fish processors or traders able to raise loans from modern financial institutions like banks, except under special external assistance to the sector. In Ghana, the Agricultural Development Bank for example, had a special revolving loan fund for fishermen but high rates of default in repayment seriously affected the operation of this facility.

People Engaged in Fisheries Along the West African Coast

The labour force in the West African region engaged in fisheries and related activities is significant. Altogether about 500,000 men along the coast from Mauritania to Cameroon are engaged in fishing, operating a total of about 120 thousand canoes (Horemans 1996). Fishermen in the region are usually between the ages of 30 and 50 years; canoe owners tend to be older. The majority of fishermen have no other occupations and have been in fishing for over 20 years. In a limited number of cases, fishermen also farm or trade. Long hours at sea and on repairs to canoe and gear, limit the amount of time fishermen have to engage in other activities. Furthermore, they often lack the bulk capital that is required to establish other businesses. Canoe owners of course have more flexibility in exploring additional income earning options if they do not go to sea. Women canoe owners tend to be quite prominent business people in their communities (*ibid.*).

It is often thought that there are as many women in fish processing and marketing as there are fishermen. Simply put, most fishermen have a female dealer with whom they trade (Akrofi this volume). The women buy and process fish and market it. In many arrangements between spouses women share the profits with men according to an agreed share system. It is equally common to find that wives purchase the fish from their husbands.

Livelihood and Human Capital in Fishing Communities

- Costs and Earnings in Artisanal Marine Fisheries
The initial capital that is needed for artisanal fishing is high in the West African sub-region, and profits appear to fluctuate dramatically, depending on the size of the catch. A fuller understanding of the economics of this enterprise is, however, hampered by the dearth of research on the profitability of the fishing operations. A study conducted in Limbe, Cameroon looking at the fixed costs and the revenue accruing from the share system revealed that the rate of profitability on purse seining (Awasha) is about 50 percent (Njifonjou 1997). In this case the revenue from the fishing was shared 50:50 between the crew and the canoe owner (*ibid.*).

Incomes from fishing vary considerably over the entire West African sub-region. It is estimated for example that in 'line fishing', fishermen in Senegal make four times more than their counterparts in Ghana. The contribution of fishing to GNP in Senegal is about 12 percent, compared to 3 percent in Ghana. Senegal of course exports more fish than does Ghana (Horemans *et al.* 1994). In addition, within each country different types of fishing carry different rates of profitability. In Njifonjou's (1997) study for example, he found that compared to the profitability rate of 50 percent for purse seining, bottom gillnet and surface gillnet fishing had profitability rates of 10 and 15 percent respectively.

Irrespective of the economic status of fishing in the sub-region, fishers are known to be highly prone to income insecurity. First, high fluctuations in income due to seasonality of income flows force fishermen to fall on their savings and working capital (Ndiaye, Y.D

1996). As part of attempts to diversify income earning sources for fisherfolk, food production activities have been introduced in places where these are possible, as in three of the project sites in The Gambia (Tujereng-Batokunku, Sanyang and Kartong). Coupled with this are high demands from family and social obligations (marriage, funerals, birth celebrations, religious ceremonies). The higher the status of the canoe owner or fisherman the more the demands on his or her resources from a large network of dependents, or those who obtain their security from this operator.

Women's incomes from fishing show similar disparities as are found among some canoe owners and their crew. On one side there are the large majority of small women processors and traders, many of whom can barely afford to subsist on their earnings. In contrast there are a relatively small number of large well-off processors and traders, some of whom own canoes and gear (Overa 1998).

Others have pointed to the consequences of the general economic malaise in many West African countries on the artisanal fishing industry. In a study of Ghana, Afful (1997) describes how the devaluation of the cedi caused a price inflation that seriously escalated investment and operational costs in this sector. The cost of an imported 40HP outboard engine increased by 850 percent between 1985 and 1994, while the cost of a locally manufactured canoe (32 m long) increased by 1,700 percent in the same period (*ibid.*). Under the circumstances many fishers seem to face difficulties in raising funds to obtain capital assets for fishing, a situation which is worsened in many places by poor loan repayment records (Bortey 1997).

It has been suggested by many studies into artisanal fishing that the lack of credit for working capital, or credit to promote investment in innovative technologies, is a major handicap for the industry. The main source of credit is very often the local well-off fish dealer (processor, wholesaler), the majority of whom are women. Fishermen have historically battled with the monopolist tendencies of the fish dealers. By virtue of their critical role as creditors, these women assume control of the sale of fish landings, often trimming off a good part of the profits from the transactions, leaving some canoe owners and crew members in a permanent state of indebtedness. Stability in the economic value of artisanal fisheries obviously carries positive impacts on the well-being of entire households, and not just the fishers who are directly involved.

- Nutrition and Health in Fishing Communities

Concern over fisheries is equally linked to the fact that fish consumption in the region which stood at approximately 15 kg per capita in the early 1980s had declined to 10 kg per capita in 1993, below the world average of 13 kg per capita. This fact is worrying considering that fish provides about 40 percent of animal protein for the majority of people in the sub-region (Horemans 1996).

A study of nutritional conditions at Limbe in Cameroon revealed that there were serious nutrition deficiencies facing the community. Nearly half of the school age children that were observed suffered from protein energy malnutrition. At the same time general environmental hygiene was poor and adults had smaller quantities of food than they required, though it was of reasonable quality (Ngo Som 1996). A detailed study of fishing communities indicate that as in many other countries, preventive diseases, mostly from poor environmental sanitation (malaria, cholera, dysentery) are widespread (Ijff 1990). Part of the problem of the poor nutrition in some cases may be linked to low educational attainment.

- Education and Training in Fishing Communities

Parents and guardians frequently withdraw their children from school early in fishing communities, to provide child labour in any of the numerous gender segregated fishing activities. Consequently, literacy rates in fishing communities are low for both men and women and children alike, ranging from 29 percent in Sierra Leone to 71 percent in Cape Verde (Horemans 1996).

On the whole, skills for fishing and related activities are more or less learned on the job through informal apprenticeships. Fathers engage their sons and other male relatives on their canoes, just as mothers involve their daughters and other female relatives in their processing and trading businesses. There are rarely any formal apprenticeship in fishing. When it does occur however, a father may negotiate with a canoe owner for his son or other male relative to join a canoe, sealing the agreement with a (Bortei-Doku Aryeetey 1995).

Various projects in the past have included training components to address the inadequacies in the skills and knowledge of artisanal fishermen. These programmes have focused mainly on use of new fishing technology, repairs and maintenance of gear, canoeing, training in book-keeping and channels of information and assistance. Such programmes have been available in some countries longer than in others. Usually the training has been geared at people who are already in the business of fishing, but in some countries such as Gabon, Cameroon, Congo, The Gambia it has been necessary to undertake fresh recruits for training, because despite their rich fisheries resources fishing is not popular. Consequently, the sector is often left to foreign fishermen, which sometimes generates a lot of conflict (Horemans 1996).

- Trends in Technological Innovation in Artisanal Fisheries

Some very significant changes have occurred in the area of fishing technology. In many places along the coast, fishermen have given up manual propelling in favour of the outboard motor. It has been pointed out, however, that in spite of the new technological innovations, the hold of tradition over the industry in terms of norms, labour organisation and business management precludes any major transformation of this industry from artisanal to industrial fishing in the near future. The rate of motorisation ranges from 7 percent in Liberia (before the war) to 90 percent in Mauritania. Two of the most active fisheries countries, Ghana and Senegal have rates of 49 percent and 67 percent respectively. The overall motorisation in the region is however, estimated at only 30 percent (Horemans 1996).

In processing there have been attempts to improve smoking ovens, the most popular of which has been the Chokor oven, a modified traditional oven originating from Chokor in Accra, Ghana. The main objective in this and other similar endeavours has been to improve fuel efficiency, and also to improve quantity processed at a time for economic and environmental reasons (Akrofi this volume).

There is little doubt that technological innovation has obvious beneficial effects on fishery activities in general, but many analysts have serious reservations about the indiscriminate introduction of new technology, arguing that there are important socio-cultural issues that need to be taken into account in the introduction of new technology. One such issue is the question of employment - the extent to which labour saving technologies are relevant in an economy marked by high unemployment in many countries. Closely linked to this is the possibility of accentuating existing social disparities by the use of more efficient technology that require fewer people. In the same way unequal access to new technologies among the fishers is likely to widen differences between fishers. From the point of view of the need for diversification in artisanal fisheries however, it may not be completely unwarranted to seek to draw people away from this industry (*ibid.*).

Gender, Processing and Marketing

Artisanal fishing is a highly segregated activity in which men and women are strictly confined to specific tasks. These tasks may differ from one country to the next, or within countries. While in some countries men dominate all activities in marine fisheries, in many other places the rigid gender segregation gives women virtual control of processing and marketing. In Cassel and Jallow's (1991) study of Gunjur in The Gambia for example, they found that men dominated the smoking and drying of fish, with women involved only on a small scale.

Fishermen and women fishmongers, often from the same kin groups, co-operate and compete through very complex networks of sellers and buyers. A large number of fishermen rely on their wives or female relatives to market or process their catch, but increasingly these transactions are becoming sensitive to market forces rather than kinship obligations. There have been instances in Ghana where fishermen have used other beaches to land their fish. By so doing they are able to sell to 'stranger' women rather than their wives and relatives, who they believe cheat them The advantage here is that depending on the availability of fish, they are able to negotiate better prices for their fish.

Despite their active involvement in most places along the coast, there are nevertheless rigidly applied cultural codes that more or less bar the extension of women's involvement beyond what is seen to be culturally permissible. For this reason traditional leadership positions in fishing communities have remained virtually closed to women, except for the non-hereditary position of women's leader. Women's status has however found other channels to grow. The few women who have been able to amass wealth in this industry have become the backbone of financial support for the purchase of fishing gear and canoes, and for financing fishing expeditions; their creditor status also gives them authority over wholesale pricing. Some attempts have been made to capture the competition and co-operation between men and women in the fishing industry in Ghana (Vercruijsse 1984; Overa 1998).

A considerable proportion of the fish traded in the sub-region is sold as smoked fish, leaving a small segment of the processed fish market to other products such as dried, cured and fried fish. Fresh fish does not attract a big market in many places due to poor cold storage facilities, though in a few instances such as in Cape Verde it constitutes 80 percent of the fish that is marketed. Fish marketing involves two fairly parallel systems of low value pelagics and high value demersal fish. It is feared that declining supply coupled with rising production costs will push prices up in the near future. Many countries in the sub-region import fish to augment local supply, even where they are exporters themselves.

A distinct advantage that enterprising women fish processors enjoy over the fishermen is being able to engage in other income earning activities, and so to plough the profits from those areas into their fish business. Further, women's control over processing and marketing gives them the opportunity to store and sell processed fish well beyond the fishing season. This enables them to stagger their incomes over the fishing season. In many places where the ecology and population density permit, during the low fishing season the women farm vegetables and cassava. Others trade in small merchandise or cooked food. Many of these are carried out simultaneously by the women. The women and men alike tend to work with mutual aid groups most of which remain informal throughout their life-cycle.

One of the ways in which development agencies have sought to mobilise women for technology and other forms of assistance has been through the formation of women only associations. In many places this appears to be appropriate because of the traditional segregation of fisheries activities between men and women. In the early 1980s the International Labour Organisation, for example, supported women's groups with the transfer of technology on fish processing.

Improved smoking ovens constitute the main items in the technology package, particularly the multiple chamber, multiple rack Chokor smoker. Other types of assistance such as credit, literacy, numeracy and marketing skills have also been introduced to women through the groups. Women's organisations are often less formal and visible compared to fishermen's associations, but they tend to experience some of the same strains on their sustainability. These include lack of credit to finance operational expenses, low access to extension services, and general alienation from decisions that affect their operations. Women's groups in Ghana often complain of lack of credit to purchase fish for smoking, while most of the assistance available to them is for technology innovation.

Compared to men's organisations women's fishers organisations have been found to be weak, and especially lacking in formal organisational procedures (Touray 1996). This has been partly blamed on the low level of education and low capital base among the members. One might also add that there is a general mistrust for any collective action among processors, particularly if it is linked to production and ownership of equipment. A study of 13 women's groups in The Gambia (Brufut and Gunjur) revealed that over 50 percent of them did not keep records on their activities. They identified their major problems as credit facilities, inadequate utensils and transport difficulties (Touray 1996).

Fishers Organisations

Based on an assumption that formalisation would be beneficial to fishers, as has been observed under certain circumstances in agriculture, numerous attempts have been made around the world to build fishers organisations. Many development assistance programmes are predicated on the existence of, or agreement that fishers' co-operatives will be introduced. The fishers' co-operative is seen as a realistic link between donor (or other outside agencies) and the artisanal fishers. The main categories of fishers working on the West African coast as fishermen and as fish processors/ allied workers include a minority of subsistence fishers who perhaps are mainly engaged in agriculture, and a significant proportion of commercial artisanal operators. As can be imagined efforts to organise them have concentrated on those running commercial operations.

The fishers' organisations can be distinguished by their main functions. Four distinct forms have emerged as follows:

- Fishermen's associations made up of boat owners and some fishermen, e.g. Senegalese Economic Groups
- Community associations which apart from issues relating to fishing are also interested in local infrastructure development, e.g. Kaback Association in Guinea
- Formal Private Enterprises, e.g. Promo-Peche in Gabon
- Women's Processing Groups

Some of these organisations enjoy political patronage as a result of which they may attract assistance from public institutions and banks. The types of fishers' organisations that are dominant differ from country to country. The private enterprise types of organisation tend to be purely economic entities organised on the model of the firm, quite distinct from the way artisanal fisheries are organised around kin groups. The enterprise, as well as many of the fishers' associations, is committed to addressing inputs and marketing needs of the members.

The community-based groups, as can be expected devote some of their attention to the development needs of the whole community (Horemans and Satia 1993).

Efforts that have been made to organise artisanal fishers into co-operatives have met with very limited results in many countries. Reporting on a study of seven fisherfolk organisations in the sub-region, Horemans and Satia (1993) observed that fishers' organisations were usually not sustainable for the following reasons:

- Over-dependence on donor / non-government agency support
- Failure to address the needs of their members
- Over burdening the organisation through multiplicity of functions / activities
- Inadequate training in co-operatives, management and book-keeping

The study further revealed that the organisations were generally subject to financial difficulties, credibility problems, capacity limitations, poor goal definition, ownership crises, and state interference. State support for co-operatives in general had probably not been strong in many countries, it appears the situation worsened during the economic difficulties of the 1970s and 1980s which was then followed by a major economic paradigm shifts to free market economy under structural adjustment programmes in the 1980s. Many non-viable fishers' co-operatives may have collapsed during this period.

With the increasing shift towards a participatory management approach as embodied in the Code of Conduct for Responsible Fisheries proposed by the Food and Agriculture Organisation, one of the expectations has been that fishers associations would form the basis for addressing issues of territorial use rights for fisheries (TURFs), as is developing in parts of Asia. Similarly, fishers' organisations are widely seen to be the channels through which co-management practices will be introduced in fishing communities. The limited activity in relation to rights based fishery is however, not entirely a result of weak fishers organisations. Fisheries scientists themselves and their social science collaborators need to undertake considerable research on the concept of property rights in the sub-region before any meaningful programmes can be introduced in that area.

Based on an assessment of the operations of the seven case studies it was evident that to be sustainable, fisheries organisations need to have dedicated leadership and day-to-day managers. In the absence of such administration, organisations become vulnerable to mismanagement and political pressure (Satia1993). Where there have been marked successes as in Shenge in Sierra Leone, this has been linked to continued efforts at institution-building, and also business development (*ibid.*).

Outside these formalised groups of fishers, there are still very active networks of reciprocal exchange, through which informal insurance transfers are channelled. O. Ndiaye (1996), looking at Senegal, classified the types of risk insurance as human and material risks, social risks and financial risks. He emphasised the burden placed on the canoe owners to mobilise resources to meet these risks, and other more broad based efforts such as fund raising through crew contributions (e.g. welfare fund), special fishing expeditions and family support (loans and grants). Ndiaye (*ibid.*) further cautioned that, informal insurance networks were themselves vulnerable to the exigencies of a fragile economy, climatic conditions and population pressures. Sometimes conflicts disrupted the smooth operations of informal networks among fishers (*ibid.*).

As part of its capacity building initiatives, IDAF introduced training in organisation as part of its activities at fisheries centres that were established in selected communities, where fisherfolk would be provided with modern amenities and improved technology, as well as literacy and numeracy training in some cases.

Migration

Migration features prominently in artisanal marine fisheries in West Africa, ranging from within country to international migration. Furthermore, the types of migration differ by the period of sojourn of the migrants, ranging from temporary seasonal migrations to long-term and quite often permanent migrations (Haakonsen and Diaw 1991).

Though principally influenced by economic factors, the motives of migration have long been recognised as multi-faceted, and include socio-cultural factors (e.g. rites de passage?). Invariably the majority of migrants are men; however it is quite common for them to be accompanied by their wives or other female relatives who hope to benefit from trading in more favourable economic climates (Haakonsen and Diaw 1991).

Horemans and Jallow (1998) found that the importance of migrants to the local fisheries varies considerably from one community to the next. In Gabon for example, 90 percent of the marine fishermen are migrants from Nigeria, Togo and Benin. There are also high concentrations of migrants in Togo (65 percent), Benin (55 percent), Cameroon (80 percent) and The Gambia (73 percent). While the flow of migrants cris-crosses the entire coast, some countries have developed reputation as sending countries (Ghana, Senegal and Nigeria), and others can be described as receiving countries as noted above. In such circumstances it has been difficult to introduce new ideas into the development of these communities because of the transient nature of the labour. The high preponderance towards migration also means that marine resources are shared between countries with direct implications for resource management. A collection of case studies (Haakonsen and Diaw 1991) from a workshop on fishermen's migrations in the sub-region (6-9 November, Kokrobite, Ghana) indicates quite clearly the diversity in migration patterns and, host country responses to migrants. Other issues concerning, security, legal status of migrants and infiltration of technology are also addressed in the collection.

Migrants invariably have to link up with local hosts and hostesses to arrange for their stay and fulfilment of obligations to the traditional authority. Many of the studies show that beyond the economic transactions with their hosts, normally migrants do not interact socially with the indigenous population. Instead they reproduce some of their own cultural arrangements back home, perhaps to compensate for their marginalisation as foreigners without a voice in the host country. Under these conditions they tend to remain vulnerable to the whims of the host communities (Haakonsen and Diaw 1991).

Even where they stay for a long time, the fishers tend to repatriate their savings to their home countries for investment in their own towns and villages. In some cases lack of access to land, and very often an irregular immigration status make it difficult for the migrants to contemplate permanent investments in the host community. As a result some migrant fishing settlements have been associated with more squalid conditions than are found among the hosts.

Institutional Arrangements for the Management of Artisanal Fisheries

Very few countries in the West African sub-region have successfully introduced a monitoring, control and surveillance (MCS) systems. The process is hampered by a weak database on the resources (stock assessments, costs, earnings and returns), as well as the difficulties with co-ordinating the several agencies whose co-operation is required for enforcement of management regimes. It has been suggested that the choice of MCS measures for most countries was often dictated by costs, vessel nationality and prevailing resource management regimes (Roberts 1995). Most developing countries do not appear to be in a position to deploy conventional surveillance equipment such as surface craft and aircraft. Different countries use the following MCS measures to varying degrees of effectiveness: licensing, registers of fishing vessels, dockside versus at sea inspections, self-reporting position and catch, on board observers, negotiation of access agreements, transit of vessels not authorised to fish, flag state responsibility, marking/identification of vessels.

There are also political considerations that restrain countries from rigid enforcement of MCS measures; for example, Roberts (*ibid.*) identifies the potential danger of job losses among fishermen resulting from efforts to restrict entry as a classic example. MCS, however, is not just about restriction. While recognising that each country must adopt an MCS regime that is contextually appropriate, Roberts argues that there are gains to be made from a certain degree of harmonisation of MCS measures in the area, including efficiency, economies of scale and negotiating power. What presently operates in most countries is a mixture of, albeit, weak traditional management systems under local fishers leaders, and national MCS under the leadership of civil, military or autonomous institutions or combinations of military and civil authorities. In Gambia for example, the Ministry of Agriculture and Natural Resources undertakes MCS activities in collaboration with the marines of the Gambia National Army, while in Guinea an autonomous body known as Centre National de Surveillance et de Protection des Pêches was created to perform MCS functions (Roberts 1995).

Traditional Management Practices

In the West African sub-region fisheries resources are more or less classified as state property, a claim supported by specified rules and regulations, but without any restriction on access to the resource (Satia 1995). Traditional institutions for artisanal fisheries management remain in force in most countries, involving a traditional leader (chief fisherman) and his council (see Brainerd 1991; Neiland *et al.* 1994). These offices are hereditary and are almost exclusively male. The chief fisherman oversees the rituals that mark the open and closed seasons, special occasions and handling of gear. In addition he is also responsible for regulating entry, labour relations, migration, etc. Though there is a general weakening of traditional institutions, the lack of alternative leadership and support structures help to maintain the influence of the chief fisherman.

In an example of traditional management practices on the Aby Lagoon in La Côte d'Ivoire, Kponhassia and Konan (1996) describe how traditionally the resources in the lagoon were covered by common property, as well as territorial rights. While fishing in the central part of the lagoon was open to all, resources along the banks, sheltered bays and shallow areas were subject to clearly defined territorial control based on boundaries with adjacent villages. Chiefs played a key role in monitoring access and violations of user rights. In 1990 they were supported by the administration to withdraw user rights from non-Ivorian fishermen, granting Ivorians exclusive use of the lagoon.

Various studies in the sub-region suggest that knowledge on fish resource management is highly mixed, with indications that fishers have slightly more incorrect facts about resource sustainability than accurate ones. Women were more likely to have incorrect facts compared to men. In a study of perceptions and practices in Cameroon it was observed that none of the men, and only one-quarter of the women agreed that the amount of fish in the sea is limited, though many fishers acknowledged that they did not always find a good catch (Demunyck and DETMAC Assoc. 1994).

Regulating Access

Various strategies have been adopted in the countries of the Gulf of Guinea to control fishing effort, or access to the shared resources. Horemans and Jallow (1998) give a detailed description of the different strategies which is summarised below.

- Types of Use Rights

Of the different types of property rights recognised by Pomeroy (1994), namely state, private, communal/common property rights and open access, it is the last type that has come to dominate entry in marine artisanal fisheries. All the same, most communities or ethnic groups purport to have customary rules governing membership and entry into fishing, though superficially these are hardly enforced in most places, partly due to population pressure and lack of employment options as noted earlier. The tendency has been to endorse open access, which perhaps helps to understand the high incidence of migration among fishers.

Existing Methods of Regulation of Access include the following:

i.   Direct regulation of exploitation
ii. Indirect control of fishing effort, including quotas
iii. Seasonal closure of fishing
iv. Mesh size regulation
v.   Traditional Measures

- Direct Regulation of Resource Exploitation

Direct regulation through a licensing system is targeted mainly at industrial vessels, though attempts have been made in some countries such as Guinea, to extend licensing to artisanal fisheries. Many of the countries have not succeeded in preparing fisheries management plans to guide the issue of licenses. A quota system of regulation was introduced in foreign coastal pelagic fisheries in Mauritania, Senegal and Guinea-Bissau, but difficulties with monitoring have made them ineffective (Horemans and Jallow 1998).

Territorial use rights in fishing (TURFs) have not taken firm root in the sub-region. In general little is known of what this means from a local perspective, and what local expectations are with regard to TURFs. Other factors affecting the creation and maintenance of TURFs have been described as natural resource attributes, boundaries, technology culture, wealth distribution and institutional factors (Charles *et al.* 1994). Far more research and documentation will be required to provide the basis for identifying appropriate frameworks for TURFs.

The issue of rights of access in fisheries is taking another turn which poses challenges for artisanal fisheries in the West African sub-region. First there is the pressure from industrial fleets which presently dominate off-shore and distant-water fishing, but are now trying to increase their access to coastal fisheries. Second, the current drive to push for the commercialisation of access rights in the form of Individual Transferable Quotas (ITQs) among fishers, which would directly limit entry only to those who can afford the quotas (SAMUDRA 1999).

Though an attractive management tool, and one with obvious economic benefit for the holders of ITQs, it has been noted that ITQs suffer from serious distributional inequities, especially in poor communities. By its nature the system makes it very difficult for crew on canoes to ever become boat owners. This is due to the disproportionate privilege enjoyed by existing canoe owners to become the first beneficiaries of ITQs in their community. Social stratification is bound to widen in fishery communities, leaving the possibility of ownership of canoes to a select group of fishermen (*ibid.*).

In terms of its value as a management tool, some doubts have been raised due to problems of quota busting which occurs because of inflexible Total Allowable Catch (TAC) limits; high grading where ITQ owners discard lower value fish; and data-fouling, which occurs when ITQ bearers under-report catch statistics, especially of high catch statistics. Critics even claim that the ITQ approach undermines the precautionary approach outlined in the Code of Responsible Fisheries adopted by member nations of the FAO. All the same the need to adopt more conservation conscious approaches in artisanal fisheries remains and has to be addressed. Concerns about the limitations of the ITQs have encouraged other ideas to flourish, such as the co-management approach.
- Indirect Regulation of Resource Exploitation
i) Reserved Zones
A popular approach to resource management within the Gulf of Guinea has been the adoption of reserved zones for the exclusive use of artisanal fisheries. The sizes of the reserved areas which vary from 2 to 8 nautical miles are largely influenced by the size of the continental shelf area. The countries apply these exclusive zones control are listed below in Table 23-1.

Position of Reserved Fishing Zone

2-4 miles	5 to 8 miles
Cameroon	Congo
Nigeria	Sierra Leone
Benin	Guinea
Gabon	
Togo	

Table 23-1 Reserved Fishing Zones in the Gulf of Guinea. (Horemans and Jallow 1998).

It appears the main objective in the adoption of exclusive zones is to separate the activities of artisanal from industrial fisheries, as a strategy to minimise conflicts between the two fleets as well as facilitate survival of the fisheries resources.

Enforcement of the fishing regulations to protect access to artisanal fisheries have been difficult in many countries, often leading to open conflicts between artisanal fishermen and trawlers as explained earlier. In Ghana, a 30-metre depth dividing line is currently being enforced by Navy patrols under the Fisheries Sub-Sector Capacity Building Project (Ghana DoF 1996) There have been reports from fishermen that the incidence of industrial fleets straying into shallow waters has reduced since the exercise was started (Hutchful n.d.).

Efforts to regulate fish exploitation through a quota system have been tried in

Mauritania, Senegal and Guinea Bissau with little success. Monitoring of allowed quotas and revenues has proved to be very difficult.

ii) Mesh Size Regulation

Besides exclusive zones, other strategies have been used such as the regulation of mesh size for both artisanal and industrial fishing to limit the catch of juvenile fish, as a strategy to minimise conflicts between the two fleets as well as facilitate survival of the fisheries resources. Occasionally a country has gone as far as to prohibit a particular type of gear, such as the ban on use of beach seining in The Gambia (Horemans 1996).

iii) Seasonal Closure of Fisheries

Sharp declines in some local fisheries have prompted governments to adopt drastic measures in an attempt to protect the species. As part of measures to reduce pressure on cephalopods for example, the government in Mauritania decided to close deep-sea fishing in the month of October in 1995 (Horemans 1996). Similar measures have been taken in Guinea (oysters), Cape Verde (lobsters) and Guinea-Bissau (marine turtles).

iv) Traditional Systems of Regulating Access

Traditional systems of regulating access are tied to religious beliefs and practices in which water and fish are deified and acknowledged through rituals on certain days. In some places such days are linked to the Christian, as well as Islamic faith. One important aspect of the ritual in almost every country is a rest day when fishing activities are customarily prohibited. In the Aby Lagoon the closures extend to the prohibition of fishing in holy places and reserves and deep waters (Kponhassia and Konan 1996). The penalties for violating this custom can be relatively heavy, involving a suspension from fishing or fine, or denial of landing rights for a specified period. In recent times, reports of violation of traditional management practices appear to be on the increase. Rest days are sometimes violated as fishermen escape to beaches away from their operational sites to land their catches. Local fishermen have also been encouraged to violate these rules where migrant fishermen have been lax in observing them (*ibid.*).

West African Sub-Regional Networks To Promote Fisheries Management

In response to the growing concern about the sustainability of the existing fisheries in the West African region, governments and development agencies are looking for ways to extract greater commitment from the various stakeholders towards sustainable fisheries. As with many such initiatives however, the successes have been modest. At the policy level various regional organisations have been formed bringing all or some of the countries along the Eastern Atlantic together. They include:

- The Fishery Committee for the Eastern Central Atlantic (CECAF) established in 1967
- The Ministerial Conference on Fishery Co-operation among African States bordering the Atlantic from Morocco to Namibia formed in 1989.
- The Sub-regional Commission for Fisheries (SRFC), comprising Cape Verde, The Gambia, Guinea, Guinea Bissau, Mauritania and Senegal established in 1985.
- This forum has enabled member states to initiate and strengthen co-operation and harmonisation of fisheries legislation of its member states.
- The Regional Fisheries Committee for the Gulf of Guinea (COREP) - Congo, Gabon,

Equatorial Guinea, Sao Tome, Principe and Zaire, established in 1984.

- The International Commission for the Conservation of Atlantic Tunas (ICCAT), formed in 1966, which includes 6 countries from West Africa.

Besides the quasi-permanent groupings there have been several workshops to address various aspects of fishing and management of resources. One of the most recent was the Second African Regional Workshop held in Malawi, November 1997 to examine case studies on co-management in the Africa Region.

Regional co-operation in West Africa faces many challenges arising from significant cultural, linguistic and geographical differences, coupled with widely varying political and fisheries management operational problems. In addition, neighbouring countries are further divided by disputes over maritime zones, posing legal complexity in attempts to enforce fisheries regulations (Roberts 1995).

Artisanal Fisheries Development and Management Policy Initiatives

Improving artisanal fisheries is now regarded as part of a multi-faceted approach to poverty reduction and food security. Sometimes in pursuit of this goal it appears as if some of the objectives that countries in the region set themselves, namely, improved productivity and better resource management are contradictory (Horemans 1996). Based upon a critical examination of management structures in the sub-region Fellizar noted for example that: "...policy instruments' effectiveness to support sustainability goals could be hampered by incompleteness, conflicts, and poor implementation. In such cases, policies are not only rendered ineffective but at times even restrict the development of the sector they are supposed to sustain" (Fellizar *et al.* 1997).

The international community, upon the recognition that centralised control of fisheries management yielded little result, has now turned to fisheries co-management in partnership with fishers. The process involves their participation in decision-making, as well as enforcement of regulatory measures. However, it is also feared that high costs associated with information dissemination, collective fisheries decision-making, and collective operational expenses could make co-management less attractive to governments and users alike (McGlade 2001).

Part of the exercise for improving the policy environment lies in clearly defining regulatory, distributive and procedural policies (Fellizar *et al.* 1997). By so doing governments would then indicate what restrictions they wish to impose on the sector; the types of services they wish to provide to different categories of fishers and who or what agencies will be responsible for pursuing the development and implementation of fisheries activities. On the whole post independence fisheries policies have tended to favour industrial fisheries at the expense of the artisanal sector. This was evidently based on a development vision of modern capital-intensive industrialisation with high levels of vertical integration (Satia 1993).

With time, failure of the newly established industrial fleets to deliver the expected outputs in many African countries has made governments turn attention, once again, to the activities of the artisanal sector. It was not only the need to improve technology, skills, and incomes, but also the desperate living conditions at this end that attracted attention. The FAO-led meetings on how to enhance artisanal fisheries recognised early that it would take an integrated approach, relying on a multi-disciplinary team of experts, to achieve the expected goals (Satia 1993).

Later, experience in working with fisheries communities through Community Fisheries Centres (CFCs) projects designed by the FAO confirmed that a participatory integrated approach was superior to conventional top-down blue print approaches to the problem (*ibid.*). This indeed, underlies, the evolution of co-management in fisheries policy today.

In spite of the growing attention being given to the management of fisheries, the extent to which basic principles of sustainable fisheries development have been incorporated into the integrated approach are not clear. Issues that are usually overlooked or not subjected to in-depth analysis include concerns over equity between crew and gear owners, problems of decreasing stock, a precautionary stance, the challenge of species conservation and biodiversity (Fellizar *et al.* 1997). This might be attributed to a vagueness in the definition of sustainable fisheries, due to the limited information on this framework at present. What is meant by sustainabilitiy and under what conditions are some of the issues that remain to be clarified. At the same time there are serious lapses in institutional and legal frameworks for supporting the development and management of the resource.

Country/Agency	Project	Assistance Package
Mauritania		
Arab Fund for Economic and Social Development	Management of the Baie du Repos, 1993-95	US$7million
Senegal		
Assistance to Artisanal Fisheries Programme (PROPECHE),	Credit and support and improvement of production, 1988	Nk
Sierra Leone		
Integrated Development Project of Shenge Region UNDP/FAO	Integrated development, 1985-95	Nk
Cote d'Ivoire		
Aby Lagoon Project IFAD/Government	Management and Monitoring, 1992	Nk
Ghana		
Yeji Artisanal Fisheries Integrated Development Project UNDP/FAO	Data collection, stock assessment, processing, 1994	US$1.4 million
Nigeria		
The Artisanal Fisheries Development Project IFAD	Credit and technical assistance for artisanal fishermen and fishsmokers, 1991-1996	US$20 million
Gabon		
Computerised Statistical Systems for Fisheries, FAO	Computerisation	US185,000

Table 23-2 Selected Socio-Economic Projects in Some West African Countries; Nk not known (after Horemans 1996).

International Programmes

- Integrated Development of Artisanal Fisheries (IDAF)

The Programme for Integrated Development of Artisanal Fisheries in West Africa (IDAF) was started in 1983 with funding from DANIDA (Denmark) and NORAD (Norway). The assistance was targeted at 20 countries between Mauritania and Angola who were willing to participate in an integrated management programme (Satia 1993). IDAF's main objective through the integrated approach was to diversify sources of livelihood through more employment options in order to raise human capital development in fisheries communities through better nutrition and improved education.   This was to be achieved among other things through attitude change and instrumental activities aimed at modernising the sector. In 1984 the World Conference on Fisheries endorsed the integrated and participatory approach by adopting an Integrated Strategy for the Development of Small-Scale Fisheries (Satia 1993).

The strategy to promote vertical integration (technology innovation) as well as horizontal integration (broad socio-economic development) arose from the recognition that fishing village communities had wide ranging developmental needs rather than fishery specific needs (Satia 1993). As a key part of the supply of the fishing inputs, credit and improved techniques were channelled through the project to the fishers. Some of the effects of these inflows were quickly realised in The Gambia, where the number of fishermen were more than doubled between 1989 and 1993, from 1270 to 3,500. Motorisation increased during this period from 55 percent to 84 percent. The fresh impetus given to fishing activities in the region provided justification for ancillary activities such as roads, water supply, communication links, health centres, schools and banking facilities. There have been major problems, however, particularly with loan repayment in some of the projects (*ibid.*).

After some initial setbacks in the first phase of IDAF (1979-87), resulting from weak participation by countries and direct users, the Project turned attention to the establishment of local institutions to strengthen project design, implementation and management. Fisheries Development Units (FDUs) were set up in participating countries and they supervised the development of local management committees in the beneficiary communities for the fisheries centres, and user groups (Meynall *et al.* 1988). Wide variations have been observed by the Project in the capacity and effectiveness of these institutions across the region (Satia 1993).

- Fisheries Co-Management

Fisheries specialists have expressed concern about the tendency to marginalise traditional fisheries management strategies in favour of central government control, resulting in mutual distrust between fishers and state, and lack of commitment among fishers to blue-print controls introduced by government (Satia 1995). African governments are responding by accepting to consider fisheries co-management, a model of joint management between users of the resource and government. It is understood that it offers both people and governments a potentially more enduring management system (Hanna 1992; McGlade 2001). For fisheries planners the question no doubt, will be how concerned should we be? A study on perceptions and practices of fishers on fishery resource management in Cameroon revealed that only 46 percent of male adults, and 36 percent of female adults could answer questions on fisheries resource management correctly. Very few people could explain the role of the government official in their area (Demuynck and DETMAC Associates 1994).

Upon the recommendation of member nations, IDAF organised a workshop as part of the 9th Liaison Officers Meeting to address the problem of participatory approaches in fisheries management (Conakry, Guinea, 13-15 November, 1995) to sensitise her partners to the need to involve stakeholders in fisheries management (IDAF 1995). Initially only a few West African countries experimented with formal attempts at co-management; they included Cote d'Ivoire and Benin. However, other countries such as Ghana later joined in the experiment with fishers-state partnerships in co-management. Currently most of the co-management projects are located in southern Africa, namely, Zambia, Zimbabwe, Malawi and Mozambique (IFM/ICLARM 1996; Jackson 1996). Studies to investigate the conditions for co-management in West Africa revealed the following among others: need for research into indigenous knowledge on fisheries management; need for studies in to fishery community organisation.

National Efforts to Support Artisanal Coastal Fisheries

Most countries in the sub-region have a specialised department or Ministry in charge of fisheries. A crops bias in many places has, however, negatively affected the public resources that are allocated to the fisheries sector. Within the sector itself a traditional focus on biological resources has shifted resources away from management and other socio-economic aspects of resource surveillance. Recent collaborative efforts between overseas centres in Norway, France and Portugal, however, are helping countries in the sub-region to change this situation (Horemans 1996).

Some of the fiscal measures that have been taken in the past to address financial difficulties of the fisheries sector are now blamed for the problem of overfishing. These include mainly subsidies on equipment and gear, fuel, soft loans, and other incentive policies. There have, unfortunately, been problems of abuse of these incentives causing policy makers to abandon them when they come under fire (Afful 1996). In Ghana, Nigeria, The Gambia and Togo fuel subsidies were introduced, withdrawn and re-introduced between 1994 and 1996 (Afful 1997).

In the general framework of structural adjustment programmes in West Africa, activities to improve research, management and monitoring, control and surveillance have been launched in various countries. Part of this process targets the participation of fishers in resource management. In Ghana the Department of Fisheries under an IDA - World Bank restructuring project known as the *Fisheries Sub-Sector Capacity Building Project (FSCBP IDA-CR 2713-GH),* is taking steps to strengthen the organisation through restructuring, staff recruitment and training, stock assessment and extension improvement. As part of this programme the Department has been encouraging the establishment and strengthening of user groups; nine have so far been identified and supported in the Central (3) and Western Regions (6), with a total membership of 203 (Ghana DoF 1996).

Managing Conflicts Involving Artisanal Fisheries

Conflicts in the artisanal fisheries have been traced to the weakening commitment to the traditional management regimes that regulate common property rights which more or less gives open access to both local and foreign fishers, overcrowding of multi-fleet, as well as flouting of regulatory mechanisms. Six case studies were conducted in Senegal, Cote d'Ivoire, Ghana and Cameroon to ascertain the extent of the problem in these countries based upon which a workshop was held on Conflicts in Coastal Fisheries, in Cotonu Benin, 1993

(IDAF 1994). In one of several types of clashes over fishing grounds, artisanal hook and line fishermen with bigger boats and powerful engines were able to fish further afield than before, quite often drifting into the zones that were previously reserved for trawlers (*ibid.*).

Artisanal and industrial fleets are frequently engaged in confrontation over entangled nets and thefts of catches in such situations. Though in many countries trawlers are restricted to operating outside a 30 meter depth to protect artisanal fishers, trawlers regularly encroached on this barrier following the lead of echosounders and fish finding devices Koranteng (1996).

Between the various fisheries departments and artisanal fishermen in each country, the main source of conflict has been over violations of fishing gear regulations such as, small mesh size, poisonous chemicals, explosives and herbs. Even the recording of catches has been a source of friction with both trawlers and artisanal fishermen. Environmental concerns have also generated misunderstanding in places where attempts have been made to restrict the use of beaches with serious problems of erosion as in Ghana (*ibid.*). Other policy measures have generated considerable tension between governments and fishers at various times in the past. In Ghana one notable example was over the withdrawal of subsidies on pre-mixed fuel introduced in 1992 and abandoned in 1994 (Afful 1997).

Channels of conflict resolution are evolving in many countries moving beyond traditional arbitration used in the communities, to legislation and enforcement procedures. Such legislation tends to rest on penalties for different offences, but many countries lack the capacity to enforce these penalties because apart from a few countries including Ghana and Nigeria, Monitoring Control Surveillance and Enforcement Units (MS) have not been established (*ibid.*). The legal aspects of undertaking this surveillance have been discussed by Roberts (1995) in detail. He concluded that in spite of obvious lapses where attempts at surveillance have been made, the signs were good that countries in the sub-region were taking measures to improve surveillance.

Perhaps more common sources of conflict are related to the observation of customary rights, which include religious practices as well as administrative processes. Many times in these cases the conflict erupts between indigenous fishers and migrant fishers. Local fishermen may for example, attempt to block foreigners' access to some of the water bodies (e.g. streams) which can cause disaffection. In other instances migrant fishermen have been accused of failing to pay the customary dues that they are required to pay to local chief fishermen in the host countries. Several factors may be a source of conflict here, for example, the amount of dues being demanded, the number of times dues should be paid for seasonal migrants, etc. Canoe owners, also sometimes, face open confrontation from their crew, and worse still, inter-crew conflicts occur sometimes leading to death (Bortei-Doku Aryeetey 1995). Between 1988 and 1992 the Fishery Protection and Monitoring Project in Senegal recorded 12 deaths and seven injured persons (*ibid.*).

Safety in Artisanal Fisheries

Efforts are also being made by IDAF to strengthen the capacity of national governments and local communities for dealing with safety at sea among artisanal fishermen. The main causes of accidents have been identified as capsizing, collision, fire at sea and entanglement of equipment and nets between fishers and trawlers. Sometimes these accidents result in the loss of life and considerable property. Estimates based on data from Mauritania to Sierra Leone suggest that along this stretch about 211 fishers die at sea in a year (Gallene 1995).

Conclusions and Recommendations

Artisanal fisheries in West Africa including the Gulf of Guinea have undergone major developments in the past two decades in relation to technology and the expansion of markets, which have had a direct impact on the socio-economic conditions of fisheries communities. Inevitably the natural resources underpinning this industry is beginning to show serious signs of stress, linked to over-exploitation. This has been traced to over-dependence on fishing and allied activities as a means of livelihood in fishing communities; expansionary policy measures in the past that encouraged more people to enter fishing such as subsidies and an open access principle in common property rights.

Areas of potential and actual conflict in fisheries policies that need to be addressed exist at different scales of operation. Part of this may be attributed to gaps in knowledge and skill among fisheries officials and other development practitioners.

With rapidly increasing populations in fishing communities the pressure on fisheries is expected to worsen, unless concerted efforts are made to regulate entry. This will require both direct measures and indirect strategies. In the first instance careful review of policies is necessary, as is indeed occurring in many countries. In addition, serious attention will have to be given to improving human capital development in fishing communities to encourage diversification in livelihood sources. Such developments should also lead to improvements in health standards, as fishers will learn improved ways of sanitation and nutrition.

Most governments desire to implement a fisheries code that is integrated, participatory and sustainability-centred. There is a strong inclination to include in this promotional instrument a strategy for achieving territorial use rights in fisheries.

There is growing pressure for countries to move towards restricted access through some form of rights-based regimes. In the long run this will help safeguard fisheries resources for future generations. The method for introducing rights-based regimes as part of a precautionary approach, however, remains highly debatable. Where does one stop in the continuum of highly privatised individual transferable quotas (ITQs) to community based co-management systems? The options have to be carefully studied together with serious research into traditional management and common property rights perspectives. This should be combined with population development programmes such as family planning, education and skills training. Ongoing programmes in fisheries co-management point in the right direction as there is ample evidence that failure to forge partnerships between fishers and the state in fisheries management will make it impossible to achieve the objectives of the FAO Code of Conduct for Responsible Fisheries.

References

Afful, K. 1996 Fiscal Policy and the Artisanal Fisheries Sector in Ghana and Senegal. Technical Report, No. 90. October.

Afful, K. 1997 The Effects of Fiscal Policy Reforms on the Artisanal Fisheries. IDAF Newsletter (FAO) No. 33: 14-16.

Akrofi, J.D. (this volume, chapter 24) Fish utilisation and marketing in Ghana: state of the art and future persepctive.

Bortei-Doku Aryeetey, E. 1995 Kinsfolk and Workers: Social Aspects of Labour Relations Among Ga-Dangme Coastal Fisherfolk. *In*: F.X. Bard and K. A. Koranteng.

Dynamics and Use of Sardinella Resources From Upwelling off Ghana and Ivory Coast. ORSTOM Editions, Paris.

Bortey, A. 1997 *Credit and Savings in Artisanal Fisheries in Ghana.*The IDAF Newsletter. September, No.35: 210-24.

Brainerd, T. 1991 *Community-Based and Traditional Fisheries Management in Africa. A Selected Annotated Bibliography.* IDRC-MR 291e.

Cassel, E.C. and A.M. Jallow 1991 Report of a Socio-Economic Survey of the Artisanal Fisheries Along the Atlantic coast in the The Gambia. IDAF/WP/41. Benin.

Charles, A.T., T.R. Brainerd, A. M. Bermude, M. M. Montalvo and R.S. Pomeroy 1994 Fisheries Socioeconomics in the Developing World. Annotated Bibliography. IDRC, Ottawa, Canada.

Demuynck, K. and DETMAC Associates 1994 The Participatory Rapid Appraisal on Perceptions and Practices of Fisherfolk on Fishery Resource Management in an Artisanal Fishing Community in Cameroon. Technical Report No. 60. IDAF, Cotonu Benin.

Fellizar, P. Jnr., R. G. Bernardo and A. P. H. Stuart 1997 Analysis of Policies and Policy Instruments Relevant to the Management of Fisheries/Aquatic Resources with Emphasis on Local Level Issues and Concerns. Philippines. SEARCA, WP No. 25. May.

Gallene, J. 1995 Intensifying the Collection of Data for a Relevant Assistance. IDAF Newsletter (FAO) No. 25:13-14.

Ghana, Department of Fisheries 1996 Fisheries Sub-Sector Capacity Building Project, (FSCBP, IDA - CR 2713-GH), Progress Report No. 1, November 1995 -July 1996. Accra, Ghana.

Haakonsen, J. M. and M.C. Diaw 1991 *Fishermen's Migrations in West Africa.* IDAF/WP/36. Cotonu, Benin.

Hanna, S. A. 1992 The Eighteenth Century English Commons: A Model for Ocean Management. *Ocean and Shoreline Management* 14.

Heinbuch, U. 1992 Relationships between Population Growth and Fisheries Development. IDAF Newsletter (FAO) No.16:19-22.

Horemans, B. 1996 *The State of Artisanal Fisheries in West Africa in 1995.* IDAF/ DIPA. Technical Report No.84.

Horemans, B. and A. Jallow 1998 Present State and Perspectives of Marine Fisheries Resources Co-Mangement in West Africa. IDAF Programme. DANIDA/FAO.

Horemans, B. and B. Satia (eds) 1993 *Report of the Workshop on Fisherfolk Organisations in West Africa* (Banjul, The Gambia, 3-5 February, 1993). Technical Report No. 45. Cotonu, FAO/IDAF.

Horemans, B., M. Kebe and W. Odoi-Akersie 1994. Working Group on Capital Needs and Availability in Artisanal Fisheries: Methodology and Lessons Learned from Case Studies. Technical Report No. 65. IDAF/FAO. Cotonu, Benin.

Hutchful, G. (not dated) The Fisheries Sub-Sector Capacity Building Project in Ghana. Department of Fisheries, Ministry of Food and Agriculture, Ghana. Mimeo.

IDAF 1994 Mobilize Efforts Against Conflicts in the Fisheries Sector. IDAF Newsletter (FAO) No. 21.

IDAF 1995 Rethinking Fisheries Management. IDAF Newsletter (FAO), No. 28.

IFM/ICLARM 1996 *Review of the Literature on Fisheries Co-management.* Red Series, No. 6.

Ijff, A.M. 1990 *Socio-Economic Conditions in Nigerian Fishing Communities.* IDAF/WP/31. Cotonu, Benin.

Jackson, J.C. (ed.) 1996 *Research Initiatives on Fisheries Co-Management in Central and Southern Africa.* Report of the Regional Workshop 20-22 November, 1995. Kariba, Zimbabwe. Red Series, No. 5.

Kotnik, A. 1981 *A Demographic and Infrastructural Profile of the Tombo Fishing Village, Sierra Leone.* Contribution of the Republic of Sierra Leone and the Federal Republic of Germany, Promotion of Small-Scale Fisheries, No.1.

Koranteng, K.A. 1996 The Marine Artisanal Fishery in Ghana: Recent Developments and Implications for Resource Evaluation. pp498-509. *In*: R.M. Meyer, C. Zhong, M.L. Windsor, R.J. Melay, L.T. Hushbak and R.M. Muth (eds). *Fisheries Resource Utilization and Policy.* Proceedings of the World Fisheries Congress.

Kponhassia G. and A. Konan 1996 *The Traditional Management of Artisanal Fisheries in Cote d'Ivoire: The Case of Aby Lagoon.* Red Series, No.12. Fisheries Project, Aby Lagoon, Adiake, Cote d'Ivoire.

McGlade, J.M. 2001 Governance and sustainable fisheries. pp307-326. *In*: B. von Bodungen and R.K.Turner (eds). *Science and Integrated Coastal Management.* 85th Dahlem Workshop, Dahlem University Press, Berlin.

Meynall, P.J., J.P. Johnson and M.P. Wilkie 1988 *Guide for Planning, Monitoring and Evaluation in Fisheries Development Units.* IDAF Field Manual, No. 2. Cotonu, Benin.

Mino-Kahozi, K.; K. Lubambala and T. Nkomko 1997 *Zaire: Results of the Frame and Socio-Economic Survey.* IDAF Newsletter (FAO) No. 33: 24-25.

Ndiaye, Y.D. 1996 *Use of Capital Income in Artisanal Fisheries: The Case of Boat Owners in Hann, Senegal.* Technical Report No. 92. IDAF/FAO, Cotonu, Benin.

Ndiaye, O. 1996 Informal Insurance Mechanisms in Artisanal Fisheries: the Case of Senegal. IDAF Newsletter (FAO) No. 29: 6-10.

Neiland, A.E and M.T. Sarch 1994 Traditional Management of Artisanal Fisheries, Northern Nigeria. Report No. 1: The Development of Survey Methodology. Research Report prepared for the Overseas Development Administration (ODA) as part of the ODA Socio-Economic Research Programme (SERP II)

Ngo Som, J. 1996 The Nutritional Condition of the Fishing Community of Limbe, Cameroon. IDAF Newsletter (FAO) No. 31: 8-10.

Njifonjou, O. 1997 Cameroon: Costs and Earnings in Marine Artisanal Fisheries in the Limbe Region. IDAF Newsletter (FAO) 33: 6-8.

Overa, R. 1998 *Partners and Competitors: Gendered Entrepreneurship in Ghanaian Canoe Fisheries.* Dissertation for the Dr. Polit. Degree, Department of Geography, University of Bergen, Norway.

Pomeroy R.S. (ed.) 1994 *Community Management and Common Property of Coastal Fisheries in Asia and Pacific: Concepts, Methods and Experiences.* Manila. ICLARM Conf. Proc. 45.

Roberts, K. 1995 Legal Aspects of Monitoring, Control and Surveillance of Fisheries and Prosecution of Offences in West Africa. Document No. 23. FAO. Dakar, July.

SAMUDRA 1999 Coastal Resources for Whom? SAMUDRA Report, No. 23. pp14-19. A summary of a keynote paper by Parzival Copes, at the founding meeting of the World Forum of Fish Harvesters and Fishworkers in New Delhi, India (18/11/99).

Satia, B.P. 1993 *Ten Years of Integrated Development of Artisanal Fisheries in West Africa.*

IDAF Programme. DANIDA/FAO. Technical Report No. 50, November. Cotonu, Benin.

Satia, B.P.1995 Research and Development Initiatives in Co-Management of Fisheries in West Africa. Presentation made at the *Workshop on Fisheries Co-Management, Hirtshals Denmark, 29031 May, 1995.*

Seynabou Sy, M. 1996 *Nutritional Condition, Food Security, Hygiene and Sanitation in the Fishing Community of Joal, Senegal.* Technical Report No. 87. IDAF/FAO, Cotonu, Benin.

Touray, I. 1996 *Study on Women's Organisations in Brufut and Gunjur Communities and the Factors that Favour or Impede their Sustainability in The Gambia.* Technical Report No. 88. IDAF/FAO, Cotonu Benin.

United Nations 1997 UN Human Development Report, New York.

Vercruijsse, E. 1984 *The Penetration of Capitalism: A West African Case-Study.* Zed Books Ltd., London

The Gulf of Guinea Large Marine Ecosystem
J.M. McGlade, P. Cury, K.A. Koranteng and N.J. Hardman-Mountford (Editors)

24

Fish Utilisation and Marketing in Ghana: State of the Art and Future Perspective

J. D. Akrofi

Abstract

Ghanaian fish production is dominated by the artisanal sector. While fish catching is the sole reserve of men, fish processing and marketing are the exclusive domains of women. The importance of fish to the Ghanaian economy, and the policies in place to sustain its exploitation are discussed. The various methods used to preserve fish in Ghana are examined and the organisation of fish marketing at the artisanal level is described. Also discussed is the new quality assurance systems introduced by developed countries and their implications for the fish export and local markets. Finally some suggestions are made to improve the fish utilisation and marketing to meet future trends.

Introduction

Background

Ghanaian fisheries comprises the marine sector and the inland sector. With a coastline of about 550 km and Exclusive Economic Zone (EEZ) waters extending 200 nautical miles from the shore, the marine sector is the main source of fish, producing 85 percent of the total catch. However, with a number of rivers, irrigation dams, ponds and lakes including the large Volta lake, (one of the largest man- made lakes in the world) the inland sector contributes substantially in making up the difference in total production.

Structure of the Industry

The marine sector has a canoe fleet (artisanal fishery), an inshore fleet (semi-industrial) and an industrial fleet (trawlers and tuna fleet). The artisanal fleet consists of about 8700 traditional wooden dug-out canoes, half of them motorised, operating out of 182 fishing villages and 296 landing beaches. (Quaatey *et al.* 1995). They operate different fishing gear, namely, beach seine, set net, drift gill net, ali/ poli/ watsa, and hook and line. Over the past ten years, total fish catches from all sources averaged approximately 400,000 tons per year. The artisanal sector produces about 70 percent of the total domestic catch, the remainder being supplied by the large industrial and shrimpers and the inshore or semi-industrial fleet of locally built trawler/ purse seines.

In the inland sector the Volta Lake is the dominant source of catches, accounting for about 50,000 tons per year which is achieved with the help of 15,000 small planked canoes operated out of nearly 2000 fishing villages along the lake. Nearly 400 tons is also produced from aquaculture.

Important Fish Species

The most important large pelagic species are the tunas while the small pelagics are the sardinellas, anchovy and the chub mackerel. The most important demersal fish species belong to the families sparidae, pomadasidae, mullidae, sciaenidae, lutjanidae, serranidae and the cephalopods. The commercially exploited resources in inland waters are the tilapias, *Chrysicthys*, *Lates, Alestes, Eutropius* and *Schilbedae* (Anon. 1995).

Fish and the Ghanaian Economy

The fisheries sub sector accounts for about 5 per cent of the agricultural GDP. Fish is a preferred and cheaper source of animal protein in Ghana, and about 75 per cent of the total production of fish is consumed domestically. The estimated per capita consumption is about 25 kg per annum, representing 60 per cent of the annual protein intake. Fish is the country's most important non-traditional export. Some 500,000 fishers, fish processors, traders, boat builders and maintenance experts are employed in the industry. These along with their family members represent about 10 per cent of the total Ghanaian population of approximately 18 million in 1998.

Fish Exports

About 40,000 tons of fish are exported annually the value of which has been increasing over the years. In 1997 the fish exports had a value of US$95.0M, which was about 28.5% of total earnings from the non-traditional export sector. High value species such as shrimp, tuna, grouper, cephalopods, snappers etc. make up the majority of the exports. The fish products from Ghana are exported as fresh, chilled with ice, canned, frozen, smoked, salted, dried and live. Tuna is the dominant export product and used to be exported as whole frozen and as loins until 1994 when canning started again, after an original start in the 1960's. Canning has increased the earnings of exports because of the value added to the product. The European Union (EU) is Ghana's largest export market for fish. Others include US and Japan.

In 1991, the EU issued two directives identified as 91/EC/439 and 91/EC/492, setting out guidelines for the harvesting, transportation, handling, storage, processing and sale of seafood and fishery products. The objectives of the directive are to ensure:
a) the application of Hazard Analysis Critical Control Point (HACCP) system in the production of improved quality, safety and value added products.
b) systematic monitoring of quality and safety of fishing grounds.
c) development of harmonised code of hygienic and safety practices for seafood and fish products industry.

The system of fish inspection currently being operated in the country is based on these objectives. The Ghana Standards Board has been appointed the competent authority to certify the export of fish and fishery products from the country to the EU and other markets abroad. Compliance by the industry will ensure continuous export of fish to the EU.

Government Policy Objectives

Considering the importance of the industry, the Ghanaian Government has always been concerned about not only the development of the industry but also the over all improvement of the standard of living of those engaged in the fishing industry. Thus, government policy objectives for the fisheries sector are as the follows:

i. Increase fish production for local consumption and export, consistent with the long-term sustainability of the resource.
ii. Alleviate poverty in fishing communities
iii. Develop resource management plans for the entire fisheries sector
iv. Integration of fishing activities in the farming system through the promotion of aquaculture
v. Privatisation of assets and operations of public bodies engaged in direct fishing or in supply of fishing gear
vi. Strengthening of the Department of Fisheries so that it can effectively carry out its mandate, particularly relating to the above task.

Objective of the Paper

In 1998, the Ghana Department of Fisheries estimated that the national fish requirement for 1997 was about 794,000.0 mt. This was based on a per capita consumption of 24kg and a population of about 17.9 million (Anon. 1998). Given this, fisheries production in 1998 only achieved 57% of the requirement. This has been the trend over the years i.e. that the fish production has been falling short of national requirement. It is, therefore, important that post-harvest losses, which are estimated to be about 20%, be reduced. This can be achieved by better utilisation of resources within current production streams.

This paper describes the status of fish processing and marketing in Ghana. The various methods in use to preserve fish in Ghana, organisation of fish marketing at the artisanal level, the new quality assurance systems introduced by developed countries and their implications for the fish export and local markets are also examined. Finally some suggestions are made to improve the fish utilisation and marketing to meet future trends.

Fish Preservation Methods

Smoking, salting, drying, icing, cold storage and canning are used in preserving fish in Ghana. The traditional methods of smoking, salting and drying are used to preserve most of the fish from the artisanal sector and the inshore fleet. In general, about 80 per cent of the fish is consumed smoked and the remaining 20 per cent consumed either fresh or salted, sun-dried or fried. The type of fish determines, to a large extent, the method used to preserve it and this is related to the preferences of consumers (FAO 1989, 1992a).

Sun-drying

This is used to dry small-sized species like the marine anchovies and the salted fish. The fish is dried on mats, cemented floors, cut grass, old fishing nets and sometimes directly on the ground.

Smoking

This is used for most species and is carried out in almost all fishing villages. Smoking combines three effects to preserve the fish. The smoke has a preservative value which kills the bacteria; it dries the fish as well as cooks the flesh. In Ghana, hot smoking is used. Smoking equipment used are normally made from oil drums, mud or clay. The traditional smoking ovens have been improved to address the problems of low capacity, high fuel consumption and the difficulty in controlling the process and constant attention required, experienced in traditional smoking plants. The popular 'Chorkor' smoking oven has been improved and has a large smoking capacity with less fuel consumption; this technology has now spread to many other African countries.

Icing

This is used to extend the shelf-life of fish landed by the inshore vessels which mostly carry ice on board. Some canoe fishermen who use hook and line for exploiting high value demersal species also carry ice to sea in insulated containers kept in special compartments in their canoes. Icing certain fish species after capture can be logistically difficult and uneconomic for the artisanal fishermen. Pelagic species such as sardinella and the mackerels do not command high prices and therefore do not warrant the cost of icing.

Freezing

This is used by distant water vessels which fish for longer periods of time and at longer distances from the habour. Different types of freezers such as plate, blast and brine freezers (applied in tuna vessels) are used to freeze fish which are then kept in cold storage rooms. Most of the frozen fish is transferred to onshore cold stores before marketing. The frozen fish is sold in the frozen state to both consumers and fish processors who process the fish before marketing. Onshore, plate freezers are only found at Tema where the main fishing harbour is located. Ice-making plants are found in the Accra-Tema area, Elmina and Apam in the Central region and Sekondi-Takoradi in the Western region.

Canning of fish

This is undertaken on a very limited scale for sardines and tunas. There are four canneries in the country, one of them Pioneer Food Cannery, is a subsidiary of Star-Kist a US based company. A substantial amount of the canned tuna is exported.

Packaging and Transportation

Wooden crates and more recently plastic fish boxes are used to carry fresh fish from the industrial and semi-industrial vessels. At the artisanal level, fresh fish is carried in wooden crates, baskets and bowls. These are also used as units of measure. Processed fish from all sources, is also packed into baskets lined with paper and jute sacks. The fish is transported mostly by public transport. There are few refrigerated vans for transporting frozen fish around the country. Most frozen fish in cartons is transported in ordinary lorries; the cartons are however, reinforced with an over wrap of thick card to insulate the fish from thawing. The cold chain however, is broken before the fish is transferred to another cold store at its destination.

Distribution and Marketing of Fish

In Ghana the private sector dominates the distribution of fish. The domestic marketing system functions fairly well, and the market signals are correctly transmitted in a timely fashion through the marketing chain. Cured fish marketing and distribution is an informal activity which operates through the "fish Mammy" system[1].

Marine Fish

The fish marketing trade is dominated by women, who are normally called fish mammies. They are deeply involved in fish marketing and can even go to the extent of financing the fishermen and therefore control the sale of the produce.

Historically in the 1960s and 1970s, when the fishing industry was still expanding there was a need to change the methods of distribution. Private entrepreneurs set up cold storage facilities to store fish from the distant water fleet.

Fish traders who are mostly women know the consumer requirements of the various ethnic groups in the country and use this knowledge to determine the marketing of the various fish product types. For example, the smoked and sun-dried anchovies, which are normally carefully dried because of their size, enjoy a very good market in the northern part of the country hence play a very important role in the cuisine of the people living within this part of Ghana. The fish traders thus send the fish to the market of their choice to ensure maximum profit; they are in fact the live wire in fish distribution and marketing in the country.

Freshwater Fish

Marketing of fish from the Volta Lake is also done by traders who are mostly women. The fish mammy system also operates here. However, as on the Lake Volta most of the villages are very remote and without good road heads leading to them, fish are taken to the lake-side markets by transport launches, mostly powered by outboard motors. Processed fish is purchased from the fishing villages by fish traders who travel on the lake to these areas while some are brought down to the markets for sale by the fishermen and their wives. There are

---

[1] This is a system in which individual female traders enter into a contractual agreement with the fisherman and processors in order to gain control over what is produced.

specific market days for the various lake-side markets and many trucks travel to these places to load the fish to the various fish markets around the country. At these lake-side markets the practice is to sell fish to regular customers. At the lake-side market, re-smoking of fish is undertaken to further reduce the water content of the smoked fish in order to prolong the shelflife of the fish. The fish are then repacked in big baskets and loaded onto the big trucks for their final destination. The packing and unpacking of fish from the producing villages through the lake-side market and to the final point of selling often causes breakage which reduces the quality and quantity of the fish.

Prices

Prices of fish and fishery products fluctuate widely throughout the year and across the country. They are determined by many factors including supply and demand situation at a particular time and of a particular species.

It has also been noted that for both marine and fresh water species the bigger species sell at higher prices. Prices of fish also increase as the distance from the source of production increases. Apart from the above, urbanisation, income levels, consumer taste and preference and prices of competing substitutes have been noted to affect fish prices in Ghana. These factors are taken into consideration by the fish mammies when choosing markets for the sale of the various species of fish. Changes in the cost of production, especially cost of fishing inputs and their availability have an effect on the price of fish. Lack of input supply definitely reduces production and pushes prices up.

Constraints

There are periods during the year when the fish produced cannot be fully utilised because the traditional processing methods cannot cope with the volume of fish landed e.g. during the main fishing season, July to September. This results in post-harvest losses. Since almost 80% of the fish landed is from the artisanal sector any post harvest losses will not only affect the protein available for domestic consumption but also the income of the artisanal fishermen. Low-income earnings and/or losses can affect the sustainable production of fish since new capital is not easy to come by. Any amount of fish landed is very valuable because it is needed to meet domestic fish requirements and to avoid the importation of fish. In addition, effective utilization of landed fish will help reduce pressure on fishery resource as there will be no real need to catch more fish.

One of the contributing factors in the decline of fish supplies is post-harvest losses caused by improper handling, processing, packaging, storage and distribution systems. Poor transport, inadequate freezing facilities, unrealistic distribution of storage capacity and the system of their operation also constitute problems, as well as poorly maintained containers which can affect total fish supply and distribution (FAO 1992b).

Limitations of traditional methods for handling and processing fish include inadequate protection of fish before, during and after processing, slow drying rates, poor quality curing salt, difficulty in controlling smoking temperatures, ineffective packing and poor storage conditions (Sefa-Dedeh *et al*. 1995). Variations in product type and quality are also common. Notable among the causes of variations in product quality are differences in freshness of the

starting material, the preparation of fish prior to curing and the methods used for curing the fish (Akrofi 1997).

In the fresh fish trade, handling and general marketing practices are highly inadequate for maintaining the fresh quality of fish due to the following reasons:
- overloading of fish into containers leading to belly burst.
- display of fish in the sun and hot humid areas.
- display of fish on the ground and over open drains.
- repeated freezing and thawing of fish.
- lack of cold / refrigerated vans to transport fish.
- display of fresh fish in dirty containers with drips from previous sale.

Drying of fish and salted fish on the ground, cut grass or old nets in unsanitary environments also degrades the products.

The problems associated with smoked fish include hygiene and sanitation, packaging, rough handling during transportation to markets and poor storage facilities. The low smoking capacities, high fuel consumption and difficulty in controlling the smoking process, also limit the efficiency of the traditional ovens. Differences in freshness of raw material also make it difficult to produce a uniform product for consumption.

Freezing facilities are required to freeze fish before storing them in cold stores. This facility is not available at many of the major landing sites. The economic returns related to investing in such a facility need to be closely investigated considering the fact that the types of fish landed in abundance during the major fishing season are not high value species (mostly sardinella and chub-mackerel). Moreover, there is no certainty about the volume and the regularity of the catch or how long the fishing season will last to warrant a huge capital outlay.

An emphasis needs to be placed on the spread of improved traditional smoking equipment such as the Chorkor oven, which has a comparatively low investment cost. The Chorkor oven can ensure the smoking of large quantities of fish effectively and at a relatively low cost. Ice can also be used to ensure a better raw material quality.

for the export market, there is a potential for developing countries to increase their share in international fish trade and in particular, gain entry into the developed markets with value added products. There are, however, various constraints to this trade in major markets such as the European Union. Tariff and non- tariff barriers created by governments and commercial constraints relating to market characteristics, quality of products and distribution channels are the three main types of constraints that have been identified (FAO 1986). The 1993 EEC sanitary regulations provide that import of fishery products from non-EEC member countries will be possible only if exporting countries design their processing facilities to meet requirements, at least, equivalent to those prevalent in the EEC countries and if the products of exporting plants are produced under the same sanitary and quality control conditions as those established for EEC plants (INFOPECHE 1993). It is not at all easy for developing countries to satisfy these conditions.

The situation is further exacerbated by the fact that the African, Caribbean, and Pacific (ACP) countries will soon be losing their preferential status on the EU market. As of now, the trade provision of the Lomé Convention, an agreement between the EU and 69 countries in the ACP countries, stipulates that fishery exports from these countries can enter the EU tariff-free. The sub-Saharan African ACP countries depend considerably on the EU market, to which they export 75 to 85 percent of their total fish exports (FAO 1995). Ghana

is one such country and relies heavily on the EU market. The General Agreement on Trade and Tariff (GATT) Uruguay round in 1994 concluded that customs regimes need to be harmonised globally. For fishery products, the main outcome was an agreement to lower customs duties by an average of 26 percent. Several of the EU tariffs for fishery products remain almost unchanged, however current concessions are subject to renegotiations and further reductions may be made in the future. For sub-Saharan ACP countries, this development will eventually mean increased competition from other developing countries.

Suggestions for Improvement

Infrastructure

- The main categories of infrastructure that require to be improved or provided are:
- landing and processing facilities at all beaches;
- all-season roads to these facilities;
- suitable means of transport for the fish separate from public transport e.g. refrigerated vans for fresh and frozen fish; trucks with custom made cargo base to transport processed fish; and
- retail facilities with adequate storage space at both landing points and markets.

These provisions are to be made with adequate attention to the following:
- adequate supply of potable water;
- provision of sanitary facilities at all beaches and markets;
- proper management of liquid and solid waste;
- provision of fly proof sheds at markets; and
- supply of ice at major landing sites.

Education

Extensive education of fishers, processors, traders and consumers on quality indices is required. There is also a need to step up the campaign to clean up the beaches and processing sites and improving the hygiene and sanitary conditions for processing.

In addition, training and educational campaigns on fish handling and quality standards will enhance the image and value of fish, resulting in greater customer satisfaction and better profits.

Access to credit

Since credit is difficult to come by, fisherfolk should be sensitised and trained to save in order to prepare them to meet the conditions of the credit institutions.

New and improved technology

More efforts should be made to spread new and improved technologies (Teutsher 1998). Examples include the use of the Chorkor smoker by providing initial credit for construction;

the use of sawdust from non-resinous wood with the LPG ovens to produce a better quality product at a lower cost; and stronger and standardised artisanal containers such as baskets and crates, which can be cleaned.

Quality Assurance

The Ghana Standards Board (GSB) has set new standards for the processing and export of fish in line with international requirements on quality based on the HACCP system. The GSB is the Competent Authority for the EU Export Fish Trade and oversees the export of fish and fishery product from Ghana.

The establishment and enforcement of quality standards is one of the important steps in orienting the industry towards more efficiency in the marketing system. The new regulations being applied to the export of fish need to be extended to the artisanal sector as well. Ghana stands to gain from implementing such quality assurance systems because by doing so Ghana will reap the following benefits:

- promotion of international trade by increasing confidence in food quality and safety;
- prevention of food borne diseases;
- reduction of quality costs, increase in economic productivity; and
- promotion of better use of resources and more timely response to problems.

Conclusion

Given the present state of the Ghanaian fishery, sustainability of production can be achieved if fish harvested at a particular time are preserved to make maximum returns in order to improve the economy.

One of the major contributing factors in the decline of fish supplies is post-harvest losses; Ghana thus stands to benefit from implementing quality assurance systems required by international markets.

To improve the present system of marketing, the government needs to play a major role at the macro-marketing level. Part of the government's strategy for achieving policy objectives is to pursue private sector oriented strategy to the utilisation of fisheries resources and divest public sector commercial assets and concerns, provide necessary services through appropriate public/private sector interface so as to create an enabling environment in which private sector activities will flourish. If this is achieved then the numerous producers who operate in the fish marketing system will see a way ahead.

References

Akrofi, J. D. 1997 *Suitability of Certain Monitoring Procedures for artisanally produced Cured Fish Product from Ghana for the European Union Market.* MSc thesis University of Hull, U.K.

Anon., 1995 Staff Appraisal report, Republic of Ghana, Fisheries Sub-Sector Capacity Building Project. World Bank Report No. 13877-GHA.

Anon., 1998 Department of Fisheries, Ministry of Agriculture, Ghana; Annual Report.

FAO (Food and Agriculture Organisation) 1986 Trade in value-added fishery products by Fish and Utilisation Marketing Service, FAO, Rome. Infofish Marketing Digest No 3/86 May/June 1986.

FAO 1989 Proceedings of the expert consultations on fish technology in Africa. Abidjan, Cote d'Ivore 25-28 April 1988. FAO Fish Rep.

FAO 1992a Proceedings of the expert consultations on fish technology in Africa. Accra, Ghana, 22-25 October 1991. FAO Fish Rep.

FAO 1992b Proceedings of Symposium on Post-harvest fish technology in Africa. CIFA Technical papers No. 19. 177p.

FAO 1995 Impact of the last Act of the Uruguay Round on the fisheries of sub-Saharan Africa. FAO Fisheries Circular No. 897. Rome; FAO/GLOBEFISH. 1995. Impact of the Uruguay Round on international Fish trade, by A. Filhol. FAO/GLOBEFISH Research Programme Vol. 38. FAO, Rome.

INFOPECHE 1993 Upgrading Fish Exports from Infopeche member countries. FIDEPRO 03/93

Teutscher, F (ed.) 1998 Report and proceedings of the sixth FAO Expert Consultation on Fish Technology in Africa. Kisumu, Kenya, 27-30 August 1996. FAO Fisheries Report, No. 574. Rome FAO. 269p.

Sefa-Dedeh, S. 1989 Harnessing traditional food technology for development. *In*: S. Sefa-Dedeh and R. Oracca-Tetteh (eds). *Harnessing traditional food technology for development*. Proceedings of a Kellogg International Fellowship Program Workshop on Food Systems (1988). Accra.

Quaatey, S. N. K., Bannerman P. O., Baddoo A. N. A., Ashong T. B. 1995 Report on the 1995 Ghana Canoe Frame Survey. Department of Fisheries, Marine Fisheries Research Division Information Report No. 29. 40p.

Institutions & Governance

The Gulf of Guinea Large Marine Ecosystem
J.M. McGlade, P. Cury, K.A. Koranteng and N.J. Hardman-Mountford (Editors)

25

How Can Collaborative Research Be More Useful to Fisheries Management in Developing Countries?

C. E. Nauen

Abstract

Most fisheries management schemes in industrialised and developing countries have not delivered sustainability, the goal usually pursued by them. This short communication takes a look at some ways in which collaborative research and the mobilisation of a wider range of societal actors, such as promoted under the ACP-EU Fisheries Research Initiative, can make a more useful contribution to fisheries management in developing countries.

Introduction

Most fisheries management schemes in industrialised and developing countries have not delivered sustainability, the goal usually pursued by them. Garcia and Newton (1997) pointed out that 70 per cent of the world's marine capture fisheries are overexploited, fully exploited or recovering. As Gréboval and Munro (1999) state in their introduction to an FAO Fisheries Technical Paper based on a Technical Working Group on the Management of Fishing Capacity, convened in La Jolla, USA from 15-18 April 1998: "Regardless of how one chooses to interpret the term 'fully exploited', there are substantial grounds for expressing deep concerns about the state of world capture fishery resources." A series of recent publications add quantitative analyses and new perspectives to the crisis in fisheries (e.g. Pauly *et al.* 1998a; Christensen 2001; Froese and Reyes, in prep.).

We therefore see more scepticism about research and applications, which confine themselves purely to the 'technical' aspects of fisheries management. These and the underlying 'reductionist' attitudes have been contributing factors to major fishery collapses and a generalised decline in ecosystem health.

On closer examination, the economics and incentive systems at work are believed to neutralise if not pervert many management attempts. Imperfect markets appear to mask scarcity effects as fish prices of previously undesirable species have increased in price relatively faster than now much scarcer 'high value' species (mostly ground fish and large pelagics) much reduced by overfishing, but which have maintained or only moderately increased their market prices (Sumaila 1999; Sumaila *et al.* 2000). The information allowing investors in fisheries to make better economic decisions is found lacking (Des Clers 1998) and this shortcoming appears to be of concern for other stakeholders as well. Most recently, Watson and Pauly (2001) showed that global marine catches have declined since reaching a peak of about 80 million tons in the late 80s instead of increasing as official world statistics have misled us into believing. Misreported national statistics were disseminated through FAO

which, as an intergovernmental organisation, relies on official national communications. This may have unwisely influenced planners and investors alike.

But perhaps equally importantly, the mechanisms are largely underdeveloped, which would allow more effective dialogue between sector stakeholders to create conditions for the success of management. The gap between the broad principles of the FAO Code of Conduct for Responsible Fisheries (FAO 1995), the demand for ecosystem-based management as incorporated in the Jakarta Mandate agreed by the 1995 Conference of the Parties to the Convention of Biological Diversity and the harsh realities of depleted resources, degraded ecosystems (Pauly *et al.* 1998a; http://www.biodiv.org/decisions/default.asp?lg=0&m=cop-02), loss of productivity and aquatic diversity (Froese and Torres 1999), remains unbridged so far. Some of the issues may be illustrated through the case of a co-operation project sponsored under the Fisheries Research Initiative between African, Caribbean and Pacific (ACP) Countries and the European Union (Anon. 1995, 1996, 1997a, 1997b).

Missing Links

The Project 'Strengthening of Fisheries and Biodiversity Management in ACP Countries' set out specifically to make FishBase and related analytical tools available to African, Caribbean and Pacific (ACP) partners enriching it through collaboration. More generally it was to contribute to a positive environment for ACP participants under the ACP-EU Fisheries Research Initiative. Interaction with ACP researchers and fisheries administrators during the years of implementation illustrate a number of obstacles in the way of effective biodiversity and resource management. With the benefit of hindsight, these seem not to be confined to the project itself but to be indicative of a more generalised set of problems in relation to fisheries management in many developing countries.

Fisheries departments are usually responsible for fisheries management in their respective countries; however, closer examination reveals a lack of means and an unsatisfactory institutional framework to carry out their mandate. Indeed, others may *de facto* influence decisions, which would be incumbent on the fisheries department or other departments of the ministry overseeing fisheries rather than those who are directly in authority. There are cases in which fisheries departments lumbered with conflicting mandates of 'extension' and 'law enforcement' are hindered in delivering either effectively. Environment and resource conservation is only gradually perceived as offering the most interesting opportunities for socio-economic development. In most cases, the human resources, institutional arrangements and budgets available for management and sector policy fall short of what is necessary. Training and retraining of fisheries officers are often found wanting when faced with the evolving nature of the task at hand.

Finally, to embed fisheries management in the wider picture, it should be mentioned that macro-economic demands are frequently put onto the sector administration without it being equipped institutionally and politically to handle them. Conversely, macro-economic decisions impact the fisheries sector, sometimes even more so than sector management (Dahou and Deme 2001). This socio-economic connectivity is, however, not further developed here.

Fishers themselves, small- and large-scale alike, though to different degrees, may not feel bound by the management scheme in place, if any, to contribute their share to long-term sustainability, when shorter-term cash flow or even survival problems loom large. Other stakeholders tend to have little or no representation on any formal scheme. Among these, fish

processors and traders, who, especially at the small-scale level are often women, may come to mind easiest. But conservationists and consumers are equally legitimate stakeholders with significant interest in long-term productivity of the resources.

Researchers working on the fisheries sector, including its ecological and socio-economic framework, should interact with all of them, though this has not always happened in the past. Research in many tropical and sub-tropical countries is insufficiently funded to make the contributions towards the overall goal of sustainable fisheries expected of them given the importance of aquatic resources and resource use in many economies. This may be aggravated by insufficient collaboration between universities, sector institutes depending on ministries in charge of fisheries, and even other research establishments or documentation centres. Each of these bodies addresses part or all of the ecological and socio-economic issues, on which decision makers need information and understanding.

The sample of stakeholders mentioned does not necessarily represent homogeneous groups, but a broad-brush picture leaving their internal subtle differences, complementarities and conflicts aside.

All too often, conservationists and fishers have identified with group interests perceived as diametrically opposed with little communication existing between them. Consumers are increasingly remote from production as indicated by the high percentage of fishery products entering global trade (FAO 2000). Since the establishment and enforcement of stricter health standards in the major import markets (Japan, European Union and North America) in the mid 1990s, worries have sometimes been aired that consumers and the veterinary inspection services in these import markets are more concerned with their own food safety than with the living conditions of the original producers (e.g. Jansen 1997).

These are but highlights on some of the fault lines. There is no shortage of attempts to establish better communication or to improve information exchange. However, the missing links enabling a reversal of the downward course of fisheries have so far been elusive.

What Could Make a Difference?

The dialogue, which laid the foundations of the ACP-EU Fisheries Research Initiative, showed the potential for relationships of voluntary co-operation based on mutual respect and shared responsibility (Anon. 1997b). It underscored needs for foresight and scientific pro-activity.

There is also increasing realisation that foresight and proactivity depend on a good understanding of the history of the ecosystem or society under study (Putman 1993; Pauly 1995; Pauly *et al.* 1998c; Nauen 1998a). Two scientific conferences at EXPO 98 under the Fisheries Research Initiative made contributions to address the historical dimension (Pauly *et al.* 1999; Pullin *et al.* 1999). This has so far attracted little attention in the context of management, concerned as it is with solving problems on a short-term scale. However, without an understanding of the history of a system no baseline exists to appreciate the range of options for decision making enabling socio-economic and ecological sustainability.

These considerations bring home yet again the need for transdisciplinary work in recognition of the complexity around us. Yet, piecing the picture together and particularly how to deal with uncertainty and risk, opens many new questions.

In the biological and ecological fields, FishBase makes available information on all 25,000 finfish in the world. The information spans many disciplines and is structured around species and their higher taxonomic aggregates as they represent current understanding of evolution (Froese and Pauly 2000). The data structure allows different 'communities' of stakeholders to draw on the database's content (and enrich it with their own specific data and knowledge, by now even through remote data entry for experts assuring quality control). It enables them to carry out their preferred types of analysis on a broader quantitative basis than they would be able to do in isolation. Experienced users thus gain new insights into their own fields and their limits, allowing them also to venture into other areas of information and analysis.

The core data are available on the Internet (http://www.fishbase.org). The site now attracts some 150,000 users per month. Many of the website visitors are from industrialised countries. The spread of the Internet in developing countries and increased emphasis on language translation is bound to increase access there beyond their current share of about 25% of users. To compensate for remaining logistical problems, several thousand conventional CD-ROMs have been shared with project partners and other interested individuals and institutions over the last few years. Froese (2001) summaries some key lessons underlying the success so far.

There are a few emerging lines, where the existing collaboration allows the development of new or complementary analytical tools responding to user needs. Extracting data and indicators relevant to ecosystems modelling such that fishery resources can be analysed in their ecosystem context enables researchers to ask new management questions (Christensen 1996; Christensen 1997). Thanks to these new concepts and outreach initiatives, managers and the public can assess fisheries in a different light.

Another opening was developed by Sumaila's analysis of price trends in fisher (1999) to probe not only the market equivalent of ecosystem changes provoked by fisheries, but also the role of protected areas as hedges against uncertainty (1998). Other areas of importance, e.g. particularly the less-developed areas of analysing the social costs of fisheries and policy research warrant future collaborative efforts. These and several other promising avenues can best be explored through collaborations, which build on the existing knowledge and develop it further under the guiding principles of the Research Initiative.

The uncertainty associated with the complexities of the large marine ecosystems, environmental forcing factors and the interconnectedness of sub-systems (eg through global scale oscillations) have implications on fisheries management (e.g. Durand *et al.* 1998; Bakun 1998). While enriching the research agenda is overdue, the inherent uncertainty in fisheries resource systems warrants *a priori* recognition that assessments cannot reasonably be expected for all resources and that consequently management approaches need careful rethinking (e.g. Walters 1998).

This also returns us to the initial question of who the legitimate stakeholders are and how to devise institutions that enable the transition from top-down fisheries management towards greater societal participation (McGlade 1994, 2001; Nauen *et al.* 1996; Adams 1997; Chakalall *et al.* 1997; Fontenelle 1997; Ruddle 1997).

Society's Increased Expectations in Relation to Research and the Role of Dialogue

There is convergence in the analyses that legitimate parties in fisheries are not just in the triangle between the fisheries administration, 'fishermen' and research. Each of these three poles is differentiated depending on the actual economic and institutional conditions and possibilities. New actors have already entered the scene (Nauen 1999), foremost in the continuum from small-scale to international interest in processing, handling and trade. The latter is a powerful, and often controversial, driving force with leverage throughout the sector and beyond. More visible for the general public, local, national and international NGOs promoting environment and biodiversity conservation have espoused reduced fishing on global and local scales.

As far as the on-going collaboration under the FishBase Project is concerned, it is acknowledged that it has not yet led to any reversal in the downward trend in fisheries and biodiversity. The project has compiled substantial amounts of relevant information, made them available through a user-friendly interface, promoted new concepts and reached many of the institutional users in ACP countries as intended (Froese *et al.* 1999). But this is not enough.

The concepts and approaches promoted under this specific project are still relatively new to many partners and need to be more fully appropriated, in order to engender significant impact (see e.g. Fontenelle 1997). Intensified capacity building of ACP institutions, ACP-EU collaboration is the natural response to seek more visible impact in terms of rehabilitated fisheries and reduced threat to aquatic biodiversity.

In general terms, research itself will have to evolve in the process not only in terms of advancing scientific knowledge but also by interacting more with society under the partnership concept for a truly 'New Deal' (Viegas 1999). Indeed, new governance institutions as advocated in Kooiman *et al.* (1999) invite the recognition of all stakeholders as providers and users of information and knowledge, new arrangements in the relations based on mutual respect.

With the recognition of the complexity of the aquatic and socio-economic systems, their spatial and temporal scales and dynamics, the role of research and knowledge can thus only increase. Approaches, such as the establishment of protected areas, are new options for managers to rehabilitate lost productivity (Roberts *et al.* 2001). Therefore, approaches to fisheries management must and can change. So great is the convergence between the general analysis and fisheries that new institutions and a 'New Deal' have also been demanded in fisheries management (Pauly *et al.* 1998b; Nauen, 1999).

Building on the good foundations of the ongoing co-operation, a strengthened ACP-EU initiative could support the wider appropriation of these new concepts and selectively help those local stakeholders, countries or regions, which have already invested in fisheries rehabilitation or will be doing so. The skills and competencies developed under the present project and other complementary ones of local and external partners would then combine to have a visible impact on fisheries. It is time to turn around some 'real life' fisheries onto a sustainable path.

Acknowledgements

The author warmly acknowledges the wide-ranging and critical discussions of the ACP-EU Steering Committee of the Project 'Strengthening of Fisheries and Biodiversity Management in ACP Countries', which contributed to the present short communication. The Steering Committee is composed of Timothy Adams, Eduardo Balguerias, Boris Fabres, Maria Luisa Cassama Ferreira/Amadu Bailo Camara, Guy Fontenelle, Thomas Maembe, Jean Calvin Njock, Helge Paulsen and chaired by the author. The Project's Principal Science Advisor,

Daniel Pauly and Senior FishBase Project staff significantly contributed to these debates, namely Rainer Froese, Maria Lourdes (Deng) Palomares and Jan Michael Vakily. The original manuscript was written in 1999 and only slightly up-dated in 2001. Any mistakes remain the author's sole responsibility.

References

Adams, T. 1997 Governance of fisheries and aquaculture in the Pacific Islands region. Annex 12 pp165-180. *In*: Anon. ACP-EU Fisheries Research Initiative. Procs. of the Third Dialogue Meeting, Caribbean, Pacific and the European Union. Belize, Belize City 5-10 December 1996. ACP-EU Fish. Res. Rep. (3).

Anon., 1995 ACP-EU Fisheries Research Initiative. Procs. of the First Dialogue Meeting. Eastern and Southern Africa, Indian Ocean and the European Union. Swakopmund, Namibia 5-8 July 1995. ACP-EU Fish. Res. Rep.(1):144.

Anon., 1996 ACP-EU Fisheries Research Initiative. Procs. of the Second Dialogue Meeting. West and Central Africa and the European Union. Dakar, Senegal 22-26 April 1996. ACP-EU Fish. Res. Rep (2):166 p.

Anon., 1997a ACP-EU Fisheries Research Initiative. Procs. of the Third Dialogue Meeting. Caribbean, Pacific and the European Union. Belize, Belize City 9-11 July 1997. ACP-EU Fish. Res. Rep. (3):180 p.

Anon., 1997b ACP-EU Fisheries Research Initiative. Procs. of the Fourth Dialogue Meeting. All ACP Regions and the European Union. Brussels, Belgium 9-11 July 1997. ACP-EU Fish. Res. Rep.(4):51 p.

Bakun, A. 1998 Ocean triads and radical interdecadal variation: bane and boon to scientific fisheries management. pp331-358. *In*: T.J. Pitcher, P.J.B. Hart and D. Pauly (eds.). *Reinventing fisheries management*. Dordrecht, Boston, London, Kluwer Academic Publishers, Fish and Fisheries Series 23.

Chakalla, B., R. Mahom and P. McConney 1997 Fisheries governance in the Caribbean. Annex 11 pp131-163. *In*: Anon. ACP-EU Fisheries Research Initiative. Proceedings of the Third Dialogue Meeting, Caribbean, Pacific and the European Union. Belize, Belize City 5-10 December 1996. ACP-EU Fish. Res. Rep. (3).

Christensen, V. 1996 Managing fisheries involving predator and prey species. Reviews in Fish Biology and Fisheries 6:417-442.

Christensen, V. 1997 Placing fisheries resources in their ecosystem context. EC Fish. Coop. Bull. /Bull. Coop. Pêche CE 10(2):9-11.

Christensen, V. 2001 100 years ICES: Where are the stocks? Poster presented at the ICES Centenary Science Conference, Oslo, Norway, 26-29 September 2001

Dahou, K. et M. Deme (éds.) 2001 *Impacts socio-économiques et environnementaux des politiques liées au commerce sur la gestion durable des ressources naturelles: Etude de cas sur le secteur de la pêche sénégalaise.* Environnement et Développement du Tiers Monde (ENDA) avec la collaboration de Centre de Recherche Océanographique de Dakar Thiaroye (CRODT). 67 p.

Des Clers, S. 1998 Information required by fishermen. Fish. Coop. Bull. /Bull. Coop. Pêche CE 11(3-4): 34-37.

Durand, M.H., P. Cury, R. Mendelssohn, A. Bakun, C. Roy and D. Pauly (eds) 1998 *Global versus local changes in upwelling areas*. Paris, ORSTOM, Série Colloques et Séminaires. 594 p.

FAO 1995 *Code of conduct for responsible fisheries.* Rome, FAO. 41 p.

FAO 2000 La situation mondiale des pêches et de l'aquaculture 2000. Rome, FAO, 142 p.

Fontenelle, G. (ed.) 1997 Activités halieutiques et développement durable. Actes du Colloque "Les Rencontres halieutiques de Rennes". Rennes, Ecole Nationale Supérieure Agronomique de Rennes/Halieutique and Association Agro-Halieutes. 168 p.

Froese, R. 2001. The FishBase information system: key features and approaches. pp36-38. *In*: E. Feoli & C.E. Nauen (eds). *Procs. of the INCO-DEV International Workshop on information systems for policy and technical support in fisheries and aquaculture.* Los Baños, Philippines, 5-7 June 2000. ACP-EU Fish.Res.Rep.(8):135 p.

Froese, R. and D. Pauly (eds) 2000. FishBase 2000: Concepts, Design and Data Sources. Los Baños, Philippines, ICLARM, 346 p. (Distributed with 4 CD-ROMs).

Froese, R. and K. Reyes in prep. *Boom and bust: Trends in world fisheries.* Univ.of Hamburg, Germany.

Froese, R., E. Capuli and D. Pauly 1999 *The FishBase Project eight years after: Progress, impact and future directions.* Http://www.fishbase.org/download/fbimpact6.zip 18p.

Froese, R. and A. Torres 1999. Fishes under threat. An analysis of the fishes on the 1996 IUCN Redlist. pp131-144 *In*: R.S.V. Pullin, D.M. Bartley and J. Kooiman (eds). *Towards policies for conservation and sustainable use of aquatic genetic resources.* ICLARM Conf. Proc.59:277 p.

Garcia, S.M. and C. Newton 1997 Current situation, trends and prospects in world capture fisheries. pp 7-27. *In*: E.K. Pikitch., E.D. Huppert and M.P. Sissenwine (eds). *Global trends: fisheries management.* American Fisheries Society Symposium 20.

Gréboval, D. and G. Munro 1999 Overcapitalization and excess capacity in world fisheries: underlying economics and methods of control. pp1-45 *In*: D. Gréboval (ed.). *Managing fishing capacity. Selected papers on underlying concepts and issues.* FAO Fish.Tech.Pap.386.

Jansen, E.G. 1997 *Rich fisheries – poor fisherfolk. Some preliminary observations about the effects of trade and aid in the Lake Victoria fisheries.* IUCN Eastern Africa Programme Report, No. 1:16 p.

Kooiman, J., M. van Vliet, and S. Jentoft (eds) 1999 Creative Governance: Opportunities for Fisheries in Europe. Ashgate, Aldershot.

McGlade, J.M. 1994 Governance of fisheries and aquaculture. Keynote address to the 1994 Statutory Meeting of the International Council of the Exploration of the Sea (ICES). Copenhagen, ICES, 26 p.

McGlade, J.M. 2001 Governance and sustainable fisheries. pp307-326. *In*: B. von Bodungen and R.K.Turner (eds). *Science and Integrated Coastal Management.* 85th Dahlem Workshop, Dahlem University Press, Berlin.

Nauen, C.E. 1998a Editorial. EC Fish.Coop.Bull./Bull.Coop. Pêche CE 11(2):2.

Nauen, C.E. 1998b New approaches to development co-operation in capture and culture fisheries. pp30-39. *In*: M.J. Williams (ed.). *A roadmap for the future for fisheries and conservation.* ICLARM Conf.Proc. 56.

Nauen, C.E. 1999 New players make a mark in ocean governance. International Journal of Sustainable Development 2(3): 382-387.

Nauen, C.E., N.S. Bangoura and A. Sall 1996 Governance of aquatic systems in West and Central Africa: Lessons of the past, prospects for the future. Annexe 11 pp147-160. *In*: Anon. *ACP-EU Fisheries Research Initiative. Procs. of the Third Dialogue*

Meeting. Caribbean, Pacific and the European Union. Belize, Belize City 5-10 December 1996. Brussels, ACP-EU Fish. Res. Rep.(3).

Pauly, D. 1995 Anecdotes and the shifting baseline syndrome of fisheries. Trends in Ecology and Evolution 10:430.

Pauly, D., V. Christensen, J. Dalsgaard, R. Froese and F. Torres Jr. 1998a Fishing down marine foodwebs. Science, 279:860-863.

Pauly, D., P.J.B. Hart and T.J. Pitcher 1998b Speaking for themselves: new acts, new actors and a New Deal in a reinvented fisheries management. pp 409-415. *In:* T.J. Pitcher, P.J.B. Hart and D. Pauly (eds). *Reinventing fisheries management.* Dordrecht, Boston, London, Kluwer Academic Publishers, Fish and Fisheries Series 23.

Pauly, D., T.J. Pitcher and D. Preikshot (with assistance of J. Hearne) (eds) 1998c *Back to the future: Reconstructuing the Straits of Georgia ecosystem.* Vancouver, UBC Fisheries Centre Research Reports 6(5):99 p.

Pauly, D., V. Christensen and L. Coelho (eds) 1999 Ocean food webs and economic productivity. Procs. of the EXPO 98 Conference, Lisbon, Portugal 1-3 July 1998. ACP-EU Fish. Res. Rep. (5):87 p.

Pullin, R.S.V., R. Froese and C.M.V. Casal (eds) 1999 Procs. of the Conference on Sustainable Use of Aquatic Biodiversity: Data, Tools and Co-operation. Lisbon, Portugal 3-5 September 1998. AC-EU Fish. Res. Rep.(6):71 p.

Putman, R.D. 1993 Democracy, development, and the civic community: evidence from an Italian experience. pp33-73. *In:* I. Serageldin, and J. Tabaroff (eds) *Culture and Development in Africa.* Procs. of an International Conference held at the World Bank, Washington, DC April 2 and 3 1992. World Bank Env. Sust. Dev. Proc. Ser.1.

Roberts, C.M., J.A. Bohnsack, F. Gell, J.P. Hawkins and R. Goodridge 2001 Effects of marine reserves on adjacent fisheries. Science 294:1920-1923.

Ruddle, K. 1997 The role of local management and knowledge systems in ICAM in the Pacific Region: a review. Annex 10 pp121-130 *In: Anon. ACP-EU Fisheries Research Initiative. Procs. of the Third Dialogue Meeting.* Caribbean, Pacific and the European Union. Belize, Belize City 5-10th December 1996. ACP-EU Fish. Res. Rep. (3).

Sumaila, U.R. 1998 Protected marine reserves as hedges against uncertainty: an economist's perspective. pp303-309. *In:* T.J. Pitcher, P.J.B. Hart and D. Pauly (eds). *Reinventing fisheries management.* Dordrecht, Boston, London, Kluwer Academic Publishers, Fish and Fisheries Series 23.

Sumaila, U.R. 1999 Pricing down marine food webs. Keynote. pp 13-15 *In:* D. Pauly, V. Christensen and L. Coelho (eds). *Procs. of the EXPO'98 Conference on Ocean Food Webs and Economic Productivity.* Lisbon, Portugal. ACP-EU Fish.Res.Rep. (5).

Sumaila, U.R., R. Chuenpagdee & M. Vasconcellos (eds.) 2000 Procs. of the INCO-DC International Workshop on Markets, Global Fisheries and Local Development. Bergen, Norway, 22-23 March 1999. ACP-EU Fish. Res. Rep. (7):115 p.

Viegas, T. 1999 Research for development: a New Deal. Int. J. Sust. Dev. 2(3):377-381.

Walters, C. 1998 Designing fisheries management systems that do not depend upon accurate stock assessment. pp 279-288 *In:* T.J.Pitcher, P.J.B. Hart and D. Pauly (eds). *Reinventing fisheries management.* Dordrecht, Boston, London, Kluwer Academic Publishers, Fish and Fisheries Series 23.

Watson, R. and D. Pauly 2001 Systematic distortions in world fisheries catch trends. Nature 414:534-536

The Gulf of Guinea Large Marine Ecosystem
J.M. McGlade, P. Cury, K.A. Koranteng and N.J. Hardman-Mountford (Editors)

26

Research and Extension Linkages in Ghana's Agricultural Development: the Case of Marine Fisheries

S. N. K. Quaatey

Abstract

Measures instituted by the Government of Ghana since 1984 and through the Ministry of Food & Agriculture (MOFA) for efficient research and extension linkages in the marine fisheries sub-sector are examined. In 1984, MOFA initiated the Medium Term Agricultural Development Programme (MTADP) to provide a focus for policy and institutional reforms to fully realize the agricultural and fisheries potential of the country.

Several projects have been implemented by the MOFA over the past decade to fulfil the goals of the MTADP. Institutions involved in marine fishery research have been strengthened through the National Agricultural Research Project (NARP) to carry out their research activities efficiently.

MOFA adopted a Unified Agricultural Extension System (UAES) in 1991 and initiated the National Agricultural Extension Project in 1992 to implement the UAES. Research – Extension Linkage Committes (RELCs) were formed through the NAEP to ensure that research work addresses the problems of the fishers and farmers. Bi-Monthly Technical Review Meetings (BMTRM) are held between researchers, extensionists, fishers and farmers to discuss technical problems in the sector.

Introduction

The importance of marine fisheries to the Ghanaian fishing industry and the national economy has been documented in many publications (Anon. 1973; Bernacsek 1986; Mensah and Koranteng 1988; Anon. 1995; Quaatey 1997). It is the backbone of the Ghanaian fishing industry, accounting for about 90% of total annual fish landings. The annual average production of fish over the last ten years is estimated as 400,000 mt.

About 10% of the population are employed in the fisheries sub-sector (Afful 1993; Anon. 1995) and the sub-sector contributes to about 3% of the national Gross Domestic Product (GDP). Fish is now the most important non-traditional export in the country. The country earned about US$ 95 million in 1997 from exporting raw and processed fish.

The fisheries sub-sector together with the crop and livestock sub-sectors constitute the agricultural sector in Ghana. Between the early 1970's and early 1980's the country experienced an economic recession (Anon. 1990). This adversely affected the performance of the agricultural sector in general and the fisheries sub-sector in particular (Anon. 1990; Afful 1993). The Government launched the Economic Recovery Programme (ERP) in 1983, which resulted in reversing the declining trend in the national economy and the performance of the agricultural sector. To sustain the gains of the ERP, the Ministry of Food & Agriculture (MOFA) embarked on a Medium Term Agricultural Development Programme (MTADP). The MTADP was to provide a focus for policy and institutional reforms in the agricultural sector for the full realization of agricultural potential in the country. According to Anon (1990), the policy for fisheries development as enunciated in the MTADP aims, among others, to:

i.   increase fish production for local consumption and export
ii.   alleviate poverty in the fishing communities
iii.   assist in the development of resource management plan for the entire fishery.

The strategy for achieving these aims, among others, included the strengthening of the Fisheries Department and related Departments and institutions so that they could support sound investments in the fisheries sub-sector and provide research and extension services effectively. Over the past ten years, MOFA has implemented many projects for the attainment of the aims of the MTADP. This paper highlights measures instituted by MOFA for effective research and extension linkages in the marine fisheries sub-sector.

The National Agricultural Research Systems

The agricultural sector was identified under the ERP as the backbone for national economic development and growth. The Government, therefore, initiated the Agricultural Service Rehabilitation Programme (ASRP) in 1987, with rehabilitation of agricultural infrastructure in the country as one of its main objectives. During the course of the ASRP it became evident that for production in all the agricultural sub-sectors to be sustained, there was a need to strengthen agricultural research. The Council for Scientific and Industrial Research (CSIR) was, therefore, commissioned in 1988 to review all agricultural research systems. The review indicated that most of the facilities at the research stations had deteriorated. There was also a lack of co-ordination among various research institutions, Universities, MOFA and other organizations conducting agricultural research in the country.

The National Agricultural Research Project (NARP) was initiated in 1991 under the MTADP to revitalize agricultural research systems in Ghana and also ensure that research findings are made available for dissemination to the farmers and fishers (Haizel, 1995). The main objectives of NARP were:

i.   to formulate research policies,
ii.   to prioritize research programmes and coordinate implementation,
iii.   to rehabilitate research infrastructures,
iv.   to improve the management of research systems (library and information services)
v.   and to facilitate research/extension linkages

In Ghana, the Marine Fisheries Research Division of the Fisheries Department is the main institution involved in marine fisheries research. The Water Research Institute of the CSIR, the Department of Oceanography & Fisheries of the University of Ghana and the Institute of Renewable Natural Resources of the Kwame Nkrumah University of Science and Technology are also involved in particular areas of fisheries research. Through the NARP, these institutions have been strengthened to carry out their research activities more effectively and minor civil works undertaken in the laboratories and stations. Equipment, vehicles, computers, books and journals have also been provided to these institutions. Some researchers and supporting members of staff of these institutions have been given further training to equip them to be able to do their jobs more effectively.

NARP now coordinates the research activities of these institutions to ensure that there is no duplication of work. The research activities of the institutions have been prioritized to reflect the national agricultural objectives. For instance, under the NARP the Marine Fisheries Research Division gave priority to the following research activities due to their importance to the national economy:

i.   studies on the small pelagics (mainly *Sardinella aurita, S. maderensis, Engraulis encrasicolus* and *Scomber japonicus*)

ii.   studies on molluscs and crustaceans.

The small pelagics are the most abundant marine resources in Ghanaian waters and constitute about 70% of marine fish landings annually (Anon. 1973; Mensah and Koranteng 1988; Quaatey 1997). The success of a fishing season in a particular year depends on the availability of the resource in that year. The molluscs and crustaceans are targeted for export to earn foreign exchange.

NARP has introduced a Management Information System (MIS) through which a database on all research activities has been created.

The Unified Agricultural Extension System

During the preparation of the MTADP, certain institutional constraints to agricultural development were identified. These included, among others, the weak and poorly developed agricultural extension systems. MOFA, therefore, adopted the Unified Agricultural Extension System (UAES) in 1991 as a means of improving the transfer of technology to fishers and farmers. A key element of the UAES is the use of a single Agricultural Extension Agent (AEA) to deliver all agricultural messages to the farmer and fishers at the fishing village and landing site level - the first point of contact. This is intended, among other things, to improve efficiency, reduce costs and coordinate extension activities in the country.

To implement the UAES, MOFA initiated in 1992 the National Agricultural Extension Project (NAEP) (Anon.1992); it is executed by the Department of Agricultural Extension Services (DAES). In addition, MOFA has vested all the planning, coordination and execution of extension activities in the agricultural sector to DAES. DAES collaborates with the departments within and outside MOFA in implementation of its tasks. The specific objectives of NAEP are:

i.   to execute the unified extension systems,

ii.   to strengthen extension services,

iii.   to strengthen the Technical Directorates of MOFA to enable them provide Subject Matter Specialists (SMS),

iv.   to assist the DAES develop human resources, and

v.   to forge linkages between research and extension

To fulfil the objectives of NAEP, twenty-two SMS centres have been established throughout the country to facilitate effective research-extension-fisher/farmer linkages and also ensure that relevant research findings are utilized by the fisher/farmer.

As a Technical Department, the Fisheries Directorate has posted personnel to some of these centres. The SMSs help to identify the technological problems facing the fisher and relay these problems to the research institutes. In cooperation with the research institutes, the SMSs participate in trials to test new technologies for their suitability and adoption.

Monthly training programme of AEAs has been instituted under the NAEP. At these training sessions the SMSs, among other things, impart the new knowledge on tested research findings and technologies to the AEAs. Problems of the AEAs in the execution of their duties and those of the fishers in adoption of any technology are discussed at these training sessions. In addition to the AEAs training session, Bi-Monthly Technical Review Meetings (BMTRM) have been institutionalised. The BMTRM had provided the forum for interaction between researchers and extensionists for formulation of appropriate solutions to the fishers' and farmers' technical problems.

During the BMTRM, the following, among others, are done:

i.   an update and upgrade of knowledge and skills of SMSs and researchers,
ii.   a review of adoption rates of previous technical recommendations,
iii.   training of SMSs on selected topics for subsequent monthly AEAs training, and
iv.   analysis of the training needs of the AEAs.

As a result of the BMTRM, stronger research and extension linkage has been forged. There has been an improvement in the SMSs message delivery techniques and methods. A number of technical leaflets on recommended technologies have also been produced.

The Research-Extension Linkage Committees (RELCs)

In order to ensure that research addresses problems of farmers and fishers, the Government has established five Research-Extension Linkage Committees (RELCs) throughout the country. The main duty of the RELCs is to oversee the functioning of research and extension systems in the regions and districts, and particularly the research-extension linkages.

The RELCs are co-ordinated by scientists from the research institutes and universities. The RELCs determine research and extension priorities, plan and promote joint training sessions, field visits, workshops, field days and on-farm trials of a technology. Farmers, fishers and NGOs participate in these sessions. The RELCs have led to greater collaboration and communication between researchers, extension agents, fishers/farmers. They have also made research development and transfer of technologies more responsive to the needs of farmers and fishers. The RELCs have been operating closely with the NARP and NAEP and this has ensured higher rate of adoption of technologies emanating from research to fishers/farmers.

Conclusion

It is clear that the initiation of the Medium Term Agricultural Development Program (MTADP) and the subsequent execution of the National Agricultural Research Project (NARP) and National Agricultural Extension Project (NAEP) have made it possible to establish an effective research-extension linkage in the marine fisheries sub-sector. The NARP and NAEP projects have sufficiently strengthened the institutions involved in research and extension systems so as to be able to perform their duties effectively. These projects have ensured that the framework for research-extension-fisher linkage has been instituted. The monthly training for AEAs and the BMTRM has ensured that the AEAs are adequately equipped with the technical 'know-how' for effective extension delivery while the SMSs message delivery techniques and methods have improved. This has contributed immensely to the transfer of technology to the fishers.

The establishment of the RELCs has also improved the interaction between fishers/farmers, researchers, policy makers, extension agents and other stakeholders under NARP and NAEP.

References

Afful, K. N. 1993 Fisheries Sector Development Strategy for Ghana (1993 – 2000). Final Report. 54p.

Anon., 1973 The Ghana fishing industry. Procs. of a symposium held at Tema, Ghana, May 4 – 5, 1972. Fishery Research Unit, Tema, Ghana. 102p.

Anon., 1990 Medium-Term Agricultural Development Programme (MTADP). Vol.1 & 2. Ministry of Agriculture, Accra, Ghana

Anon., 1992 National Agricultural Extension Project. Report on Orientation Workshop, December 14 – 18, 1992. Winneba, Ghana.

Anon., 1995 Staff Appraisal Report. Fisheries Sub-sector Capacity Building Project: 56pp. Fisheries Department, Ghana.

Bernacsek, G. M. 1986 Profile of the marine resources of Ghana. CECAF/TECH/86/71: CECAF Programmes, Dakar, Senegal. 71p.

Haizel, K. A. 1995 The Development of the National Agricultural Research Strategic Plan: pp 1 – 12. *In*: Procs. of the National Agricultural Research Strategic Plan. November 1994. Accra, Ghana.

Mensah, M.A. and K.A. Koranteng 1988 A Review of the Oceanography and Fisheries Resources in the coastal waters of Ghana, 1981-1986. Mar. Fish. Res. Rep. 8. Fishery Research Unit of Ghana. 35p.

Quaatey, S.N.K. 1997 Synthesis of recent evaluations undertaken on the major fish stocks in Ghanaian waters. A working document for the eleventh session of the CECAF Working Party on Resources Evaluation, October, 1997. Accra, Ghana.

V

Index